International Conference on Information Technologies in Business and Industry 2016

Journal of Physics: Conference Series Volume 803

Tomsk, Russia
21 – 23 September 2016

ISBN: 978-1-5108-3774-4
ISSN: 1742-6588

Printed from e-media with permission by:

Curran Associates, Inc.
57 Morehouse Lane
Red Hook, NY 12571

Some format issues inherent in the e-media version may also appear in this print version.

Printed by Curran Associates, Inc. (2017)

For permission requests, please contact the Institute of Physics
at the address below.

Institute of Physics
Dirac House, Temple Back
Bristol BS1 6BE UK

Phone: 44 1 17 929 7481
Fax: 44 1 17 920 0979

techtracking@iop.org

Additional copies of this publication are available from:

Curran Associates, Inc.
57 Morehouse Lane
Red Hook, NY 12571 USA
Phone: 845-758-0400
Fax: 845-758-2633
Email: curran@proceedings.com
Web: www.proceedings.com

Table of contents

Volume 803

International Conference on Information Technologies in Business and Industry 2016

21–23 September 2016, Tomsk, Russian Federation

Papers

Scheduling algorithms for automatic control systems for technological processes

A S Chernigovskiy, R Yu Tsarev, D V Kapulin

Siberian Federal University, 79, Svobodny ave., Krasnoyarsk, 660041, Russia

E-mail: achernigovskiy@sfu-kras.ru

Abstract. Wide use of automatic process control systems and the usage of high-performance systems containing a number of computers (processors) give opportunities for creation of high-quality and fast production that increases competitiveness of an enterprise. Exact and fast calculations, control computation, and processing of the big data arrays – all of this requires the high level of productivity and, at the same time, minimum time of data handling and result receiving. In order to reach the best time, it is necessary not only to use computing resources optimally, but also to design and develop the software so that time gain will be maximal. For this purpose task (jobs or operations), scheduling techniques for the multi-machine/multiprocessor systems are applied. Some of basic task scheduling methods for the multi-machine process control systems are considered in this paper, their advantages and disadvantages come to light, and also some usage considerations, in case of the software for automatic process control systems developing, are made.

1. Introduction

Nowadays an enterprise cannot exist in the competitive environment and even function at all if processes and operations control of production are not carried out in an automatic mode by control systems [9]. An automatic process control system (APCS) is an integral part of production at the large modern enterprises. Application of APCS not only allows reducing usage of human hand labor, but also promotes improving production quality [2, 6]. As a result, requirement and necessity for use of APCS are obvious. Also, it is not necessarily to use APCS in production everywhere; it is possible to carry out automation step by step, starting with the labor-consuming processes, and processes requiring high accuracy.

With high rates of technical progress in the field of computing, microprocessors become more and more powerful, and their cost becomes less. In this regard, the modern APCS the multiprocessor and multicomputer systems begin to be used more often. They allow increasing reliability of APCS on the whole, because in case of a failure of one unit, the system will not cease to function because of components redundancy [14].

Redundancy is especially necessary in case of real-time operation, as the organization of concurrent and multithreaded processes is possible because of it [1]. But not only correctly organized architecture of hardware is necessary for achievement of the maximum productivity. As well the software developed for APCS, using multi-machine and multiprocessor computing systems, has a great influence on the speed of receiving results and correct controlling actions, on the speed and quality of input data processing [7]. For this purpose it is necessary to select correctly the concurrent processes proceeding in APCS and to assign to their available computing devices of system so that to receive minimum possible time of result receiving [11]. For this purpose, it is possible to use task scheduling algorithms for the distributed computer systems. These algorithms are used to create such task schedule or operations execution which in case of execution on APCS gives minimum time of result receiving [16]. In this paper, basic task scheduling algorithms, which can be used in case of design of automatic process control systems, are considered.

The algorithms given below are intended for the one problem solution: making of such plan in which execution time of all tasks will be minimum (it is denoted as $C_{\max}$ in a scheduling problem definition). In such case, considered algorithms execute it in different methods and using different models. Scheduling is carried for identical machines, which are machines that have identical productivity. We will carry out the review of some algorithms and will show their applicability boundaries, as well as advantages and disadvantages.

2. Scheduling algorithms' inputs and outcomes

In this section we consider well-known scheduling algorithms that have been proven to solve effectively different practical scheduling problem and can be applied for automatic control systems for technological processes.

2.1. Hu's algorithm

First of all, it is necessary to consider that many algorithms use the precedence constraints among tasks to show dependence of one task on others and impossibility of their execution, the prior will not be executed yet. Usage of the oriented graphs is represented as the most convenient method. Hu's algorithm is not an exception here; it is applicable for the acyclic oriented graphs in which each vertex has only one immediate successor, except finite vertex (intree) [8]. It is possible to call such graph as a «reversed» tree, its difference is in that a tree graph has one vertex without the entering arcs and some leaving arcs (root); in this model, the reverse case is used - there is one vertex without the leaving arcs (sink) with a set of the entering [3]. Also restriction on the processing time of each task is introduced, they need to have equivalent duration in one or in p of time units ($p_j = p$ or $p_j = 1$). Thus, it can be written a scheduling problem definition for Hu's algorithm as follows, $P|intree, p_j = p|Cmax$ (or $P|intree, pj = 1|Cmax$) [8] that means: making of such task execution plan that concerns precedence, processing times are identical, that gives minimum completion time of all tasks set for finite number of identical machines (processors). It is proved that the plan produced by Hu's algorithm is optimal for the above described problem definition. Time complexity of algorithm is $O(n)$.

It is very convenient to use this algorithm for such processes which have a large number of input data and afterwards are processed in parallel and then as result give only one value or the controlling action, because the structure of such process completely matches structure of a graph which each vertex has only one immediate successor. Also, it is very convenient to apply it in that case, when there is no opportunity to measure runtime of each separate task and it is easier to take each of them for a time unit if the schedule received such methods afterwards does not arrange with productivity results. It is possible to perform further optimization or to use other algorithms. Hu's algorithm is easy to use and does not require big computations; it can be applied directly under system design to provide satisfactory software modules placement result. Sometimes, during the design, developers while planning placement of the software modules use the methods explained in Hu's algorithm unconsciously. It is possible to tell that this algorithm is result of natural attempt of computation scheduling and system optimization for the multi-machine (multiprocessor) computing system.

As a disadvantage, of course, it is necessary to consider that this algorithm does not provide preemptive schedules (with interruptions), but in certain cases it is not necessary or even technically impossible. But, nevertheless, Hu's algorithm is considered one of the simplest and basic algorithms for task scheduling with the described «reversed» tree structure. One more disadvantage is equal duration restriction of all system tasks, it can lead to some system idle time, but as it is mentioned, sometimes it is simply impossible to avoid.

2.2. Coffman-Graham algorithm

The optimality of this algorithm is proved for $P2|prec, p_j = p|C_{\max}$, ($P2|prec, p_j = 1|C_{\max}$) [4, 5, 15], all designations match with those designated above, the difference only in $P2$. Here $P2$ is an optimality for two machines (processors). As it is possible to note, one immediate successor restriction (*intree* absence in a problem definition) for each graph vertex is lifted in this algorithm (Fig. 1), but essential restriction is introduced, which is optimal only for two machines. However, it is proved that it solves not only the minimum makespan problem, but also finds minimum value of the tasks (operations) completion moments problem ($\sum C_j$), having solved both of these tasks $C_{\max}$ and $\sum C_j$, it is possible to receive the so-called *ideal* schedule. It means that this algorithm one of the best in case of the given restrictions. Time complexity $O(n^2)$ is higher than Hu's algorithm have, because of two stages of scheduling: labeling each task and following tasks assignment for each machine according to received labels. Each task can use the first labeling stage for obtaining the task list for the following use in algorithms which are intended for task list scheduling which will be considered further.

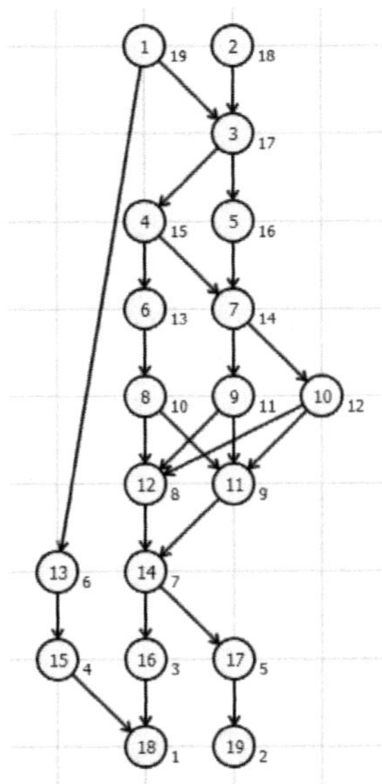

Figure 1. The graph example acceptable for Coffman-Graham algorithm.

As it was already told, this algorithm is one of the best for two-machine systems, the schedule received with its help is considered *ideal*, and also, unlike Hu's algorithm, it can be applied to any of the acyclic directed graphs. *Ideal* means that the result schedule will have not only the minimum length, but also ending time of each task in the schedule will be as minimal as possible (Fig. 2). Also, as well as Hu's algorithm, it is useful while it is impossible to estimate task duration and it is possible to use unit time restriction for tasks.

The main delimiter is non-optimality for $m \geq 3$ machines and not non-preemptiveness; some advantages of non-preemptive schedules were mentioned above.

	0–1	1–2	2–3	3–4	4–5	5–6	6–7	7–8	8–9	9–10
1	1	3	5	7	10	8	11	14	17	19
2	2	13	4	6	9	15	12		16	18

Figure 2. A schedule derived by means of Coffman-Graham algorithm from the graph in fig. 1.

2.3. LPT (Largest Processing Time) algorithm

The LPT algorithm solves the $P \,||\, C_{max}$ problem. From a problem definition, it is visible that the precedence relation is not considered, and some independent tasks list is processed. The task list can be created by Coffman-Graham's labeling algorithm, considering dependences, or any other possible way, while analyzing the structure of the graph vertexes precedence. The LPT algorithm is seldom applied in a pure form; more often it is used for combinatory optimization. With optimizations and using reasonable tasks choice, which scheduling gives the best result, this algorithm is capable to show good efficiency results. Time complexity is $O(n \cdot log(n))$.

The LPT algorithm is very simple and evident in use that is an important advantage. Also, unlike the first two considered algorithms, it can be applied to any duration tasks, without $p_j=p$ type restrictions, and it allows avoiding those idle times which can be received as a result of algorithms with identical task durations. The LPT algorithm plays an important role in further optimization. It is used in many modified algorithms. Therefore, it should not be treated as a non-optimal solution. It can be applied in the way that the system productivity will be increased on the whole.

An important restriction is the impossibility of the direct algorithm use for graph structures, but only for independent tasks lists which need to be obtained independently. Also, its relative optimality is equal to $t_{LPT}\ (TS, m)\ /\ t_{opt}\ (TS, m) \leq 4/3 - 1\ /\ (3m)$ [4, 5], where to $t_{LPT}\ (TS, m) = C_{max}$ of the dchedule received by LPT algorithm for the independent tasks list TS for m of machines, $t_{opt}\ (TS, m) = C_{max}$ of the optimal schedule for the same TS and m values. It means that the non-optimal decision at worst equals $4/3 - 1\ /\ (3m)$, where m number of machines (handlers) in certain cases can be received.

2.4. McNaughton's algorithm

McNaughton's algorithm is optimal for the $P|pmtn|C_{max}$ problem [10]. For a start, it is worth mentioning such concept as preemptive scheduling. Preemptive scheduling considers the possibility to interrupt a task and to continue its performing later using another machine (processor). In this case, it is also necessary to consider the information transfer time subsequently, as well as some delays at interruptions, but nevertheless, scheduling algorithms consider ideal cases (Fig. 3). Here, *pmtn* means what it was told earlier, it indicates that it is possible to split (interrupt) tasks and then to carry out them using other machine. Of course, it should be taken into account, whether there is a possibility of tasks interruption in that industrial control system for which the software is projected, at technical or program impossibility of interruptions this algorithm won't manage to be applied. Time complexity is $O(n^2)$.

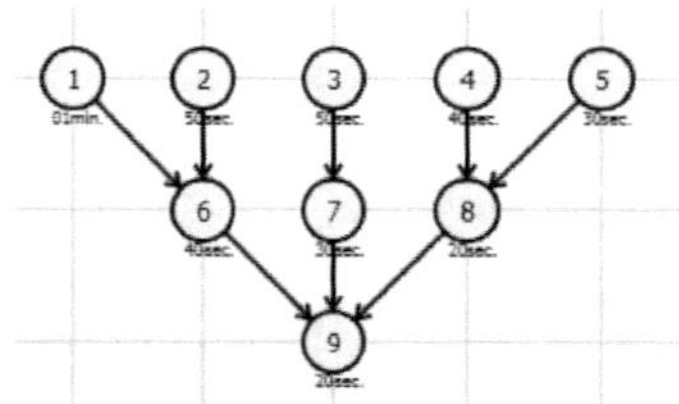

Figure 3. The graph example acceptable for McNaughton's algorithm.

The undoubted advantage of this algorithm is preemptiveness, it is the possibility accounting tasks division and the subsequent possible execution using another machine. Also, this algorithm can use non-unit task durations, therefore, the plan received with its help will not contain idle times, which can arise because of restrictions on task durations (Fig. 4). Therefore, in contrast to the LPT algorithm, this algorithm considers a possibility of tasks interruption and, also, allows avoiding the idle times caused by impossibility of task splitting into parts and, as a result, of emergence of an idle time of longer task processing.

This method has the same disadvantages as LPT, i.e. it cannot be applied for a graph structure directly, although it is proved to be optimal for an independent tasks list. It is recommended to follow the same procedure to construct the task list of, as for the LPT algorithm.

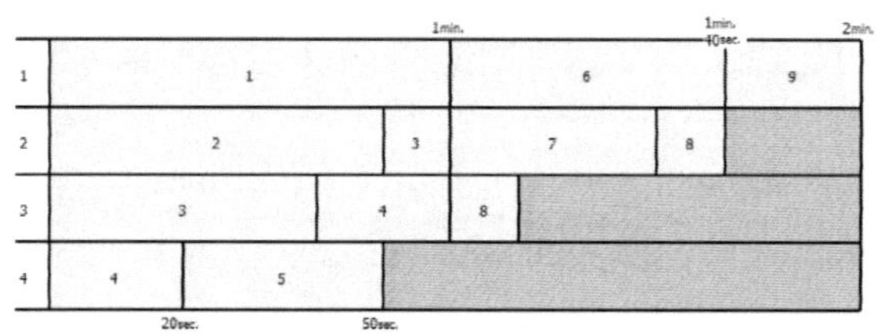

Figure 4. A schedule derived by means of McNaughton's algorithm from the graph in fig. 3 for four processors.

2.5. Muntz-Coffman algorithm

The algorithm is intended for $P|pmtn, intree|C_{max}$ and $P2|pmtn, prec|C_{max}$ problems solution [12, 13]. It means that it is optimal for a graph where each vertex has only one immediate successor (Fig. 5). Thus, it is possible to split task execution and then continue with another computer (processor). In addition, it is optimal for two machines for any acyclic kind of a graph.

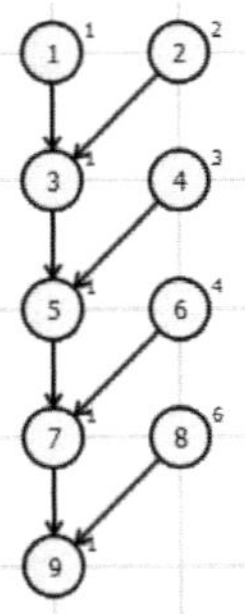

Figure 5. The graph example acceptable for Muntz-Coffman algorithm

Moreover, it is possible to receive the execution task schedule with processor resource sharing a based on preemptive schedule, if such opportunity is available. In this case, it is not important what kind of duration vertexes have – number of time units or normal time – the algorithm can work with any of them.

This algorithm contains all advantages of the previous algorithms, it is optimal for intree (as well as Hu's algorithm), and also for any kind of the acyclic graph for two machines (as well as Coffman-Graham's algorithm). It operates with unit time tasks and with usual time durations (as well as LPT algorithm); it allows creating preemptive schedules with interruptions (as McNaughton's algorithm) (Fig. 6), as well as machine (processor) resource sharing schedules (Fig. 7).

The same delimiter, as in the first two algorithms have is impossibility of application to any kind of the directed graph structure. The first restriction – the graph needs to be a «reversed» tree type; the second – it can be any kind of graph, but the schedule will be optimal for a two-machine system.

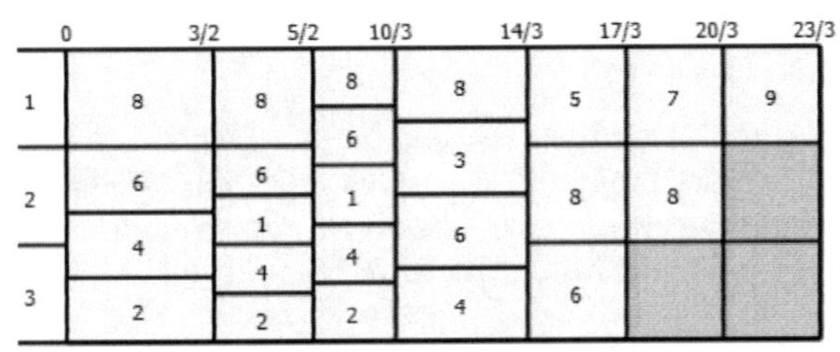

Figure 6. A preemptive schedule derived by means of Muntz-Coffman algorithm from the graph in fig. 5.

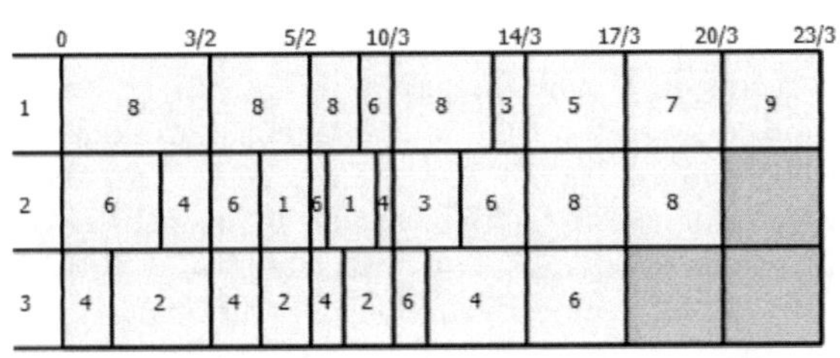

Figure 7. A resource sharing schedule derived by means of Muntz-Coffman algorithm from the graph in fig. 5.

In Table 1, it is possible to find distinguishing features of observed scheduling algorithms. Table 1 covers these features to facilitate the selection of the proper algorithm to solve the scheduling problem.

Table 1. Distinguishing features of scheduling algorithms

	Hu's	Coffman-Graham	Muntz-Coffman	LPT	McNaughton
Nonpreemptive	+	+		+	
Preemptive			+		+

Unit processing times	+	+	+		
Any kind of processing time				+	+
Simple schedule structure	+	+		+	
Machine resources sharing scheduling			+		

3. Conclusion
In this paper, the main task or operations scheduling algorithms of the software of APCS for execution on the distributed multiprocessor or multi-machine system are considered. In the paper, the review of basic algorithms of scheduling is made, advantages and disadvantages of each of them are revealed; some recommendations about application of algorithms for design of the software of APCS are given. Boundaries of applicability of each of the algorithms, as well as time complexity of their execution are shown in the article.

Algorithms features considered in this article give sufficient information for software developing of APCS, which optimally uses computing resources, for the purpose of receiving a minimum result receiving time. However, eventually, a necessary scheduling algorithm choice, as well as the design need, is to be applied reasonably by experts and system designers. In the making of the optimum task execution plan, it is necessary to consider the structure of processed data, its transmission flows and many other features of APCS functions program implementation. The set of hardware and the software features metering is a ticket to success of the best APCS composition and the minimum time of receiving the final product.

Acknowledgements
This work was supported by the Ministry of Education and Science of the Russian Federation in the framework of the Federal target program «Research and development of priority directions of development of the scientific-technological complex of Russia for 2014-2020» (agreement № 14.578.21.0116, unique ID project RFMEFI57815X0116).

References
[1] Al-Asaad H 2014 Real time scheduling of multiple executions of tasks to achieve fault tolerance in multiprocessor systems *AUTOTESTCON Proc. (St. Louis, United States, 15-18 September 2014)* 323–328
[2] Cai Z, Hong R and Cui J 2015 Control system design for induction hardening machine based on PLC and HMI *Heat. Treat. Met.* **40** 224–227
[3] Chernigovskiy A S, Tsarev R Y and Knyazkov A N 2015 Hu's algorithm application for task scheduling in N-version software for satellite communications control systems *Int. Siberian Conf. on Control and Communications, SIBCON 2015: Proc. (Omsk, Russia, 21-23 May 2015)* 1–4
[4] Graham R L 1966 Bounds for certain multiprocessing anomalies *Bell Syst. Tech. J.* 45 1563–1581
[5] Graham R L 1969 Bounds on multiprocessing timing anomalies *SIAM J. Appl. Math.* 17 263–269
[6] Hamm C and Schaefers E 2015 Numerical control for ultra-precision machining *Proc. of the 15th Int. Conf. of the European Soc. for Precision Eng. and Nanotechnology, EUSPEN 2015 (Leuven, Belgium, 1-5 June 2015)* 3–7
[7] Hashimoto K, Tsuchiya T and Kikuno T 2000 New approach to fault-tolerant scheduling using task duplication in multiprocessor systems *J. Syst. Software* **53** 159–171
[8] Hu T C 1961 Parallel Sequencing and Assembly Line Problems *Oper. Res.* **9** 841–848
[9] Grigorenko O, Stratan D and Sedykh J 2015 Current trends in the strategy of innovative development of industries in Russia *Mediterr. J. Soc. Sci.* **6** 364–370
[10] McNaughton R 1959 Scheduling with deadlines and loss functions *Manag. Sci.* 6 1–12
[11] Mertzios G B, Shalom M, Voloshin A, Wong P W H and Zaks S 2015 Optimizing busy time on parallel machines *Theor. Comput. Sci.* **562** 524–541
[12] Muntz R R and Coffman E G Jr 1969 Optimal preemptive scheduling on two-processor systems *IEEE Trans. Comp.* C-18 1014–1020
[13] Muntz R R and Coffman E G Jr 1970 Preemptive scheduling of real-time tasks on multiprocessor systems *J. ACM* 7 324–338

[14] Simevski A, Kraemer R and Krstic M 2014 Investigating core-level N-modular redundancy in multiprocessors *Proc. 2014 IEEE 8th Int. Symp. on Embedded Multicore/Manycore SoCs, MCSoC 2014 (Wakamatsu, Japan, 23-25 September 2014)* 175–180

[15] Sinnen O 2014 Reducing the solution space of optimal task scheduling *Comput. Oper. Res.* **43** 201–214

[16] Valckenaers P, Van Brussel H, Verstraete P, Saint Germain B and Hadeli 2007 Schedule execution in autonomic manufacturing execution systems *J. Manuf. Syst.* **26** 75–84

Segmentation of anatomical structures of the heart based on echocardiography

V V Danilov[1], **I P Skirnevskiy**[1], **O M Gerget**[1]

[1] Tomsk Polytechnic University, 30, Lenina Ave., Tomsk, 634050, Russia

E-mail: viacheslav.v.danilov@gmail.com

Abstract. Nowadays, many practical applications in the field of medical image processing require valid and reliable segmentation of images in the capacity of input data. Some of the commonly used imaging techniques are ultrasound, CT, and MRI. However, the main difference between the other medical imaging equipment and EchoCG is that it is safer, low cost, non-invasive and non-traumatic. Three-dimensional EchoCG is a non-invasive imaging modality that is complementary and supplementary to two-dimensional imaging and can be used to examine the cardiovascular function and anatomy in different medical settings. The challenging problems, presented by EchoCG image processing, such as speckle phenomena, noise, temporary non-stationarity of processes, unsharp boundaries, attenuation, etc. forced us to consider and compare existing methods and then to develop an innovative approach that can tackle the problems connected with clinical applications. Actual studies are related to the analysis and development of a cardiac parameters automatic detection system by EchoCG that will provide new data on the dynamics of changes in cardiac parameters and improve the accuracy and reliability of the diagnosis. Research study in image segmentation has highlighted the capabilities of image-based methods for medical applications. The focus of the research is both theoretical and practical aspects of the application of the methods. Some of the segmentation approaches can be interesting for the imaging and medical community. Performance evaluation is carried out by comparing the borders, obtained from the considered methods to those manually prescribed by a medical specialist. Promising results demonstrate the possibilities and the limitations of each technique for image segmentation problems. The developed approach allows: to eliminate errors in calculating the geometric parameters of the heart; perform the necessary conditions, such as speed, accuracy, reliability; build a master model that will be an indispensable assistant for operations on a beating heart.

1. Introduction

Currently, there are many different kinds of medical research studies and medical imaging modalities, closely related to image processing problems, such as angiography, echocardiography (EchoCG), cell magnetic resonance therapy and computed tomographic scanning. Today, these types of imaging modalities have been actively developing and they are the main diagnostic tools in their fields. The main disadvantage of the above-described methods of diagnosis is a limited software bundle of existing facilities. Thus, the effectiveness and validation of automated diagnostic methods directly depend on the level of proficiency of a researcher. In addition, image processing methods used in medical diagnostic belong to the static methods, i.e. working with each individual image and not taking into account the dynamics. Analysis shows that in the diagnostic studies, for example in angiographic and echocardiographic diagnosis, an application of dynamic segmentation methods of image sequences is the most efficient.

It should be noted that image segmentation is a preliminary stage of image processing and analysis of any system. The quality of image segmentation has a significant impact on the quality of the final stage of image processing. For example, one of the key features is the accuracy in the task of searching for the boundaries of anatomical structures of the heart by echocardiography. Thus, the main aim of our research is to develop algorithms for automated dynamic imaging and segmentation of two-dimensional and three-dimensional echocardiographic diagnosis in medical challenges. The main hypothesis, considered in the study, is that the medical images of anatomical structures of the heart can be classified according to their geometrical properties and the properties of their environment; a segmentation algorithm can be defined for each class of objects, which allows getting a result that satisfies the subsequent usage in the scientific work.

2. Materials and methods

Methods of digital processing of signals and images, mathematical analysis and mathematical modeling are used as the theoretical methods of the research. For solving posed tasks we also used: the methods of computer vision, methods of conversion and analysis of images, neural networks, statistical analysis, expert evaluation and decision-making. As for epicardial 2D echocardiography, datasets were acquired on a porcine animal model using an X7-2 transducer on a Philips iE33 machine and a PMS5.1 Ultrasound software (Philips Healthcare, Andover MA). The experimental protocols were approved by the Boston Children's Hospital Institutional Animal Care and Use Committee (IACUC). All animals received humane care in accordance with the 1996 Guide for the Care and Use of Laboratory Animals recommended by the US National Institutes of Health. The datasets were then processed off-line on the machine equipped with Intel Core i7-4820K 3.7GHz CPU using MATLAB (MathWorks, Natick MA).

The preparatory stage of image analysis is connected with processing digital noise, reduction of the noise and interferences, and assessment of the resulted image. Digital image noise is an image artifact presented as random deviation of brightness or color from its original values. Noisy pixels may be erroneously classified as edge ones. An algorithmic mistake in segmentation by the threshold method is shown in Figure 1, where the noisy area of the blood pool was classified as an endocardium.

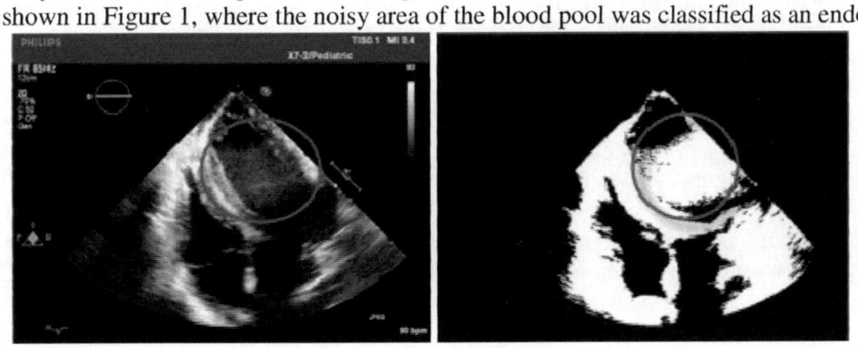

a b

Figure 1. Results of thresholding: a – reference image; b – the mistake of the thresholding method.

As for the segmentation stage, the Laplacian automated segmentation method was used. Using this approach allows one to recognize edges with high accuracy. However, Laplacian and Gaussian segmentation methods strongly depend on digital noise. In order to improve segmentation accuracy, the pre-processing stage using the median filter was implemented. The results are shown in the Figure 2.

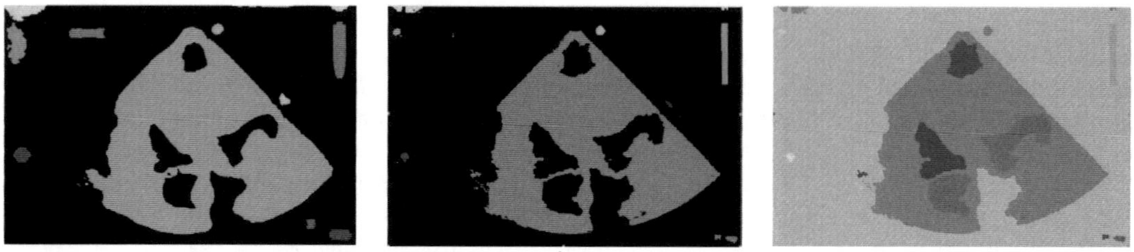

a b c

Figure 2. Automated segmentation: a – Gaussian method with median filtering; b – Laplacian method with median filtering; c – Final results of segmentation.

3. Results of pre-processing and segmentation of 2D EchoCG images

This section is dedicated to the study of algorithms of echocardiographic image processing. The main results received during the study are described below.

Procedure for conducting EchoCG:

1. During conducting EchoCG for a patient, it is possible to obtain up to 357 slices in four key views.
2. Under the existing manual processing technology of echocardiographic images, obtained by ultrasound, the average time of image analysis that the doctor manually does for one patient takes about 30 minutes. The doctor usually analyzes from 1 to 10 slices of the examination. It is not possible to trace the change dynamics of cardiac parameters from the entire set of frames, which does not allow concluding about a disorder of the cardiac activity of the heart.

Segmentation and processing of EchoCG images:

1. Existing algorithms for detecting the boundaries of the heart chambers, based on texture segmentation and morphological transformations, do not take into account the angle of inclination of the heart chambers and discontinuous contours. Algorithms, using active contours, require a template - the initial approximation of the chamber border that iteratively manually adjusted to the real border. But due to the characteristics of each slice, it is difficult to find contour parameters, such as flexibility, elasticity, step, that give the acceptable results. Furthermore, these methods are iterative. There are no EchoCG image processing algorithms on the long axis.
2. Many filters have been used to suppress speckle phenomena and noise, for instance median filters, Kuan filter [1], [2], sigma filter [3], Ramponi filter [4] et al. The values of the standard parameters, such as root mean square deviation (RMSD) [5] and a peak signal to a noise ratio (PSNR), were assessed [6]. The studies were conducted based on the real images in various positions (276 images were processed). Results are shown in Table 1. According to the considered filters, it is more efficient to use the Ramponi filter to remove speckle noise, as it provides a minimum value of RMSD and a maximum value of PSNR.

Table 1. The values of the root mean square deviation and a peak signal-to-noise ratio for different filters

Filter type	RMSD	PSNR, dB
Kuan filter	8.0	14.0
Median filter	12.0	13.0
Sigma filter	10.0	15.0
Ramponi filter	5.0	16.0

3. In this work, the possibility of using parallel processing methods to increase the speed of the pre-processing and segmentation algorithms was also explored. For example, one of the classic digital

nonlinear algorithms for noise removing is a median filter implemented using NVIDIA CUDA technology. The results are shown in Figure 3.

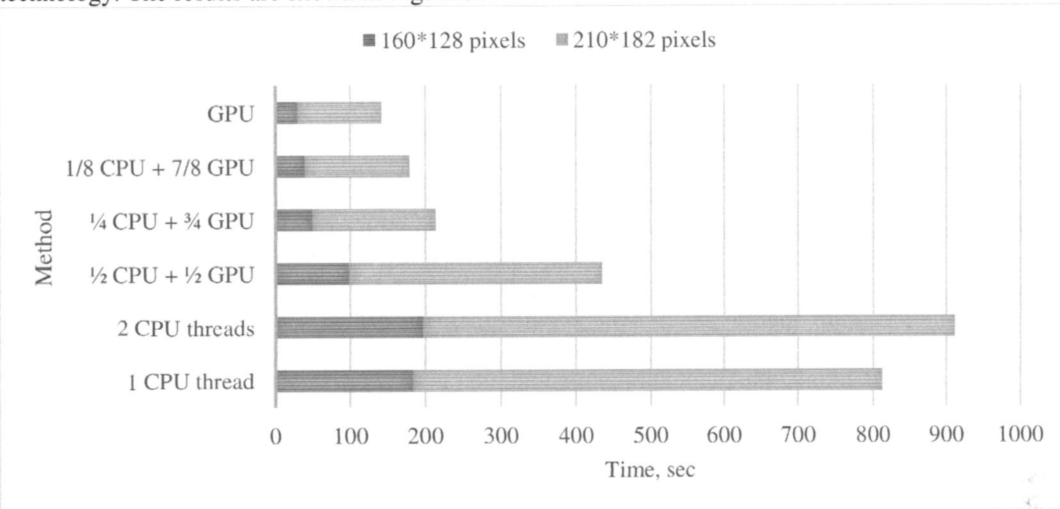

Figure 3. Results of the median filter, implemented on NVIDIA CUDA technology.

Basing on the represented data, we can conclude that the non-linear algorithms with high complexity can be implemented using parallel computing technology for increasing performance and reducing the operating time, which is one of the crucial characteristics of the implementation of the system with automatic pre-processing and segmentation of medical images.
4. Research of the developed algorithm, using different types of segmentation and an active contours method, was carried out on the basis of EchoCG test images of the heart in the apical view. The basic characteristics of the left ventricle, such as the area, length and volume, were calculated for each image. The fastest algorithm with high accuracy determination of the length, area and volume is an algorithm based on the sector segmentation.

4. Conclusion
As a result of the research study, the medical images of anatomical structures of the heart can be classified according to their geometrical properties and the properties of their environment; the segmentation algorithm can be defined for each class of objects, which allows getting the most accurate result. To increase the processing speed of complicated medical images, that is one of the crucial characteristics of the implementation of the automatic segmentation systems and medical image pre-processing, it is necessary to use parallel computing technologies.

5. Discussions
The existing manual EchoCG image processing technology does not allow one to complete full analysis of registered frames of each patient because of the complexity and the limited time of an appointment. Also, it is not possible for the doctor to estimate the change in the parameters of the dynamics from slice to slice in key positions that causes the loss of large amounts of information. In this regard, great attention is paid to the algorithm using different types of segmentation, as well as the method of active contours that were performed on the test EchoCG images.

6. Acknowledgments
This work was in part supported by the Russian Federation Governmental Program "Nauka" N: 4.1660.2016.

References

[1] Kuan D T, Sawchuk A A, Strand T C and Chavel P 1985 Adaptive noise smoothing filter for images with signal-dependent noise *IEEE Trans. Pattern Anal. Mach. Intell.* **7(2)** 165–177

[2] Kuan D T, Sawchuk A A, Strand T C and Chavel P 1987 Adaptive Restoration of Images with Speckle *IEEE Transactions on Acoustics, Speech, and Signal Processing* **35(3)** 373–383

[3] Lee J S 1980 Digital image enhancement and noise filtering by use of local statistics *IEEE Trans. Pattern Anal. Mach. Intell.* **2(2)** 165–168

[4] Ramponi G and Moloney C 1997 An Adaptive Rational Operator *IEEE Signal Processing Letters* **4(3)** 68–71

[5] Jähne B 2005 *Digital image processing 6th ed.* (New York:Springer Berlin Heidelberg) p 654

[6] Gonzalez R C and Woods R E 2008 *Digital Image Processing 3rd ed.* (New Jersey: Pearson Hall) p 977

Prospects and features of robotics in russian crop farming

B D Dokin[1], A A Aletdinova[2], M S Kravchenko[2]

[1] Siberian Research Institute of Mechanization and Electrification of Agriculture, Krasnoobsk, Krasnoobsk settlement, 630501, Russia
[2] Novosibirsk State Technical University, 20, Karla Marksa Ave., Novosibirsk, 630073, Russia

E-mail: sibime@ngs.ru, aletdinova@corp.nstu.ru, lint_5@mail.ru

Abstract. Specificity of agriculture, low levels of technical and technological, information and communication, human resources and managerial capacities of small and medium Russian agricultural producers explain the slow pace of implementation of robotics in plant breeding. Existing models are characterized by low levels of speech understanding technologies, the creation of modern power supplies, bionic systems and the use of micro-robots. Serial production of robotics for agriculture will replace human labor in the future. Also, it will help to solve the problem of hunger, reduce environmental damage and reduce the consumption of non-renewable resources. Creating and using robotics should be based on the generated System of machines and technologies for the perfect machine-tractor fleet.

1. Introduction

The European Association of Agricultural Engineering forecasts the need to increase agricultural production by 2050 to 70%, considering the growth in the number of hungry population of the planet [1]. Maybe in 35 years, the situation will be smoothed over by the decrease of the number of women of the reproductive age, the aging world population, urbanization [2], but the problem of hunger remains urgent. Food and Agriculture Organization of the United Nations (FAO) calls it the global food challenge [1]. Production of agricultural products is not unlimited. Currently, the total arable land area of the Earth is about 1.3 billion hectares, which represents 9% of the land. This indicator has remained stable in recent years [3]. In all cases, the expansion of cultivated areas have the following consequences: changes in gas and water cycle, climate change, change of river flows, geographical, ecological, biocenotics change, possible natural habitat and population destruction, etc. According to K. Moebius, the change of at least one condition results in the changes in the number of species, i.e. changing biocenosis.

The crop production in Russia is connected with the global restrictions, as well as slow intensification, impoverishing of farms, neglected farmlands [4, 5]. To solve these problems, FAO proposes considering the intensification of crop as a strategy sustainable production. This allows getting a better crop yield from the same area of the land, while reducing its negative impact on the environment [4].

For over the last sixty years, the robots have been playing a fundamental role in improving industrial efficiency and reducing production costs. At the same time, automated production operations have been implemented in agriculture, and, during the last decade, there has been a tendency of robotics applications. Transition to the precise agriculture required application of new technologies.

The term "smart agriculture economy" has already been traced in foreign publications. It is regarded as a high-tech cluster intended to introduce innovative technologies, in particular, smart generation of agricultural machinery [6]. Market Analysis of the European agricultural machinery indicated that 70% of the machines that perform operations on fertilization and spraying, are already using smart technology [7]. It is also remotely possible to adjust the depth of tillage and seeding, to create interactive maps of fields, embedded smart irrigation systems, etc. Thus, new technology led to the creation of a large number of innovations that can afford to move towards sustainable development.

Relevant questions are: whether the Russian agriculture is ready to adopt, in particular crops, the use of agro robots; whether the robots meet farmers' needs; whether it is currently effective in Russia to replace labor mechanics with agronomists robotic systems.

2. Materials and methods

The development of agricultural robotics is influenced by the following research areas: artificial intelligence; the Internet of things; Big data; cloud technologies; 3 D-printing; the mobile Internet; new generation materials; biotechnology; bionics; neuroscience; microelectronics; mechanization and automation; nanotechnology; cognitive science.

The characteristic features of crops impose certain requirements on the robotics:

– dependability on the weather and climate conditions;
– use of agrolandscape zoning;
– territorial dispersion of fields and farms;
– seasonal work;
– use of crop rotation;
– soil and plants handling.

Creating machines that do not require constant attention and their administration is the most important task of modern technology [8]. Livestock production is the most developed field in the world in terms of robots implementation in agriculture. In 2013, 4790 robots for milking were sold; in 2014 – 5180. For crops the number was significantly smaller. Meanwhile, the demand for them will increase.

The interest of scientists and researchers in development and use of robots in crop farming is increasing (Table 1).

Table 1. The number of articles on Google Scholar (2010–2016) that match the search query about agricultural robotics.

Search queries \ Period	2010	2011	2012	2013	2014	2015	till august 2016
Agricultural robotics	2960	3320	4140	4290	4010	4280	2500
Automated agricultural plant growth monitoring	1350	1510	1580	1890	1870	2140	1250
Automated plant species classification and maturation determination	203	188	231	265	281	324	168
Automatic vehicle guidance	780	960	1020	1020	969	1130	625
Autonomous seeding, phenotyping and weed removal	17	11	13	12	26	35	25
Automatic mowing plants	600	650	790	792	764	776	490
Agricultural harvest robots	1970	2140	2580	2820	2960	3120	1870
Robotic tillage	65	89	89	94	115	116	62

In 2010, the number of articles about it on Google Scholar site has reached 2960, by the end of 2015, interest has increased 1.45 times (the authors note that the approach used, the results obtained might have an error, but it can be used). Considering separate groups of operations, it is necessary to emphasize that the greatest academic interest of Russian and foreign scientists lies in the harvest, and automated monitoring of the plant growth. In their article, Z. Ying, Y. Wang, Z. Li explored the successful model of agricultural products e-commerce on the basis of analysing the characteristics of agricultural products circulation, and put forward the idea of developing agricultural products e-commerce [9]. In their research, Armstrong L.J., D. Diepeveen, R. Maddern aimed to assess these new data mining techniques and to apply them to a soil science database to establish if meaningful relationships can be found [10]. In Russia, the first work on the use of robot technology in agriculture appeared in 1984 [11]. The development of robotic technical devices for crop production in Russia is carried by the All-Russian Institute of Mechanization (VIM). First prototypes and preproduction models were created there [12].

The Russian agriculture sector has started using foreign and domestic autonomous operation system models, machines with multi-functional chassis, modern combines, unmanned aerial vehicles, special vehicles. Satellite positioning system technology, which has its disadvantages, such as problems with the GPS quality of coverage, is replaced or supplemented by the creation of an interactive map technology, using active RFID systems of radio-frequency identifiers, infrared, ultrasonic positioning and laser-guiding. Remote or autonomous means of controlling robotic equipment are currently used in crop production. The examples of robotic devices are autonomous systems (Autonomous Tractor, Kinze Manufacturing ect.); multifunctional chassis (Spirit, Cleorpath Robotics, Lynex, Hortibat ect..); harvesters (Argobot, Energid, Romobility Yoto ect..); unmanned aerial vehicles (Agribatix, SenseFly, Precision Howk etc.); specialized vehicles (VIM-ELEC 2.0, BoniRob, HortiBot, Vibro Crop Robottii, Prospero, Spider Mini etc.). The application of these devices accumulated both positive and negative experience.

Mankind is on the way of solving the problem of hunger. Major trends in the creation of robots for agriculture are aimed at: transition from technical operations to the implementation of the entire production process; minimization of human labor costs; preservation of human health; a reduction of the consumption of natural resources; a reduction of environmental damage.

3. Discussion.

The expert-analytical PBK report gave an assessment of the impact of progress in various scientific fields on the development of industrial machinery in different sectors [13]. Grouping areas of economic activity allowed identifying three groups: with a high level of technological development, e. g., the army, care for the disabled, work at home, leisure and play; with a medium level of technology development , e. g. logistics, construction, health care. Currently, the areas with the lowest level of technological development are computer vision, speech understanding, sensory systems, bionic and navigation, power supplies, hardware miniaturization in creating robots in agriculture, industry, education, the extraction of minerals (Table 2).

Table 2. The group of applications with low level of technology.

Technologies	Application area			
	Industry	Mining	Agriculture	Education
Computer vision	Medium	High	High	High
Speech understanding	Low	Low	Low	High
Sensor networks	Medium	Medium	Medium	Medium
Bionic system	Low	Low	Low	Low
Navigation systems	Medium	High	High	Low
Power supplies	Low	Low	Low	Low
Microrobots	Low	Low	Low	Low

Currently, the lowest level of technological development is observed in computer vision, speech understanding, sensory systems, bionic and navigation, power supplies, hardware miniaturization in creating robots in agriculture, industry, education, the extraction of minerals. This can be explained by the size of these areas, their characteristics as well as by the wide range of tasks. According to the forecast made by RBK, the greatest demand on the domestic Russian market can be expected for the purchase of military and special robotics for mining, operation of the infrastructure, logistics and health.

In March, 2016, during the meeting of the Presidium and the Board of scientific and technical council of the Ministry of Agriculture of the Russian Federation, the main directions of the Forecast for scientific and technological development of the agro-industrial complex of the Russian Federation for the period of up to 2030 were defined. The exhaustion of long-term effects of "green revolution" in 1960-1980 requires a transition to a new technological paradigm – the technology which is called «low external input sustainable agriculture» (LEISA). It includes biotechnology, precision agriculture, robotics, composite fertilizer, integrated biological protection, resource efficient local agriculture. There has been a growing contribution of platform technologies for multi-sectoral purposes [14]. As noted by V.I. Kiryushin in 2004, high technologies require modern software, including computer-aided design, electronic control and mapping [15]. But are Russian agricultural enterprises ready for them?

The analysis of the current situation in the Russian agriculture has shown that large agricultural holdings have a high potential for the introduction of robotic systems (Table 3).

Table 3. The potential for introduction of modern technologies which are interconnected with robotization of technological processes in crop farming (compiled on the basis of data [14]).

Modern technology	Economy type			
	APH (subsistence farming)	K(F)/SP (semicommodity economy)	Average agricultural enterprises, agricultural production cooperatives (commodity production)	Large agricultural holdings (trade, export-oriented economy)
Precise agriculture	Low	Low	Medium	High
Automation and computerization	Low	Low	Medium	High

Medium and small agricultural enterprises still need to build up their technical and technological, information and communication, human resources and management potentials.

According to A.N. Ananiev, the application of robots in agriculture is profitable, provided that it replaces at least two employees with its full depreciation for over three years [16].

A low demand for agricultural robots intended for crop production can be explained, on the one hand, by their high price and, on the other hand, by the need for their further development, providing their compatibility. Farmers show great interest in robotics, but they are wary of their implementation. Most often their questions concern their efficiency, the cost of the technical systems, required personnel training, information and communications support, etc.

Apart from the readiness of agricultural produce, technical re-equipment of the agricultural machinery industry is required. Russia has an extensive network of such plants, but, as noted by Y.N. Blynsky, S.G. Shchukin, a lot of highly qualified scientific and engineering personnel left the country in the 90s, the indirect production costs increased and now there is a technological gap between Russian and foreign manufacturers [17]. The lag of Russian robotics is noticeable in basic control devices, drives, training and software systems [18]. At the same time, on the Russian market there are low import duty rates for foreign agricultural machinery. According to the All Russian Institute of mechanization, tractors of 78 modifications produced by 13 foreign companies are imported to the European part of Russia. According to the data from the State Technical Authority (Gostekhnadzor), 128 makes of tractors produced by 34 companies and 65 modifications of harvesters, produced by 23 companies, are imported to Novosibirsk Region. A number of farms has already purchased foreign sowing complexes, which prove to be applicable in real production. According to the State of supervision in the Novosibirsk region supplied 34 tractors firms 128 brands, harvesters – 23 firms 65 modifications. A number of farms already purchased foreign sowing complexes, which proved to be inapplicable in real production.

One of the main reasons for this situation is the absence of federal and regional programs for technological and technical modernization of agriculture; an information gap among managers and farms specialists; deceptive advertising on the part of manufacturers and resellers. Mindless acquisition of foreign technological machinery (including robotics) creates a delayed-action bomb in the form of problems connected with operation of multi-brand equipment.

In addition, agriculture in comparison with other areas of economics can be called an outsider in terms of implementation of data support, engineering, animal veterinary, agronomical and other services, to say nothing of executive decision-making. Accountancy systems and legal information services are most commonly used in organizations.

Thus, in our view, the introduction and development of robotics in the Russian crop farming has the following problems:

– under-funded research and development in the robotic sphere;
– lack of the regulatory and methodological framework for measuring, testing and monitoring, as well as evaluation of the quality and safety of robotics;
– necessity of technological modernization and re-equipment of agricultural enterprises
– shortage of a highly qualified personnel to ensure the introduction, development and implementation of robotics, and eventually its management; the necessity to increase information and communication technology literacy among population;
– necessity of cost-effectiveness (which can be achieved by mass production, the replacement of human labor, the development of new materials, etc., that lower the costs);
– a low level of automation and data support of technological and control processes on the farms;
– the question of how networks will cope with the load of billions of connected devices still remains open; there is a growing problem of information security;
– necessity to create advanced technological and machinery System for agriculture, taking into account the introduction of robotics in manufacturing;
– development and implementation of government programs to support the entire cycle of the research work before putting prototypes into production and providing a unified intelligent control.

Due to the development of Russian information and communication technologies, we should expect the growth of computing power in the short term, reduction of components cost, standardization of platform technologies in robotics [19].

In our opinion, it is quite promising to expect robotic machines with electronic tracking systems to appear on Siberian fields performing the complex mineral fertilizing based on yield maps or maps of crops which need fertilizing elements.

The most demanded and compatible equipment will be robotic machines with the remote control system in the unit with a train in agricultural machinery, drills, sprayers for low-growing crops.

There is no possibility to create advanced robots without further development and improvement of the mobile Internet; artificial intelligence; the Internet of Things; cloud computing; new generation materials; storage, accumulation and use of renewable energy. It requires the state program to support the development of robotic systems and management systems.

4. Conclusion.

Thus, the creation of robotics applied to crop requirements of plant features: depending on weather and climate conditions; agrolandscape use zoning; territorial dispersion of fields and farms; seasonal work; the use of crop rotation; operations with soil and plants. Advanced robotic systems have to minimize human labor costs, go to the realization of the whole technological process, reduce harm to the environment, the consumption of natural resources. In the future, new technologies will enable obtaining high yields of crops with a minimum total cash costs and the balance between the preservation of human activity and the biosphere. But it is necessary to solve a number of serious problems: to provide funded research and development activities in the field of robotics; to create a regulatory and methodological framework for measuring, testing and monitoring, evaluation of the quality and safety of robotics; to conduct technological modernization and re-equipment of enterprises of agricultural engineering; to improve personnel literacy in terms of use of information and communication technologies and robotics; to reduce the cost model (engineering); to create a modern information and communication structure in rural areas. It will also enable creating a System of progressive technologies and machines based agrolandscape zoning and the use of robotics, which is only possible with the support of the state. In the near future, in Siberia robotic machines will be of great demand in the sphere where fertilizers and aggregates with a train in agricultural machinery, such as drills, sprayers for low-growing crops, will be highly required.

References

[1] The Global Food Challenge. – URL: http://www.cema-agri.org/page/global-food-challenge (Accessed 09.09.2016)
[2] Antonov A.I. 2006 Dynamics of the population of Russia in the XXI century and the priorities of the demographic Publishing House "Key-C" 192
[3] Geography of the main branches of agriculture. – URL: http://www.geographyofrussia.com (Accessed 09.09.2016)
[4] Aletdinova A.A., Kurcheeva G.I. 2014 Factors innovative development of agro-industrial complex of Russia *Cluster Structure of Industrial Economics* Publishing House of SpbSTU 192-214
[5] Pimentel D., Hurd L.E., Bellotti A.C., Foster M.J., Oka I.N., Sholes O.D., Whitman R.J. Food 1973 Production and the Energy Crisis *Science* **182–4111** 443–449
[6] Enabling Smart Farming in Europe – URL: http://www.cema-agri.org/page/enabling-smart-farming-europe (Accessed 09.09.2016)
[7] Armağan Z. E. 2016 Global trends in agriculture and technological solutions *Fifth World Summit on Agriculture Machinery* 28
[8] Elyashiv N.N. 2016 *Fundamentals of Robotics: A Training Manual* (RGPPU p. 49
[9] Ying Z., Wang Y., Li Z. 2005 Research on circulation and e-commerce of Chinese agricultural products *ICEC '05 Proceedings of the 7th international conference on Electronic commerce* 846-850
[10] Armstrong L.J., Diepeveen D., Maddern R. 2007 The application of data mining techniques to characterize agricultural soil profiles *AusDM '07 Proceedings of the sixth Australasian conference on Data mining and analytics* **70** 85-100
[11] Vasaynin V.I. 1984 *Agricultural robots* (Kolos) p. 224
[12] Hort D.O., Filippov R.A., Kutyrev A.I. 2015 Multifunctional robotic vehicle with vision *Innovations in agriculture* **4 (14)** 115-121
[13] RBC. – URL: http://www.rbc.ru/ (Accessed 09.09.2016)
[14] 2016 Forecast of Russian scientific and technological development of the agro-industrial complex for the period till 2030: global challenges *Newsletter "Agrarian heart rate of a great country"* **4** 14-21
[15] Kiryushin V.I. 2004 Exact agricultural technology as the highest form of intensification of adaptive-landscape agriculture *Agriculture* **6** 16-20
[16] A.N. Ananiev Problems of application of robots in agriculture. – URL: http://www.biz-for.ru/robotics/robotsprobl/robotsprobl.php (Accessed 09.09.2016)
[17] Blynsky Y.N. Schukin S.G. 2013 Agricultural Engineering Western Siberia *Innovation and food security* **1** 67-76
[18] Maksimov P.L., Ivanov A.G., Mokhov A.A., Petrov V.A. 2015 Study of the possibilities of automation of farm work *Bulletin of the Izhevsk State Agricultural Academy* **3 3 (44)** 32-38

[19] Potential of innovations Russian market automation systems and robotics. – URL: http://www.slideshare.net/tuknov/otchet-robotfinal291014 (Accessed 09.09.2016)

Recognition of a person named entity from the text written in a natural language

A V Dolbin[1], **V L Rozaliev**[1], **Y A Orlova**[1]

[1] Volgograd State Technical University, 28, Lenina ave., Volgograd, 400131, Russia

E-mail: vladimir.rozaliev@gmail.com

Abstract. This work is devoted to the semantic analysis of texts, which were written in a natural language. The main goal of the research was to compare latent Dirichlet allocation and latent semantic analysis to identify elements of the human appearance in the text. The completeness of information retrieval was chosen as the efficiency criteria for methods comparison. However, it was insufficient to choose only one method for achieving high recognition rates. Thus, additional methods were used for finding references to the personality in the text. All these methods are based on the created information model, which represents person's appearance.

1. Introduction

Named entity recognition belongs to the category of information retrieval tasks. Nowadays, there are many methods for extracting knowledge from the text. Despite the fact that this problem is relatively new, it is already one of the most important directions in the field of computer technology. A large amount of information in the text is described implicitly. For example, if extracted information relates to the elements of human appearance, maybe it becomes possible to conclude about person's professional activities or habits. However, this task is not always easy as it seems because of the features of natural languages. Usually, the text is written in a free form and, therefore, it is quite difficult to interpret the meaning of words. [1]

This paper presents an investigation which method is better for recognition of the appearance elements of a man from the text, composed in a natural language. These two methods were selected for investigation since they are very common and show very good results in practice. In particular, the study was carried by the example of the Russian language. However, it is necessary to bear in mind that latent semantic analysis and latent Dirichlet allocation can be used for a large variety of natural languages.

2. The information model of person's appearance

First of all, it was necessary to create an information model for providing a research basis. This model should met the following requirements:

- Scalability. Requirements could be changed after experiments.
- Model should fit easily into one of the common programming paradigms.
- Visibility. It should be easy to find necessary information.

- Fullness. The model should contain detailed description of person's appearance so that it was possible to visualize a human.

The frame representation language (FRL) was chosen as a formal description of person's appearance since it has all necessary parameters. Figure 1 shows the information model of person's appearance in FRL. Each slot of the frame has a gender determination procedure, which starts after the slot has been filled. The gender determination process relies heavily on morphological aspects of slot's value. "M" is a predefined set of possible parameters for the appearance model.

```
(frame Human_Appearance
(Height      ( value  (M)  )  ( IF_ADDED(detect_gender)  ) )
(Body        ( value  (M)  )  ( IF_ADDED(detect_gender)  ) )
(Head        ( value  (M)  )  ( IF_ADDED(detect_gender)  ) )
(Hair        ( value  (M)  )  ( IF_ADDED(detect_gender)  ) )
(Face        ( value  (M)  )  ( IF_ADDED(detect_gender)  ) )
(Forehead    ( value  (M)  )  ( IF_ADDED(detect_gender)  ) )
(Eyebrows    ( value  (M)  )  ( IF_ADDED(detect_gender)  ) )
(Eyes        ( value  (M)  )  ( IF_ADDED(detect_gender)  ) )
(Eyelashes   ( value  (M)  )  ( IF_ADDED(detect_gender)  ) )
(Nose        ( value  (M)  )  ( IF_ADDED(detect_gender)  ) )
(Lips        ( value  (M)  )  ( IF_ADDED(detect_gender)  ) )
(Chin        ( value  (M)  )  ( IF_ADDED(detect_gender)  ) )
(Teeth       ( value  (M)  )  ( IF_ADDED(detect_gender)  ) )
(Neck        ( value  (M)  )  ( IF_ADDED(detect_gender)  ) )
(Shoulders   ( value  (M)  )  ( IF_ADDED(detect_gender)  ) )
(Chest       ( value  (M)  )  ( IF_ADDED(detect_gender)  ) )
(Back        ( value  (M)  )  ( IF_ADDED(detect_gender)  ) )
(Legs        ( value  (M)  )  ( IF_ADDED(detect_gender)  ) )
(Arms        ( value  (M)  )  ( IF_ADDED(detect_gender)  ) )
)
```

Figure 1. The person entity information model.

3. Latent semantic analysis and latent Dirichlet allocation

Latent semantic analysis (LSA) is a method for processing text information in a natural language, which examines the dependency of keywords on collection of documents (or sentences). The main purpose is to find a vector of documents, which is as close as possible to the specified keyword. [2]

The algorithm of latent semantic analysis is as follows:

- Select specified keywords. These keywords should describe the appearance of a person.
- Construct frequency matrix A, in which columns are documents and rows are keywords. Each element of the matrix should represent the occurrence of the keyword in the document.
- Apply a TF-IDF method on the basis of a frequency matrix to ensure that results are relevant. [3]
- It is necessary to use singular value decomposition on the basis of a resulting matrix. SVD separates the frequency matrix into a matrix of documents (U), a matrix of keywords (Vt) and a diagonal matrix (S).

$$A = U * S * V^t \quad (1)$$

- A number of rows from matrix Vt and columns from matrix U can be dropped. This statement follows from the feature of singular value decomposition. Thus, less significant values will not take part in future computations. According to experiments, the best variant is to use Vt with 2 rows and U with 2 columns.
- Matrix U contains the coordinates of keywords and Vt – coordinates of documents. It is now possible to obtain the nearest documents, which has the same semantic meaning as a specified keyword. [4]

Latent Dirichlet allocation (LDA) serves the same purpose as LSA – it finds the dependency between the specified keyword and related documents, but in another form. LDA provides the information about how likely each keyword refers to each topic. On the other hand, the result contains probability of the document belonging to one of the topic. [5]

LDA takes at least three parameters as an input:

- Number of iterations.
- Number of topics.
- Text in a natural language.
- Set of keywords.

The algorithm of this method is as follows:

- Compose the list of keywords. These keywords should describe the appearance of a person.
- Assign topic for every word in each document. This assignment can be arbitrary or based on any external signs (in this case, it can be the same frequency matrix as in the LSA method).
- Calculate the percentage of topics, which relates to specified document p(t|d) and the probability that the word relates to current topic p(w|t). A specified word relates to the topic with probability:

$$p = p(t \mid d) * p(w \mid t) \qquad (2)$$

- Repeat the previous iteration of a predetermined number of times.

Both methods have much in common. The performance of these methods is low for the large document collection. Latent semantic analysis treats the text as a set of words that do not have any semantic links; so does latent Dirichlet allocation. However, this approach has an advantage in partial disambiguation from words with several meanings.

4. Person entity recognition

A person is one the most difficult named entity to recognize in the text. It is very easy to recognize the person if the text contains the name or surname of the entity. Nevertheless, such cases are extremely rare and references to an individual occurs differently. It is necessary to perform pre-processing of the text to increase the accuracy of latent semantic methods. LSA and LDA should perform searching of person's appearance only in documents, which contain such description. [6, 9]

One common approach is to use contextual rules. Every rule represents a standard regular expression, which is constructs from the training sample as follows:

- Any mention of a person should be replaced with a special word {PERSON}.
- If the word in a training sample is an element of a human appearance, it should be set in its initial form.
- All other words should be replaced with its parts of speech.
- After processing the entire training sample, similar contextual rules should be combined using special characters. The '?' symbol means that this position can be omitted, the '+' symbol means that the position can be repeated one or more times in a row and the '|' symbol represents logical "or".
- Optionally, some phrases can be listed at the end of the training sample, which indicates that the sentence excludes the possibility of containing the entity.

Thus, the outputs are kind of regular expressions, which are applied to the text to define a named entity with it.

The next step is to resolve the reference of pronouns in the third person. Reference resolution of pronouns in the third form is one of the most common, but at the same time the simplest case, but completeness of information retrieval receives a great boost. This task can be considered as a problem of binary classification. Therefore, it is a good opportunity to use support vector machines (SVM). The simplest case of SVM was programmed using the linear classifier. Thereby the training sample must be compiled carefully to avoid the case of linear inseparability. [7, 8]

The list of parameters for support vector machines which were used for training:

- Number of sentences between antecedent and anaphora.
- Whether the antecedent is in the nominative.
- Position of the anaphora in the sentence.
- Position of the antecedent in the sentence.
- Number of nouns and pronouns, which are located in sentences with antecedent and anaphora.
- Is the antecedent and anaphora case matches.
- Is the antecedent and anaphora genus matches.
- Is antecedent and anaphora both in a plural or singular form.

5. Latent semantic methods comparison

All methods, which were mentioned in this paper, were written in Python 3. Pymorphy2 library was used for any morphological operations. It uses the "OpenCorpora" dictionary that counts about million and a half tokens at this moment.

A training set for support vector machines consists of five hundred samples. The training set for contextual rules also counts five hundred sentences. Latent semantic analysis and latent Dirichlet allocation do not need any training data.

Let N be a total number of person's appearance elements, and N_r – the number of items found by one of the method. These two methods are compared by calculation of the completeness of information retrieval, which is as follows: [10]

$$CIR = \frac{N_r}{N} \tag{3}$$

Experiment results are shown in table 1. Natural language texts used in the experiment were taken from areas that had been used to compile OpenCorpora, such as imaginative literature, blogs, Wikipedia and legal texts. Table 1 shows that if the search is performed only based on text passages that contain person references, the accuracy of recognition will increase greatly. In turn, the latent semantic analysis shows slightly better results for the recognition of human appearance in comparison with latent Dirichlet allocation regardless of the volume of the input information.

Table 1. The results of the comparison of latent semantic analysis and latent Dirichlet allocation.

Number of documents	The approximate number of words per document	LSA without person recognition	LDA without person recognition	LSA with person recognition	LDA with person recognition
1	100	0.7	0.68	0.85	0.83
2	100	0.69	0.67	0.84	0.82
5	200	0.69	0.67	0.84	0.82
7	200	0.67	0.65	0.83	0.81

10	500	0.65	0.62	0.81	0.79
12	500	0.64	0.62	0.80	0.78
15	500	0.64	0.61	0.79	0.78

6. Acknowledgments
The study was financially supported by RFBR research projects №15-07-05440, 15-07-07519, 15-37-70014, 16-07-00453.

7. Conclusion
Completeness of information retrieval on the subject of human appearance has been investigated using the methods of latent semantic analysis and latent Dirichlet allocation. The completeness of information retrieval was chosen as the efficiency criteria for methods comparison. Both methods have shown quite good results in practice. However, LSA is slighter better than LDA in case of recognition the appearance of a person. Preprocessing the text with finding references to the person greatly increases the accuracy of both methods.

The program was written in Python 3 with the usage of pymorphy2 library for morphological analysis. OpenCorpora was chosen as a main dictionary, with the help of which words were put in their normal forms.

References

[1] Dmitriev A, Zaboleeva-Zotova A, Orlova J and Rozaliev V 2013 Automatic Identification of Time and Space Categories in the Natural Language Text *App. Com.* 187-90

[2] Soloshenko A, Orlova J, Rozaliev V and Zaboleeva-Zotova A 2015 Establishing Semantic Similarity of the Cluster Documents and Extracting Key Entities in the Problem of the Semantic Analysis of News Texts *Mod. Appl. Sci.* **9(5)** 246-68

[3] Ramos J 2003 Using TF-IDF to Determine Word Relevance in Document Quieries *Dep. Of Comp. Sci.*

[4] Deerwester S, Sumais S, Furnas G, Landauer T and Harshman R 1990 Indexing by Latent Semantic Analysis *Jour. Of the Amer. Soc. For Inf. Sci.* **41** 391-407

[5] Blei D, Ng A and Jordan M 2003 Latent Dirichlet Allocation *The Jour. Of. Mach. Lear. Res.* **3** 993-1022

[6] Ritter A, Clark S, Mausam and Etzioni O 2011 Named Entity Recognition in Tweets: an Experimental Study *Proc. of the Conf. on Emp. Met. In Nat. Lan. Proc.* 1524-34

[7] Keerthi S, Chapelle O and DeCoste D 2006 Building Support Vector Machines with Reduced Classifier Complexity *Jour. Of Mac. Ler. Res.* **7** 1493-515

[8] Lee H, Chang A, Peirsman Y, Chambers N, Surdeanu M and Jurafsky D 2013 Deterministic Coreference Resolution Based on Entity-Centric, Precision-Ranked Rules *Com. Lin.* **39(4)** 885-916

[9] Minkov E, Wang R and Cohen W 2005 Extracting Personal Names From Email: Applying Named Entity Recognition to Informal Text *Proc. of the Conf. on Emp. Met. In Nat. Lan. Proc.* 443-50

[10] Zhou B and Yao Y 2010 Evaluating information retrieval system performance based on user preference *Jour. of Int. Inf. Sys.* **34(4)** 227-48

Automation of testing modules of controller ELSY-TMK

A E Dolotov[1], R G Dolotova[1], D V Petuhov[2], A P Potapova[1]

[1] Tomsk Polytechnic University, 30, Lenina ave., Tomsk, 634050, Russia
[2] EleSy Company, 161 A, Altayskaya ave., Tomsk, 634021, Russia

E-mail: dolotovae@tpu.ru, dolotova@tpu.ru

Abstract. In modern life, there are means for automation of various processes which allow one to provide high quality standards of released products and to raise labour efficiency. In the given paper, the data on the automation of the test process of the ELSY-TMK controller [1] is presented. The ELSY-TMK programmed logic controller is an effective modular platform for construction of automation systems for small and average branches of industrial production. The modern and functional standard of communication and open environment of the logic controller give a powerful tool of wide spectrum applications for industrial automation. The algorithm allows one to test controller modules by operating the switching system and external devices faster and at a higher level of quality than a human without such means does.

1. Introduction

The ELSI-TMK controller is designed for the following purposes: measuring continuous signals presented by direct current voltage and (or) direct current, gathering and processing the data from primary sensors, forming control signals according to a specified algorithm, receiving and sending information through serial communication channels in systems of measurement, controlling and management of the oil and gas industry facilities, power facilities and facilities of other branches of industry, including energy consumtion audit both for commercial and technological purposes and development of security systems [2].

The controller is a restorable, multipurpose, multichannel item with a variable structure of functional modules and ability to back up the power supply and the central processor unit. Its special feature is an ability to apply the whole set of input-output modules and also to apply one or several interface modules connected by a switching panel. The TA715 module which is a part of the ELSY-TMK controller is designed for measurement of direct current voltage or for direct current.

The operation algorithm of the controller is defined by an operating program developed by the user according to requirements to the control system created by means of the controller. The basic area of application is the systems of automatic and automated control of manufacturing processes.

2. Materials and methods

The controller represents a set of modules (Fig. 1) connected by a switching panel which is designed for mechanical integrating of the controller modules, for organising the electrical connections between modules, and also for installating the controller on an installation site. Depending on the number of installed modules, the panel has some modifications. For the function of power supply back-up, some special variants of modification are provided. The object-to-be-tested is the controller of the TA715 module designed for measurement of direct current voltage or for direct current. The TA715 module is a part of the controller programmed by the ELSY-TMK [3-6].

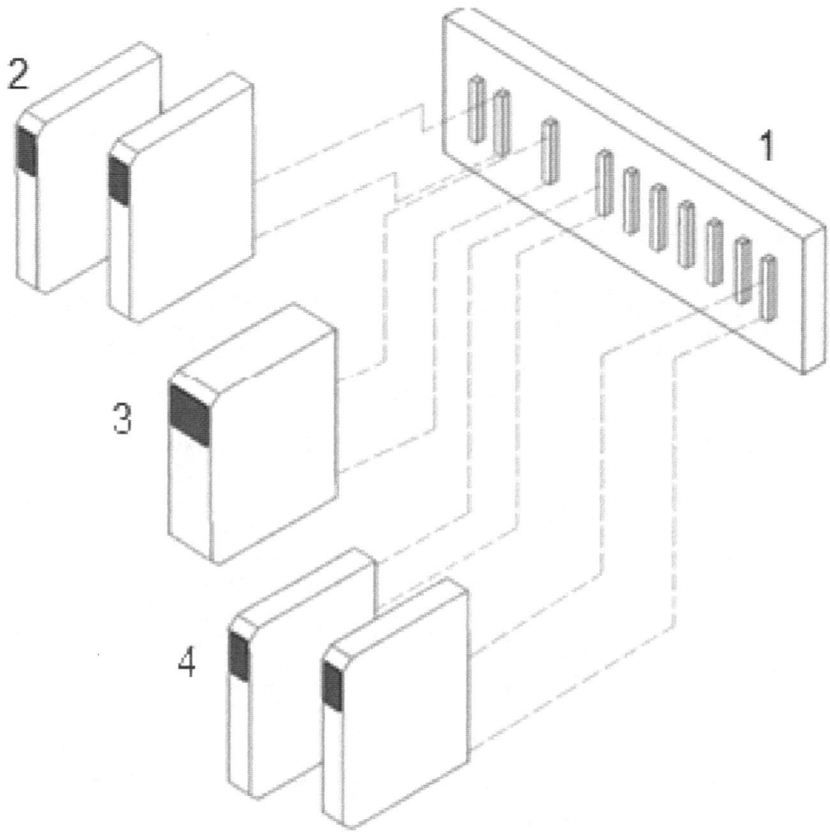

Figure 1. A controller. A hardware structure: 1 - switching panel, 2 - power supply, 3 - processor module, 4 - functional modules.

The controller is realised on a modular approach on the basis of a parallel highway. The controller highway contains the following parts: an address bus, a data bus, a control bus, an interrupt bus and a power bus. Controller modules are connected to the bus in parallel and they exchange data with the processor by means of the highway and also are connected to the power voltage. The controller power supply transforms voltage of power from *220* B/*50* Hz or *24* in direct current to a supply voltage of modules: plus *5*, plus *12*, minus *12* B.

The controller software is based on the *CoDeSys* development system by the company "3S-Smart Software Solutions" designed for controller programming in languages according to the IEC 61131-3 standard. The base structure of the *CoDeSys* complex includes two systems: a development system and an execution system. The development system runs on a PC and represents a tool for designing, system configuring and operating program code development for the programmable logic controller. The execution system (ES) runs on the controller and provides the code loading of the applied program in the controller, execution of the operating program and executing of debugging functions. The base version of *CoDeSys* is specially adapted for operating on the ELSI-TM controller. In addition to available tools of the complex, the add-in components of the controller support are developed for the most effective development of application programs.

3. Automation of voltage tests of a TA75 module

For automation purposes, the connection and control of an industrial computer were established with the object-to-be-tested and the calibrator on the test bed (Fig. 2).

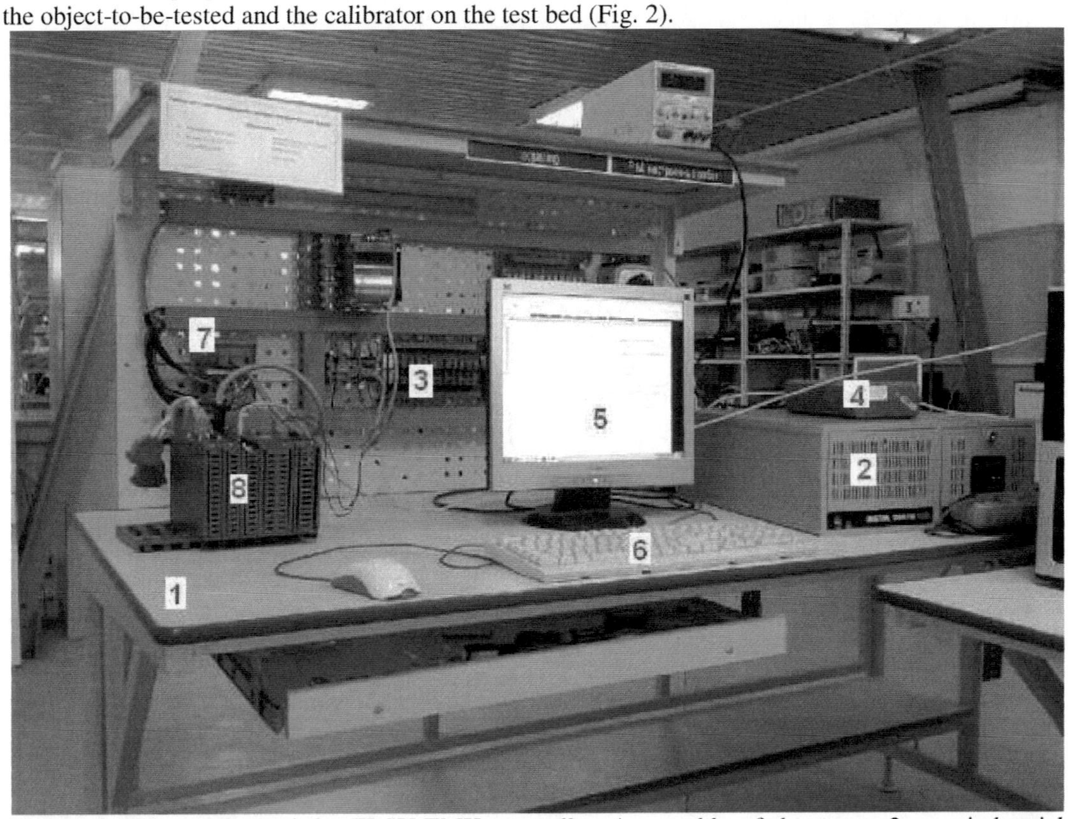

Figure 2. The test bed of the ELSY-TMK controller: 1 - a table of the tester, 2 - an industrial computer, 3 – a system of signal switching, 4 – a unit for specifying the external voltage (SA100calibrator), 5 – a monitor, 6 – a keyboard, 7 – a special demountable connector of the switching system, 8 – an ELSY-TMK controller with extension modules as an object-to-be-tested.

The test bed is designed for automatic tests of extension modules of the TA715 analogue input terminals, the TD711 digital input terminals, the TD712 digital output terminals of the ELSY-TMK controller. The control of the controller is carried out with the Modbus tcp/ip interface. The signals from the switching system pass to the extension modules of the controller. The switching system represents sets of relays controlled by means of the industrial computer through the DIO144 expansion card. There are controlling and individual relays. By means of controlling relays the signal wire commutes with one of the external devices that set the voltage: the Advantage card, TMA, the CA100 calibrator. By means of controlling relays the signal is inverted, and also the signal wire is closed with the 680 Ohm and 6.8 kOhm resistors. Individual relays commute the signal wire to one of channels of a specified extension module.

For example, for testing of the TA715 extension module for the current, the signal from the external setting device is commuted by the system to one of the input terminals of module analogue channels. Each channel of this module is checked in series. Also all other channels must not have any connection to the signal wire. The test process control is carried out by the industrial computer with installed software. The algorithm of the given program allows conducting tests of three types of extension modules. External setting devices, such as a SA100 calibrator and a TMA calibrator, are controlled through a COM-port. The Advantage card is controlled through a demountable ISA connector.

All TA715 analogue channels should be tested for the given admissible relative error of voltage value. If any channels of the module do not pass the test for the given error, the information about the damage of specified channels is displayed along with the information about the given relative error at the specified voltage level. Further, the tester removes the module from the switching panel, prepares the report and starts testing of the next module.

After the start of the executable file of the test software, there is a readjustments of basic connection elements with the object-to-be-tested and testing tools. By pressing the button "Connect", the tester starts the procedures that establish connection with the controller through modbus tcpip [7]. To operate the modbus interface, the free library "libmodbus" is used. From the ComboBox1 element a peripheral device is chosen, from which the analogue power will pass: the Advantach card, TMA or the CA100 calibrator. After the start of "TA715 voltage test", the main algorithm function of testing of the module of analogue input terminals is executed. In the boot record, the initial values of all main function variables of the testing algorithm are defined.

4. Conclusion

The developed software allows conducting tests of extension modules of the ELSY-TMK controller. The tests of the TD712 module of digital output were carried out by means of the verified module of digital input terminals - TD711 module. The automatic mode of tests of the TD712 module output channels is carried out by connecting the buses of the digital input terminals of the TD711 module with digital output of the TD712 module and by starting the "Test TD712" program on the industrial computer.

The testing means software sequentially shorts the TD712 module channels one by one by changing the value of the register bit which is responsible for the state of this channel. The register value is changed by sending the write instruction under the modbus tcp/ip protocol. If the bit is equal to 1, the current passes through the channel. It can influence the corresponding channel of the TD711 module of digital input terminals. This can be checked by reading the corresponding register with the modbus tcp/ip. The current information about the testing of module channels is displayed on the screen.

References

[1] Dolotov A, Petuhov D and Dolotova P 2016 Automation of test unit controller TA715 ELSY-TMK *International scientific conference* (Tomsk: Tomsk Polytechnic University) pp 532-536
[2] Child G 2001 *Programmer's Reference for C/C, 2nd ed H Shild* (Moscow : Williams) p 448
[3] Guide 2005 *The programming of the controller ELSI-TM* (Tomsk: EleSy Company) p 2
[4] Guide 2015 *The programming of the controller ELSI-TM*. Specifications (Tomsk: EleTim Company) p 95
[5] Guide 2005 *The programming of the controller ELSI-TM. Test procedure* (Tomsk: EleTim Company) p 25
[6] Guide 2005 *The programming of the controller ELSI-TM* (Programming system CoDeSys) (Tomsk: EleSy Company) p 125
[7] Stroustrup B 2008 *The C++ programming Language* (Moscow : Binom-Press) p 53-56

Parametric identification of the controlled object based on transfer characteristics

S V Efimov[1], M I Pushkarev[1], P Rapisarda[2], S V Zamyatin[1], V M Marukyan[1]

[1] National Research Tomsk Polytechnic University, 30, Lenina ave., Tomsk, 634050, Russia
[2] University of Southampton, University Road, Southampton SO17 1BJ, United Kingdom

E-mail: efimov@tpu.ru

Abstract. Classical identification techniques are shown; their strengths and shortcomings are analyzed. The correlations associating direct indexes of transfer characteristic quality of the controlled object and its pole-zero configuration are received. The identification technique of the linear dynamic object based on correlations is developed. The numerical model is given.

1. Introduction

According to the classification criteria automation objects identification techniques are divided into several groups – in accordance with scope of initial information, type of experiment, type of object of research, mathematical tool, etc.

The most widespread and commonly occurring identification techniques are used to be a frequency method and a method with the help of transfer function [1-10].

As a rule, the frequency identification technique is used in laboratory setting. It implies a possibility of artificial influence on the controlled object by means of a sinusoidal signal of different frequencies. It is assumed that the object structure is known, it is necessary to define its parameters, b_m, b_{m-1},… b_1, a_n, a_{n-1},… a_0:

$$W(s) = \frac{b_m s^m + b_{m-1} s^{m-1} + \ldots + 1}{a_n s^n + a_{n-1} s^{n-1} + \ldots + a_0}. \quad (1)$$

Based on (1), we obtain complex harmonic locus as a sum of the real frequency response and the imaginary frequency response:

$$W(j\omega) = \frac{b_m (j\omega)^m + b_{m-1} (j\omega)^{m-1} + \ldots + 1}{a_n (j\omega)^n + a_{n-1} (j\omega)^{n-1} + \ldots + a_0} = P(\omega) + jQ(\omega). \quad (2)$$

In order to receive test values, real frequency response $P(j\omega_1)$ and imaginary frequency response $Q(j\omega_1)$ influence the controlled object by means of sinusoidal signal with frequency ω_1. Equating the theoretical real frequency response and the imaginary frequency response (2) and test values, we obtain two equations

$$P_{th}(\omega_1) = P_{ex}(\omega_1);$$
$$Q_{th}(\omega_1) = Q_{ex}(\omega_1).$$

IOP Conf. Series: Journal of Physics: Conf. Series **803** (2017) 012036 doi:10.1088/1742-6596/803/1/012036

However, in (1), and correspondingly in (2), $m+n$ of unknown parameters is contained, for that reason to define all the parameters of system transfer function, it is essential to conduct $k/2$ experiments, where $k=m+n+1$, m, n are the order of a numerator and a denominator, correspondingly. Thus, having solved a set of k equations, all the parameters of transfer function are defined.

Together with the frequency technique identification method based on transfer characteristics is widely used. The controlled object is influenced with unit step function and the object reaction is obtained - transfer characteristic in the form of continuous curve or values array with incremental sampling. Approximating the transfer function by analytic expression and differentiating this expression, we shall obtain impulse response. In engineering grapho-analytical methods have become a frequent practice, thus allowing to calculate the parameters for the given controlled object structure.

In accordance with the form of the transfer characteristic curve the additional constructions for approximation are provided and approximation formulas for calculation are applied. So, for identification of the controlled object described by differentiation equation $T\frac{d}{dt}y(t)+y(t)=kx(t)$, it is essential to define the parameters k, T. Let us consider fig. 1 showing determination methods of system parameters.

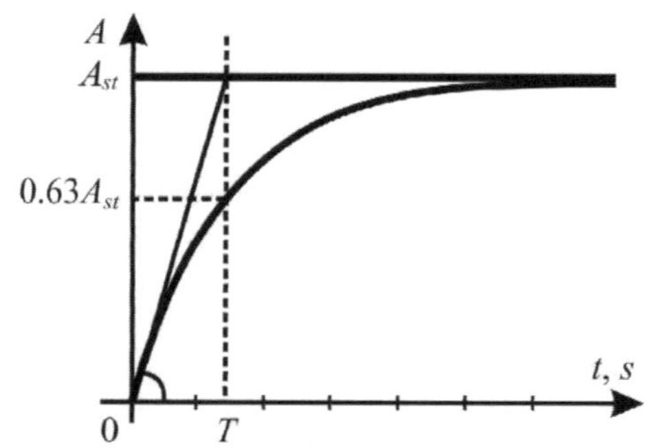

Figure 1. Transfer characteristic of the first-order object.

As the transfer characteristic is the reaction of the controlled object to the step function, the coefficient of object transfer $k=\frac{A_{st}}{x}$, where A_{st} is a stationary value, x is step impact on the object.

Time constant T can be defined in two ways. While defining T by the first method, a tangent to the curve of transfer characteristic in the point $t=0$ is plotted, from a cross point of tangent and direct $A=A_{st}$ a vertical to time axis is drawn. Using the second method, it is assumed that T corresponds to the time of achievement by the curve of transfer characteristic 63% from value A_{st}, fig. 1.

In a similar way, delay systems and vibration systems are identified, fig. 2.

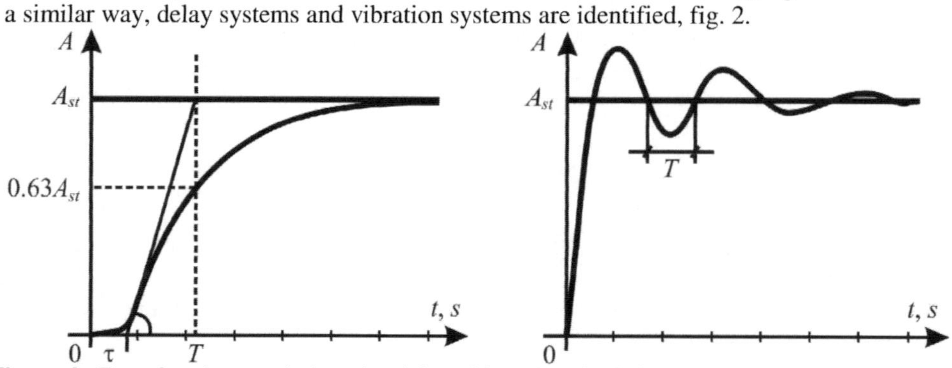

Figure 2. Transfer characteristics: the delay object (on the left); the vibration object (on the right).

Suggested identification techniques have their own strengths and shortcomings. The frequency technique is suitable for identification of not high order objects; otherwise it leads to a large number of experiments that is not always acceptable in virtue of the controlled object specific character. The method gives reasonably exact results in the case when the controlled object structure is a priori known, but in this connection a large number of experiments and calculations for identification process are required.

In its turn, identification technique based on transfer characteristic identifies the objects less accurately by reason of approximation and approximate formulas application. The main shortcoming of that technique is that it does not provide the occurrence of zeros in the controlled object, i. e. the transfer function of such object shall be $W(s)=\dfrac{ke^{\tau s}}{a_n s^n + a_{n-1}s^{n-1}+...+a_0}$. This technique is fairly simple in calculations and it is recommended to apply for the systems of not higher than the second order.

It is known that system transfer characteristic $h(t)$ on unit step function is described by the Heaviside formula:

$$h(t)=\frac{G(0)}{H(0)}+\sum_{k=1}^{n}\frac{G(s_k)}{s_k H'(s_k)}e^{s_k t}, \tag{3}$$

where $G(s)$, $H(s)$ are a nominator and a denominator of the controlled object, s_k is a k-pole of the object.

However (3) can be presented as:

$$h(t)=\frac{G(0)}{H_0(0)}+\frac{\prod_{i=1}^{n}|s_i|}{\prod_{j=1}^{m}|N_j|}\sum_{k=1}^{n}\frac{\prod_{j=1}^{m}(s_k-N_j)}{s_k\prod_{i=1,i\neq k}^{n}(s_k-s_i)}e^{s_k t}, \tag{4}$$

where s_k and N_j are the poles and zeros of the object.

Relationship (4) connects the pole-zero configuration of the object with transfer characteristic curve $h(t)$.

Thus, the problem solution of the pole-zero configuration, corresponding to the given curve of transfer characteristic $h(t)$ is of great interest.

2. Problem statement

Let us assume that unit step impact X is given to the input of the controlled object. In the output, the reaction of the object to this impact is received, fig. 3.

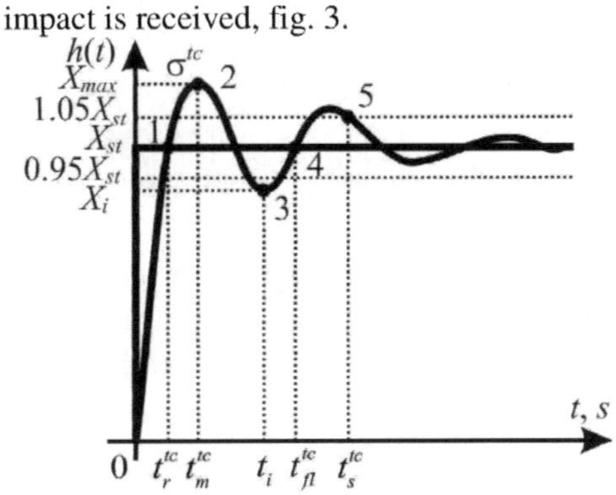

Figure 3. The controlled object reaction to the unit step impact.

It is necessary to identify the transfer characteristic curve according to a probable pole-zero configuration of the system on the root flat.

3. The solution of the identification problem

It is known that the system dynamics is influenced both by all system poles and zeros [11]. However, a degree of impact of one or another pole and zero is different. The type of transfer function depends on the system pole-zero configuration: steady, aperiodic, vibrational.

The first step in identification is the choice of the controlled object structure. The designer has to define a number of poles and zeros. The transfer function object order can be predetermined by an iterative method sophisticating it until we arrive at the satisfactory result. Taking into account the form of the transfer function curve, it is possible to reduce calculations. For example, the object with the transfer function curve, fig. 3, as vibration has minimum two complex conjugate poles, we can exclude versions with one or two real poles, as a steady-state process will take place. Then, if the obtained results of identification with the given structure turn out to be unsatisfactory, the transfer function object order increases, poles and/or zeros are added.

For the given object, the field equation is derived:

$$H(s,s_1,s_2...s_n)=(s-s_1)(s-s_2)...(s-s_n), \tag{5}$$

where s is a Laplace operator; s_i are the object poles. The zeros form equation $G(s)=k(s-N_1)(s-N_2)\ldots(s-N_m)$, where k is a constant factor of the transfer function, N_j are predetermined zeros.

Those equation poles (5) which influence the summation curve are called dominant. The information on the choice and predetermination of dominant poles is given in [11].

Let us assume that for the transfer function curve, fig. 3, the dominant poles will be $s_{1,2}=\delta_1\pm\omega_1 j$. Then (5) is as follows:

$$H(s,\delta_1,\omega_1,s_3...s_n)=(s-(\delta_1+\omega_1 j))(s-(\delta_1-\omega_1 j))(s-s_3)...(s-s_n), \tag{6}$$

Then, on the transfer function curve we define main direct indexes of quality: overregulation σ^{tc}; overregulation time t_s^{tc}; characteristic time t_r^{tc}.

For overregulation the following relationship is true:

$$\sigma^{tc}=\frac{\xi}{X_{st}}, \tag{7}$$

where overshoot $\xi=X_{max}-X_{st}$; X_{st} is a steady state value; X_{max} is maximum.

In its turn, for the vibration system overshoot, ξ is defined in accordance with [8] by the expression:

$$\xi=2A_1\frac{\omega_1}{\omega_0}\exp\left[\frac{\delta_1}{\omega_1}\left(\pi+\sum_3^n\varphi_k-\sum_1^m\Phi_j\right)\right]+\sum_3^m B_k e^{s_k t_m}, \tag{8}$$

where $W(s)=\dfrac{G(s)}{H(s)}$; $s_{1,2}=\delta_1\pm\omega_1 j$ are dominant poles; φ_k and $\Phi_j\in[-\pi;\pi]$ are angles formed by dominant pole s_1, other system pole or zero and positive direction of an X-axis of the root flat; $\omega_0=\sqrt{\delta_1^2+\omega_1^2}$; $A_1=\left|\dfrac{G(s_1)}{s_1H'(s_1)}\right|$; $B_k=\left|\dfrac{G(s_k)}{s_kH'(s_k)}\right|$; $t_m=\dfrac{1}{\omega_1}\left(\pi+\sum_3^n\varphi_k-\sum_1^m\Phi_j\right)$.

Having substituted expression (8) into (7), we shall obtain:

$$\sigma^{tc}=\frac{2A_1\frac{\omega_1}{\omega_0}\exp\left[\frac{\delta_1}{\omega_1}\left(\pi+\sum_3^n\varphi_k-\sum_1^m\Phi_j\right)\right]+\sum_3^m B_k e^{s_k t_m}}{X_{st}}. \tag{9}$$

According to the cosine theorem, the angle formed by dominant pole $s_1=\delta_1+\omega_1 j$ of the root flat, one of the zeros N_j and positive direction of the X-axis is defined by formula [12]:

$$\Phi_j = \operatorname{sign}\left(\delta_1 - \operatorname{Re}(N_j)\right)\cdot \arccos\frac{l^2\left(s_1, N_j\right) + l^2\left(N_j, \operatorname{Im} N_j\right) - l^2\left(s_1, \operatorname{Im} N_j\right)}{2l\left(s_1, N_j\right)\cdot l\left(N_j, \operatorname{Im} N_j\right)}, \tag{10}$$

where l is a distance between the points on the root flat. For example, $l(a,b)$ is a distance from point a to point b.

In the same way, for the angle formed by point $s_1=\delta_1+\omega_1 j$, one of non-dominant poles of the system s_k and positive direction of X-axis of the root flat:

$$\varphi_k = \operatorname{sign}(\delta_1 - \operatorname{Re}(s_k))\cdot \arccos\frac{l^2(s_1, s_k) + l^2(s_k, \operatorname{Im} s_k) - l^2(s_1, \operatorname{Im} s_k)}{2l(s_1, s_k)\cdot l(s_k, \operatorname{Im} s_k)}. \tag{11}$$

Having substituted (10) and (11) into (9), we obtain:

$$f\left(\delta_1, \omega_1, s_3 ... s_n, N_1 ... N_m, k\right) = \sigma^{tc}. \tag{12}$$

Relationship (12) connects overregulation σ^{tc} estimated on the transfer function curve, with the predetermined by designer poles $s_1, s_2 ... s_n$, zeros $N_1, N_2 ... N_m$ and constant factor k of transfer function of the controlled object.

The Heaviside formula presented in (4) allows setting actual values of output signal at the particular period of time according to values of the controlled object zeros and poles.

Let us set peak time X_{max} by variable t_m^{tc}. Overregulation is calculated from point 2 (fig. 3) of the transfer function curve, where the output signal reaches value X_{max}, so with regard to (4) we obtain:

$$h(t_m^{tc}, \delta_1, \omega_1, s_3 ... s_n, N_1 ... N_m, k) = X_{max}. \tag{13}$$

On the other hand, the point of the transfer function curve is the extreme point, thus

$$\frac{dh(t_m^{tc}, \delta_1, \omega_1, s_3 ... s_n, N_1 ... N_m, k)}{dt} = 0. \tag{14}$$

In a similar way to (13), at time period t_s^{tc} the Heaviside formula will be as follows:

$$h(t_s^{tc}, \delta_1, \omega_1, s_3 ... s_n, N_1 ... N_m, k) = X_{st} \pm 0.05 X_{st}. \tag{15}$$

For time period $t = t_r^{tc}$:

$$h(t_r^{tc}, \delta_1, \omega_1, s_3 ... s_n, N_1 ... N_m, k) = X_{st}. \tag{16}$$

In its turn, it is obvious that for the constant factor of the transfer function of the controlled object it is true:

$$\left|\frac{k\prod_{i=1}^{m} N_i}{\left(\delta_1^2 + \omega_1^2\right)\prod_{j=1}^{n} s_j}\right| = X_{st}. \tag{17}$$

For the controlled object, the complementary equations can be predetermined, which define species of predetermined zeros and poles (real or complex conjugate). Let us assume that the designer predetermines a couple of complex conjugate zeros, then

$$\begin{cases} \operatorname{Re} N_k = \operatorname{Re} N_{k+1}, \\ \operatorname{Im} N_k = -\operatorname{Im} N_{k+1}, \end{cases} \tag{18}$$

and for the real zero or pole it is true:

$$\begin{cases} \operatorname{Im} N_j = 0, \\ \operatorname{Im} s_i = 0. \end{cases} \tag{19}$$

In accordance with (4), when the output signal of the controlled object reaches values X_i at time periods t_i (for example, point 3 in fig. 3), the following relationship takes place:

$$h(t_i, \delta_1, \omega_1, s_3 ... s_n, N_1 ... N_m) = X_i. \tag{20}$$

Let us consider the flex point 4 of the transfer function curve in fig. 3. It is known that the second derived function at the flex point of the curve is equal to zero, so if we assume that at point 4 $t=t_{fl}^{tc}$, we obtain

$$\frac{d^2h(t_{fl}^{tc},\delta_1,\omega_1,s_3...s_n,N_1...N_m)}{dt^2}=0. \tag{21}$$

Let us connect (12)–(21) into the set of equations:

$$\begin{cases} f\left(\delta_1,\omega_1,s_3...s_n,N_1...N_m,k\right)=\sigma^{tc}, \\ h(t_m^{tc},\delta_1,\omega_1,s_3...s_n,N_1...N_m,k)=X_{max}, \\ \dfrac{dh(t_m^{tc},\delta_1,\omega_1,s_3...s_n,N_1...N_m,k)}{dt}=0, \\ h(t_s^{tc},\delta_1,\omega_1,s_3...s_n,N_1...N_m,k)=X_{st}\pm 0.05X_{st}, \\ h(t_r^{tc},\delta_1,\omega_1,s_3...s_n,N_1...N_m,k)=X_{st}, \\ \left|\dfrac{k\prod\limits_{i=1}^{m}N_i}{\left(\delta_1^2+\omega_1^2\right)\prod\limits_{j=1}^{n}s_j}\right|=X_{st}, \\ \operatorname{Re}N_k=\operatorname{Re}N_{k+1}, \\ \operatorname{Im}N_k=-\operatorname{Im}N_{k+1}, \\ \operatorname{Im}N_j=0, \\ \operatorname{Im}s_i=0, \\ h(t_i,\delta_1,\omega_1,s_3...s_n,N_1...N_m)=X_i, \\ \dfrac{d^2h(t_{fl}^{tc},\delta_1,\omega_1,s_3...s_n,N_1...N_m)}{dt^2}=0. \end{cases} \tag{22}$$

The set of equations (22) connects predetermined by the designer zeros, poles and the constant factor of the controlled object transfer function with direct indexes of quality: overregulation, characteristic time, setting time, and with values X_{max} and X_{st} defined on the transfer function curve obtained while signaling to the object input in a stepped configuration.

Solution (22) allows defining zeros $N_1...N_m$, poles $s_{1,2}=\delta_1\pm\omega_1 j$, $s_3...s_n$ and the constant factor of object transfer function k relating to defined direct indexes of quality on the transfer function curve.

By reason of complexity and the nonlinear nature of the set of equations (22), the solution is performed numerically.

The equations (18) and (19), the sets (22) allow finding a solution with predetermined zeros and poles, thus reducing calculations.

In its turn, relations (20) and (21) are not basic and are added to system (22) by the designer who wants to improve identification accuracy under the given transfer function object order. The number of given relations as in (20) and (21) is defined by the designer.

4. Accuracy evaluation of identification

After identification, it is very important to evaluate the error of the obtained results. On calculated poles $s_{1,2}=\delta_1\pm\omega_1 j$, $s_3...s_n$ and zeros $N_1...N_m$, the transfer characteristic graph is constructed. To evaluate accuracy, it is supposed to discretize the predetermined and the obtained on roots transfer characteristics at time intervals Δt. Let us set values of the given transfer function at time periods in $h^{tc}(t_i)$, and calculated – $h^{cc}(t_i)$. Then at each time step t_i error is estimated. Absolute or ratio error, root-mean-square error, etc. can be taken as an assessment criterion.

Let us take a ratio error as the assessment identification criterion. So,

$$\delta_i=\frac{\left|h^{tc}\left(t_i\right)-h^{cc}\left(t_i\right)\right|}{h^{cc}\left(t_i\right)}100\ \%.$$

In its turn, identification results are considered to be satisfactory under condition of

$$\max\left(\delta_i\right) \le \delta_{req},$$

where $\max(\delta_i)$ is a maximum ratio error at time period t_i, δ_{req} is the given ratio identification error.

5. The example

The transfer function curve of the controlled object is given, fig. 4a. Find its transfer function.

As there is oscillation transient, at first, the object structure is given as $W(s) = \dfrac{k}{\left(s-(\delta_1+\omega_1 j)\right)\left(s-(\delta_1-\omega_1 j)\right)}$. For the given object structure, the set of equations (22) was produced containing relationships which connect overregulation, characteristic and setting time with the predetermined poles and constant k, the derivative at the maximum point and equations setting the species of poles. However, identification with such structure gave unsatisfactory results, fig. 4b, thus the object order increased. Satisfactory results were obtained with the transfer function of the fifth-order object with zeros $N_{1,2}$=–9.412±0.944j, N_3=–9.81, poles $s_{1,2}$=–0.281±2.414j, $s_{3,4}$=–2.167±0.337j, s_5=–17.194, and constant factor k=3.34. The maximum ratio error reached 4.88%.

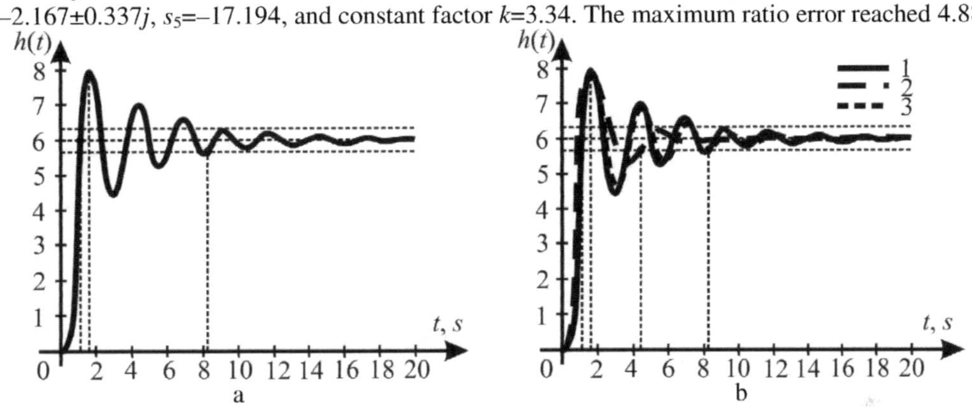

Figure 4. Step response: a – basic identification data; b – the identification results: 1) the given one, 2) the object of the 2nd order, 3) the object of the 5th order.

6. Conclusion

The analysis of classical identification techniques such as frequency identification and with the help of transfer function has been carried out.

The equations connecting quality indexes of transient with the values of zeros, poles and the constant factor of the controlled object transfer function have been obtained.

The object identification technique on the basis of transfer characteristics has been presented.

References

[1] Konovalov V I 2010 *Identifikacija i diagnostika system* [*Identification and diagnostics of systems*] (Tomsk: TPU) p 156

[2] Ljung L and Glad T 1994 *Modeling of dynamic systems* (Englewood Cliffs, N.J.: Prentice Hall) p 368

[3] Chen L, Li J and Ding R 2011 Identification for the second-order systems based on the step response *Mathematical and computer modelling* **53 (5–6)** 1074–1083

[4] Semenov A D, Artomonov D V and Brjuhachev A V 2003 *Identifikacija ob'ektov upravlenija* [*Control object identification*] (Penza: PSU) p 211

[5] Goodwin G C and Payne R L 1977 *Dynamic system identification: experiment design and data analysis* vol 136 (New York: Academic Press) p 291

[6] Mohamed A 2010 *Modelling, simulation and identification* (Sciyo) p 354

[7] Figueroa J L, Biagiola S L, and Agamennoni O E 2008 An approach for identification of uncertain Wiener systems *Mathematical and Computer Modelling* **48 (1–2)** 305–315

[8] Liu X G and Lu J 2010 Least squares based iterative identification for a class of multirate systems *Automatica* **46 (3)** 549–554

[9] Efimov S V, Kurgankin V V and Zamyatin S V 2014 Designing transfer function with the required direct performance measures based on the Laplace transform *Optoelectronics, Instrumentation and Data Processing* **50 (4)** 348–353

[10] Voevoda A A, Troshina G V 2014 Active identification of linear stationary dynamic objects on base of the Fisher information matrix: The steady state *Proc. 12th Int. Conf. on Actual Problems of Electronic Instrument Engineering, APEIE-2014* 7040785 84-86

[11] Efimov S V and Pushkarev M I 2011 Determining direct measures of performance based on the location of zeros and pole of the transfer function *Optoelectronics, Instrumentation and Data Processing* **47 (3)** 297–302

[12] Uderman E G 1972 *Metod kornevogo godografa v teorii avtomaticheskih sistem* [*Root-locus method in automatic system theory*] (Moscow: Science) p 448

Comparison of different spatial mathematical models of the extruder screw channel

S V Ershov[1], **N M Trufanova**[1]

[1] Perm National Research Polytechnic University, 29, Komsomolsky Ave., Perm, 614990, Russia

E-mail: Ershov_sv@bk.ru

Abstract. One of the important problems in the processing of polymeric materials by an extrusion method is the occurrence of local overheating. The aim of this work is to identify these areas and finding ways to reduce their influence on the finished product. The paper discusses the heat and mass transfer processes, which occur in the extruder screw channel. In the work a comparison of the results, obtained by various mathematical models, was made: a flat channel model (full-scale), and the screw channel models with regard to the gap between the screw crest and the inner surface of the body and without it. The investigation of the possibility of using the inverse movement principle in the study of the problems, associated with the cable insulating on the extrusion equipment was made. As a result of research, temperature fields at the outlet of the extruder channel were obtained.

1. Introduction

To date, extrusion is the most common method of cable insulation. Local overheating resulting from the processing can negatively affect the quality of the insulation. As the processable polymer has a high viscosity, friction between particles of the material reaches a large value, whereby a polymer is heated up to high temperatures that can cause thermal degradation of the polymer [1]. The aim of this work was to study flow and heat transfer processes in the extruder screw channel in the production of plastic insulation and comparison of different geometric models.

Currently, there are a number of mathematical models describing the flow and heat transfer in extruders screw channels [2-5]. All known approaches, based on the laws of conservation, use the principle of inverse movement and turns the screw channel on a plane, that is an assumption.

A schematic illustration of the extruder is represented in figure 1. It consists of the screw, body, heating elements and a forming device. The extruder is divided into four zones: the feeding zone, a delay melting zone, a melting and metering zone. In this paper, a study was conducted with the last 5 turns of the extruder, because this is where the greatest overheating takes place.

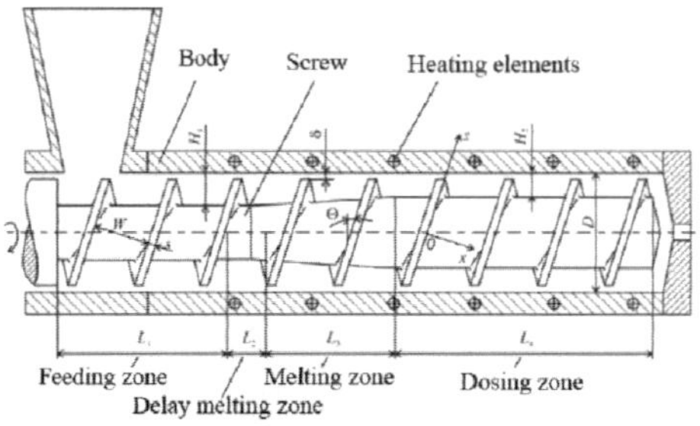

Figure 1. A schematic illustration of the extruder.

2. Mathematical statement and initial data

The mathematical representation of the motion and heat transfer processes of polymers in a screw channel of the plasticizing extruder is based on the laws of conservation of mass, momentum and energy [4,6]:

$$\nabla \cdot \mathbf{V} = 0; \tag{1}$$

$$\rho(\mathbf{V}\cdot\nabla)\mathbf{V} = -\nabla P + (\nabla\cdot\tilde{\tau}); \tag{2}$$

$$\rho C(T)(\mathbf{V}\cdot\nabla)T = \nabla\cdot\lambda\nabla T + (\tilde{\tau}:\nabla\mathbf{V}), \tag{3}$$

where ρ and C – the density and specific heat capacity of the polymer; λ – thermal conductivity; V – velocity vector; T – temperature; P – hydrostatic pressure; $\tilde{\tau}$ – stress deviator tensor; $(\tilde{\tau}:\nabla V)$ – An irreversible increase in internal energy per unit volume due to the energy dissipation in the viscous flow.

The connection between the deviator stress tensor and the strain rate tensor is given by:

$$\tilde{\tau} = \mu_э \tilde{\dot{\gamma}}, \tag{4}$$

where $\tilde{\dot{\gamma}}$ – deformation rate tensor; $\mu_э$ – effective viscosity which is a function of shear rate and temperature[1,7,6].

The dependence of the effective viscosity on the shear rate of the investigated polymer is determined by the power law [6]:

$$\mu_э = \mu_0 \dot{\gamma}^{\frac{n-1}{2}} \tag{5}$$

Viscosity dependence on temperature was determined by the Reynolds equation:

$$\mu_0 = \mu_{0|_{T_0}} \exp(-\beta(T - T_0)) \tag{6}$$

To study the processes in the extruder channel, models presented in Figure 2 were built.

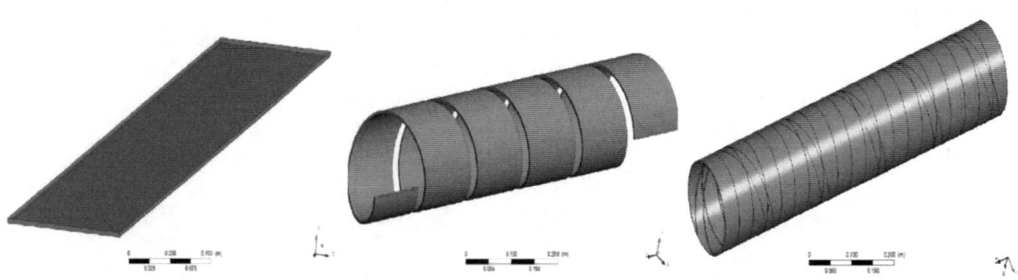

a b c

Figure 2. A geometric model of the investigated screw: a – Flat model; b – Cylindrical model without a gap; c – Cylindrical model with a gap.

IOP Conf. Series: Journal of Physics: Conf. Series **803** (2017) 012037 doi:10.1088/1742-6596/803/1/012037

The geometry of the channel is presented in Table 1. The body temperature was set at 200 °C, the temperature of the screw is 210°C, the temperature of the input material is 225°C . Calculations were carried out for the screw rotation speed of 60 revolutions / min.

Table 1. The geometry of the channel

Parameters	Value
Inner diameter of a barrel, mm	160,0
Outer diameter of a screw, mm	159,4
Twist, mm	160,0
Channel width, mm	137,3
Flight Wight, mm	15,3
Length of the dosing geometric zone, turn	5
Channel depth in the metering zone, mm	4
Helix angle, °	17°39′
Radial clearance between the flight and the barrel, mm	0,3

The properties of the material are presented in Table 2. The values of coefficients for rheological models have been determined experimentally on rotating rheometer DHR-2.

Table 2. Polymer properties

Material	Density, ρ (kg/m^3)	Specific heat capacity, C $(J/kg{\cdot}K)$	Thermal conductivity, λ $(W/m{\cdot}K)$	μ_0, $(Pa{\cdot}s)$	B, $(1/K)$	n	T_0, (K)
PE 153-02K	779	2500	0,182	23336.9	0.0124	0.34	433

3. Results and Discussion

As a result of the study, temperature fields at the outlet of the extruder were obtained. Since processable polymer has a sufficiently high viscosity, the friction between the particles of the material heats it to a high temperature. Figure 3 shows the temperature field for the flat channel model using the principle of inverse movement and without it. For visualization of the obtained results, there was a 3-time increase of the field.

a

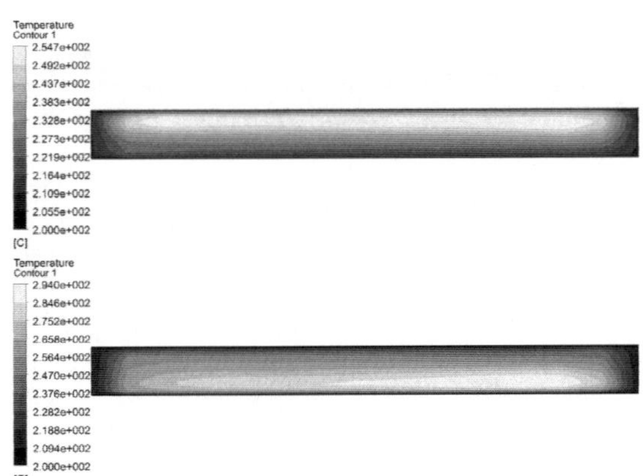

b

Figure 3. Temperature fields for flat channel models: a – using the principle of reversed motion; b – without using the principle of reversed motion.

When using the principles of inversed motion, the maximum shear rate occurs near the movable wall. Therefore, in investigation of the heat and mass transfer processes in the channel, using this principle, the character of the flow is changing. The maximum temperature is displaced to the movable walls.

Figure 4 shows the temperature field for three different spatial models: the flat channel model, the cylindrical channel model with a gap between the crest of the screw and the inner surface of the body and without it.

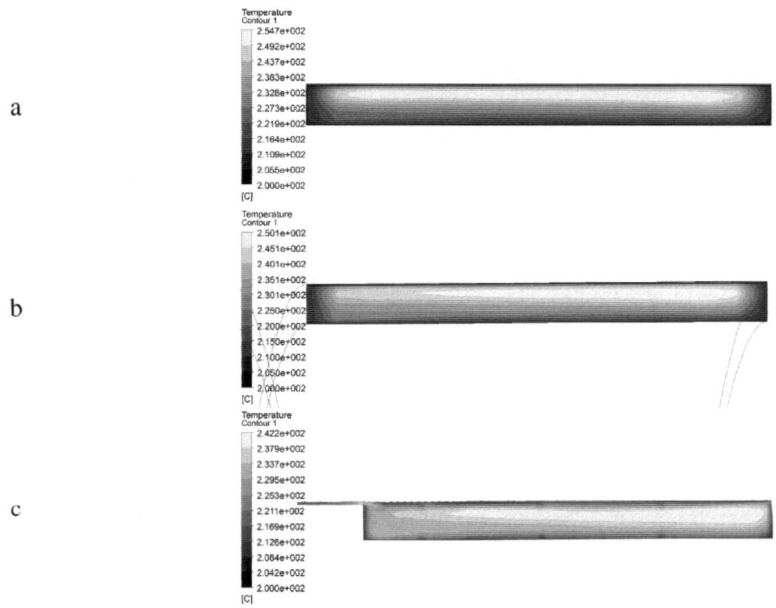

Figure 4. Changing of the temperature field for the inverse motion using the power law, at the outlet of the channel: a – flat channel; b – helical channel without a gap; c – helical channel with a gap.

In the transition from the flat model to the actual screw channel with a gap, the character of the polymer flow is changing, so that the maximum and average values of temperature decrease.

4. Conclusion

The largest difference in the maximum temperature between the flat model and the helical channel with a gap amounts to 5.1%, and in the average temperature - to 1.45%. As we approach the geometry of the real form (from the flat to the screw channel with a gap), the maximum value of the temperature decreases.

Changing the temperature field, the viscosity and shear rate fields along the channel in all three geometrical models, when using the inverted motion, has the same pattern for investigated rheological law. Even though, the power law operates in a limited range of shear rates.

In all cases of the movable screw model (unconverted movement), the average temperature is higher than in case of inverted motion. This is caused by the changing character of the polymer flow, in which the maximum value of shear rates is displaced closer to the moving surface.

References

[1] Trufanova N M, Shcherbinin A G and Yankov V I 2009 *Melting polymers in extruders* (Moscow- Izhevsk: SIC " Regular and Chaotic Dynamics ", Institute of Computer Science) p

336
[2] Rauwendaal C 2008 *Extrusion of polymers*.(St. Petersburg: Profession) pp 295-353
[3] Rauwendaal C 2004 Finite element studies of flow and temperature evolution in single screw extruders *Plastics, Rubbers and Composites* **33 (9/10)** 390–369
[4] Torner R V, Gudkov A F and Nikolaev I K 1965 Rectilinear–parallel steady motion of abnormally viscous fluid between two parallel walls *Mechanics of polymers* **6** 138–145
[5] Manab Kumar Das, Ghoshdastidar P S 2002 Experimental validation of a quasi three-dimensional conjugate heat transfer model for the metering section of a single-screw plasticating extruder *Journal of Materials Processing Technology* **120 (1–3)** p 397
[6] Tadmor Z, Gogos C 1984 *Theoretical Foundations of polymer processing*. (Moscow: Chemistry) pp 96-159
[7] Malkin A Y, Isaev A I 2007 *Rheology: concepts, methods, and applications*. (St. Petersburg: Profession) p 184

The architecture of the management system of complex steganographic information

O O Evsutin[1], R V Meshcheryakov[1], A S Kozlova[2], T M Solovyev[1]

[1] Tomsk State University of Control Systems and Radioelectronics, 40, Lenina Ave., Tomsk, 634050, Russia
[2] Tomsk State University, 36, Lenina Ave., Tomsk, 634050, Russia

E-mail: eoo@keva.tusur.ru

Abstract. The aim of the study is to create a wide area information system that allows one to control processes of generation, embedding, extraction, and detection of steganographic information. In this paper, the following problems are considered: the definition of the system scope and the development of its architecture. For creation of algorithmic maintenance of the system, classic methods of steganography are used to embed information. Methods of mathematical statistics and computational intelligence are used to identify the embedded information. The main result of the paper is the development of the architecture of the management system of complex steganographic information. The suggested architecture utilizes cloud technology in order to provide service using the web-service via the Internet. It is meant to provide streams of multimedia data processing that are streams with many sources of different types. The information system, built in accordance with the proposed architecture, will be used in the following areas: hidden transfer of documents protected by medical secrecy in telemedicine systems; copyright protection of online content in public networks; prevention of information leakage caused by insiders.

1. Introduction

At present time, the rapid increase of multimedia traffic proportion in data networks is caused by the high popularity of different kinds of audio and video interactive streaming applications that are used by the Internet users for video conferences, training, remote medical services, television broadcasting, etc.

So, there is a need for solving problems such as the prevention of confidential information leakage, assurance of impossibility of unauthorized multimedia content changes, copyright protection for multimedia content, medical confidentiality protection (for medical applications).

One of the most effective approaches for solving such problems is based on the steganography methods (which are the practice of transferring files, messages, images, or video within another file along with the concealing of the transmission fact) by inserting the hidden embeddings to the protected content for different purposes [1].

Steganography methods are classified as follows:

- Methods of hidden information embedding into digital objects to ensure its confidentiality during transmission and/or storage [2–5].

- Methods of digital watermarks embedded into the multimedia content to protect information from unauthorized copying and use [6–8]. Here, by the digital watermark we mean an invisible tag embedded in digital objects for confirming the authenticity of the protected objects as well as a conventional watermark.
- Methods of identification numbers embedded into the multimedia content to protect it from unauthorized copying and replication [9, 10]. Identification numbers can be considered as a kind of digital watermarks. The difference is that digital watermarks are not unique while an identification number is unique for each copy of the software or multimedia content. These methods often do not separate from each other.
- Methods of headers embedding. These methods are similar to the methods of the first type; they are not intended to provide privacy, but designed for hidden addition of auxiliary public information into the multimedia content.

Thus, a wide range of problems related to information security and digital content processing can be effectively solved by using steganographic methods.

The analysis of the US Patent Office Database [11] showed that the number of patented technical solutions relating to steganographic devices and methods is annually increasing. In particular, about 150 patents relating to the subject area were issued in 2014. World's leading IT companies such as Microsoft Corporation, Google Inc., Sony Corporation, etc. are engaged in such developments.

However, the analysis of the collected patent documents published between 2000 and 2015 and the literature review allowed us to establish the following fact: the existing solutions for multimedia data protection using digital steganographic methods are intended for private use and for the solution of individual information security problems and often designed for work with the content of a specific type (i.e. they do not have the complexity and flexibility).

2. Application of steganographic methods for scientific and practical problems

2.1. The main applications of steganography methods

The following principal scientific-practical problems, for which a management system of complex integrated steganographic information can be applied, are considered:

- copyright protection for video content, including materials of video conferencing and video-reportage recordings;
- video authenticity protection in surveillance systems;
- medical data protection;
- information leakage detection and prevention.

The ability of watching the same video on any number of mobile or stationary devices leads to the emergence of its multi-format multiple copies with various levels of quality. This different video content representation allows unauthorized copying of video, which leads to copyright infringement and inability to register a significant number of views legally. One way to confront the unauthorized video replication and transcoding in the film industry is the use of formats such as "Ultra Panavisio", "IMAX" and others. The materials of video conferencing and video-reportage recordings are special cases of video content.

At present time, the most common applications that use video conferencing for business purposes are the following: online meetings and meetings between employees, webinars, online interviews in recruitment, marketing activities and online press conferences, presentation of services and products, video consultations, telemedicine. One of the major trends of modern market development is a consequence of the cloud computing spread, transferring the entire infrastructure to service providers and the use of videoconferencing as a service. The users of videoconferencing software application, including a large number of corporations and government organizations, are increasingly demanding copyright protection and confidentiality.

Steganography technology provides the ability of embedding the additional information about video, facts about its views and transcoding facts for manufacturers and rightholders of video conferencing recordings.

The processing and protection of the surveillance system data have its own specificity. Traditionally, video surveillance systems are used for the security of the protected site perimeter, but now there is a tendency to expand the functionality of these systems.

The intelligent video surveillance system allows accumulating data for making decisions, for example, the data about customers: their numbers, shop routs, etc. The video combined with the technological process data increases the quality control level at an assembly line.

Thereby, video surveillance data become a source of additional information for the effective production and business organization.

In this case, steganography can provide authenticity protection in the video surveillance system with automatic embedding of digital watermarks (DVRs, cameras, external and internal monitoring, etc.).

Telemedicine is another specific and rapidly developing area where the multimedia content presence is caused by the use of different systems of monitoring of a personal health status, registration and transfer of the data produced by embedded and carried medical devices, by the use of medical video conferencing and local or cloud medical information storage.

Such kind of multimedia content a priori has to be protected in respect of medical confidentiality and protection of personal information in the long run, often for life.

Today, the video recording of clinical and outpatient study and medical operations is introduced in many medical organizations. Normalization, unification of data, using specially designed formats, recording of medical attributes, data authentication (including the place and time) – these are the factors relevant today for the entire world medical community. So, in this case the problems of information protection and video content protection against modification occupy leading positions. Steganography applications are an effective instrument providing personalization and a possibility to proof tampering of original data. It should be noted that the steganographic methods are used not only by legitimate users and information owners to ensure its protection or to solve some auxiliary problems. Hackers also can use them for the organization of confidential information leakage through insider channels.

A significant part of the insider incidents is carried out with the aid of steganographic techniques when confidential information is taken away from company by means of the hidden embedding to multimedia with innocuous content. Steganalysis of such content type in the presence of the hidden embedded data embedding allows one to identify and prevent such incidents.

2.2. Existing systems for steganographic data protection and steganalysis

There are many systems for embedding hidden information into digital images, audio and video data. Here are the most well-known: Mp3Stego (for embedding information into MPEG-1(2) Layer 3 audio); ImageSpyer (for embedding data into images); S-Tools (allows one to place small files into the GIF and BMP images or WAV audio); MSU Stego Video (for hiding data in AVI files).

Complex systems of network traffic steganalysing in real-time are MBEGA – an optimized complex steganalysis method for searching stego images in the flow of network traffic, which was proposed by S. Geetha and N. Kamaraj (India) [12], and the StegAlyzerRTS real-time steganography scanner developed by SARC (Steganography Analysis and Research Center, Backbone Security.Com, Fairmont, USA) [13].

The algorithm implemented in the MBEGA system includes many different techniques for automated result obtaining. Each of these techniques is looking for the set of statistical image characteristics (about 140 characteristics: the study of the LSB correlation of statistical properties, cross-correlation of structure properties, Fridrich's features, higher-order statistical features, etc.) and compares them to average network traffic patterns or statistics in an image database. The accelerated modification of the genetic algorithm is used for the search optimization.

StegAlyzerRTS is the world's first commercially available network traffic analysis tool for detecting of hidden embeddings in real time. The latest version of this application is able to operate in networks with the capacity of 1Gbit/s.

The StegAlyzerRTS algorithm is based on the comparison of potential stego container features with the file fingerprints from the database which contains multiple file profiles associated with 1225 steganografic and watermark applications. In addition, the algorithm allows retrieving hidden information from suspicious files and does not affect network performance.

Thus, all existing solutions are aimed either at embedding hidden data into files or at network traffic steganalysis.

According to the results of the literature review and a patent-governmental research, it can be stated that at the moment, there is no complex software solution for steganographic information management, which includes both the embedding/extracting of hidden information and the traffic steganalysis in local and global networks.

3. Description of the architecture of the distributed steganography management system

To solve the problems that we examined above, we offer to use a distributed management system of steganography information.

The system has a modular architecture and is oriented to solve a wide range of steganography problems including real-time processing. Apart from the features of existing embedding/extracting and analysing algorithms, during the system design, we paid much attention to scalability and the possibility of using it in distributed computing. It is connected to the huge multimedia data volumes to be processed and the high level of computing complexity of the existing steganography algorithms.

The scheme of modules interaction of the management system is presented in figure 1.

The media content transmit/receive module was designed for exchanging data between different multimedia sources. In particular, the ability to work in local networks, in the Internet, and also using different analog and digital television receivers is assumed. The components of the module which are serving for data transmitting/receiving could be divided logically or physically.

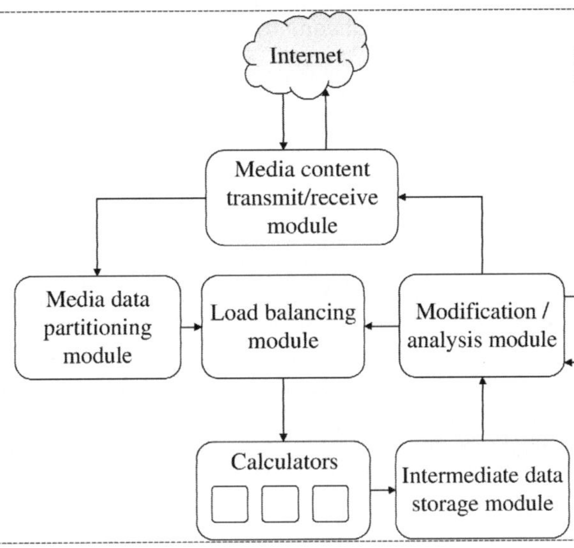

Figure 1. The architecture of the proposed system.

The processing of the resulting content consists of two main steps: stream fragmentation into structural elements and modified data analysing. Structural elements imply the logically divided parts of a data stream which are specific for the specific media content type and its representative format.

Usually, media data are sent in the compressed mode. Consequently, the choice of structural elements is defined by the syntax feature of the compression standard. The structural elements for digital images could represent a palette, a block of pixels of a specific size, and some of specified pixels. In case of video content, it could be frames, slices, macroblocks, motion vectors, etc. Audio content could be fragmented into channels, frames, samples [14]. Apart from the media data format, the type of fragmentation into structure elements is defined by algorithms of modification/analysis implemented in the system. Within the developed system, the problem of the structural element selection of the analysing content is solved by the module of media data fragmentation.

The module that implements the management of stenographical information is the module of modification/analysis. The module implements a set of algorithms of embedding/extraction of stenographical data to work with trusted containers, and also detection algorithms of hidden attachments in untrusted containers. It should be noted that the described module works not directly with the streams of media data, but with its structural elements that were selected at the stage of fragmentation.

The algorithm of steganographic embedding detection may use previously formed special steganographic signatures during operation. The steganographic signature means the complex of characteristics (properties) or patterns (templates) associated with the media containers containing hidden embeddings. It is proposed to store these signatures in a specialized database for the efficient operation. Such kinds of databases are filled during the system creation and its operation in an automatic or semi-automatic way.

The result of the modification/analysis module (depending on the objectives) is either a multimedia container with/without extracted hidden embeddings or a report about presence/absence of hidden embeddings in the suspicious container.

In the first case, the data are sent to the transmit/receive module of media content. In other cases, the results are written in the dedicated reporting database. The report access can be obtained either from a local network or from the Internet through a web server interacting with the reporting database after the authentication and authorization procedures.

The algorithms implemented by the modification/analysis module and the media data partitioning module can be characterized by great computational complexity, and as a consequence, the proposed system architecture is focused on the use of distributed computing. Organization and control of computing resources are performed by the load balancing module. This module receives the tasks from the data partitioning module and the modification/analysis module in the form of four objects (S, A, D, M). S – the set of the input data identifiers, by which the input data can be received requesting either the intermediate data storage module or the media content transmit/receive module; A – the algorithm identifier to be executed; D – a set of output data identifiers, M – metadata including information about the algorithm complexity and the requirements to the type and speed of calculation.

Then, the load balancing module decides to delegate the task to the specific calculator depending on the system load, types of available calculators and the task. After the task is completed, the output data are sent either to the intermediate data storage module or directly to the module that set this task.

The intermediate data storage module provides optimal information distribution over the available devices in terms of access speed. At the same time, particular qualities of algorithms are taken into consideration for using this information in subsequent stages of system operation.

The proposed system can be deployed on a single server or on a cluster. Moreover, the system architecture is focused on cloud technologies as one of the most effective ways of scaling. The distributed character and the possibility of using heterogeneous calculators allow solving a wide class of steganography problems in real time.

4. Conclusion

Within the limits of the given paper, the literature review and patent research were conducted in the area of steganography methods and information protection. It was established that the currently existing scientific and technical decisions are intended for private use, aimed at solving only a limited range of information security problems and do not possess flexibility and scalability.

Therefore, the development and implementation of the information system which allows embedding hidden information into multimedia for different purposes without reference to the data type and data stream analysing for the presence of unauthorized embeddings is relevant. In this paper, the architecture of such system is introduced for the use in real time by means of the distributed computing. System algorithmization as regards to embedding of hidden information into multimedia content and extracting information has to be built on the basis of classical methods, their combinations and newly proposed approaches to stego container construction, which is based on multimedia data information sequences. In turn, the algorithms of determining the embeddings in multimedia content must be hybrid and combine mathematical statistics and computational intelligence.

5. Acknowledgments

This work was supported by the Ministry of Education and Science of the Russian Federation within the 1.3 federal program «Research and development in priority areas of scientific-technological complex of Russia for 2014-2020» (grant agreement № 14.577.21.0153 of November 28, 2014; identifier RFMEFI57714X0153).

References

[1] Fridrich J 2010 *Steganography in digital media: principles, algorithms, and applications* (Cambridge: Cambridge University Press)

[2] Shanableh T 2012 Data hiding in MPEG video files using multivariate regression and flexible macroblock ordering *IEEE T. Inf. Foren. Sec.* **7** 455–464

[3] Kanan H R and Nazeri B 2014 A novel image steganography scheme with high embedding capacity and tunable visual image quality based on a genetic algorithm *Expert Syst. Appl.* **41** 6123–6130

[4] Evsutin O O, Kokurina A S, Shelupanov A A and Shepelev I I 2015 An improved algorithm for data hiding in compressed digital images based on PM1 method *Computer Optics* **39** 572–581

[5] Evsutin O O, Kokurina A S, Mescheryakov R V and Shumskaya O O 2016 An adaptive algorithm for the steganographic embedding information into the discrete Fourier transform phase spectrum *Advances in Intelligent Systems and Computing* **451** ed A Abraham, S Kovalev, V Tarassov and V Snášel (Springer International Publishing) 47–56

[6] Glumov N I and Mitekin V A 2011 A new semi-fragile watermarking algorithm for image authentication and information hiding *Computer Optics* **35** 262–267

[7] Lusson F, Bailey K, Leeney M and Curran K 2013 A novel approach to digital watermarking, exploiting colour spaces *Signal Process.* **93** 1268–1294

[8] Ali M, Ahn C W and Siarry P 2014 Differential evolution algorithm for the selection of optimal scaling factors in image watermarking *Eng. Appl. Artif. Intel.* **31** 15–26

[9] Li Y N 2015 Robust content fingerprinting algorithm based on sparse coding *IEEE Signal Proc. Let.* **22** 1254–1258

[10] Thanh T M and Iwakiri M 2016 Fragile watermarking with permutation code for content-leakage in digital rights management system *Multimedia Systems* **22** 603–615

[11] United States patent and trademark office. URL: http://www.uspto.gov/

[12] Geetha S and Kamaraj N 2010 Optimized image steganalysis through feature selection using MBEGA *Int. J. of Computer Networks & Communications* **2** 161–175

[13] SARC – Steganography Analysis and Research Center: Home. URL: http://www.sarc-wv.com/

[14] Salomon D 2007 *Data compression: the complete reference* (London: Springer–Verlag)

IOP Conf. Series: Journal of Physics: Conf. Series **803** (2017) 012039 doi:10.1088/1742-6596/803/1/012039

Music Tune Restoration Based on a Mother Wavelet Construction

A S Fadeev, V I Konovalov, T I Butakova, A V Sobetsky

Tomsk Polytechnic University, 30, Lenina Ave., Tomsk, 634050, Russia

E-mail: fas@tpu.ru

Abstract. It is offered to use the mother wavelet function obtained from the local part of an analyzed music signal. Requirements for the constructed function are proposed and the implementation technique and its properties are described. The suggested approach allows construction of mother wavelet families with specified identifying properties. Consequently, this makes possible to identify the basic signal variations of complex music signals including local time-frequency characteristics of the basic one.

1. Introduction

Discrete and continuous wavelet transforms are becoming an indispensable part of modern mathematics and other human activities. Furthermore, owing to the growth of computing technologies in recent years, hardware and software contributed to solution of many mathematical problems connected with pattern recognition. So, modern high-performance computing allows solving a number of problems including speech recognition, graphical objects, processing of seismic data, cardiograms, etc. A lot of them require using wavelet transforms. However, some problems in musical pattern recognition and their implementation in the automated information systems have been insufficiently studied [**Ошибка! Источник ссылки не найден.**, 10]. One of such problems is a problem of identifying a separate note in one-voice and polyphonic melodies performed on certain musical instruments.

2. Music signal model

A simple mathematical model of any music melody consists of a set of notes played in different times on a certain musical instrument [**Ошибка! Источник ссылки не найден.**]:

$$F(t)=A_1n_1(t-\theta_1)+A_2n_2(t-\theta_2)+\ldots+A_Nn_N(t-\theta_N)+h(t),$$

where $n_i(t)$ is the amplitude-time characteristics of a single note voice; θ_i is the temporal shift determining initial time of each note sounding; A_i is the sound volume of the separate note; $h(t)$ is the signal of disturbance introduced by sound-recording equipment; t is time.

The majority of musical instruments possesses the property of self-similarity which allows obtaining the temporal function of any note $n_i(t)$ from the same musical instrument from the temporal function of one note $n_0(t)$ of a certain musical instrument by scaling function $n_0(t)$ along time axis:

$$n_i(t)=n_0\left(\frac{t}{m_i}\right),$$

where m_i is the scale factor, i is the position of note $n_i(t)$ by height relative to note $n_0(t)$. For uniformly tempered pitch of European music, m_i is represented as $m_i = 2^{\frac{-i}{12}}$, $i = 0, \pm 1, \pm 2...$ [**Ошибка! Источник ссылки не найден.**].

For example, note «C» of the second octave is situated 12 semitones higher than note «C» of the first octave, and has a pitch frequency two times higher than the pitch frequency of note «C» of the first octave [**Ошибка! Источник ссылки не найден.**]. The time function of note «C» of second octave $n_2(t)$ is two times compressed relatively function $n_1(t)$:

$$n_2(t) = n_1\left(\frac{t}{2^{\frac{-12}{12}}}\right) = n_1(2t).$$

Any note in the range of a certain instrument may be selected as base note $n_0(t)$ regardless of being used in concrete musical signal $F(t)$. For example, the temporal function of note «A» of the first octave with pitch frequency υ =440 Hz may be taken as $n_0(t)$.

This property is implemented in all modern musical synthesizers using the method of wave table [**Ошибка! Источник ссылки не найден.**]. Such synthesizers use the data bank of voices of one (basic) note for each musical instrument. When forming a tune of one instrument, the basic note signal is always scaled by a pitch by value m_i. Each note is shifted in time by value θ_i and scaled by amplitude by value A_i according to this instrument part [**Ошибка! Источник ссылки не найден.**]. All formed functions for musical instrument are summarized. As a result, the model of a musical signal formed by a synthesizer for a definite musical instrument has the following form:

$$F(t) = \sum_i A_i \cdot n_0\left(\frac{t-\theta_i}{m_i}\right).$$

Similarly, for k-different musical instruments simulated by the synthesizer, a musical signal may be represented by a sum of signals of all notes played at different moments of time with different amplitude:

$$f(t) = \sum_k \sum_i A_i \cdot n_0^k\left(\frac{t-\theta_i}{m_i}\right),$$

where $n_0^k(t)$ is the time-dependent function of a basic note signal for a k-musical instrument; θ_i is the time interval shift of note $n_0^k(t)$; m_i is the scale of note $n_i^k(t)$ in regard to basic note $n_0^k(t)$ specifying frequency of the main tone; A_i is the magnitude of note $n_i^k(t)$.

The above-mentioned function $f(t)$ represents an idealized model of a musical signal obtained similarly to recording several musical instruments in an orchestra or an ensemble [**Ошибка! Источник ссылки не найден.**].

According to the model of musical signal $f(t)$, the task of musical signal identification may be presented as the task of identification of note amplitudes with certain scale m_i and time shift θ_i for all k musical instruments presented in analyzed signal $f(t)$. The task of identification of a singly recorded musical instrument melody or a melody generated by a musical synthesizer may be considered as a particular case of this task [**Ошибка! Источник ссылки не найден.**].

3. Continuous wavelet transform

Selecting the mathematical apparatus for time-frequency analysis of signal $f(t)$ which is nonstationary and nonperiodic one, it was determined to apply a continuous wavelet transform (CWT):

$$Wf(\tau, s) = \int_{-\infty}^{\infty} \frac{1}{\sqrt{s}} f(t) w\left(\frac{t-\tau}{s}\right) dt,$$

where $w(t)$ is the mother wavelet function; s is the coefficient of wavelet scaling; τ is the coefficient of wavelet shift.

IOP Conf. Series: Journal of Physics: Conf. Series **803** (2017) 012039 doi:10.1088/1742-6596/803/1/012039

One of the features of CWT is the formation of wavelet family $w_{s,\tau}(t)$ by shifts τ and scaling s of mother wavelet $w(t)$. The wavelet family formation is similar to the system of formation of note family $n_i^k(t)$ of one basic note $n_0^k(t)$ by shifts θ_i and scaling m_i. Therefore, CWT usage in this task is rather restricted.

The procedure of selecting a mother wavelet function is empirical for each definite task and reduced to searching functions of mother wavelets in CWT until the achievement of a desired result. The research of wavelet function properties [**Ошибка! Источник ссылки не найден.**] showed that the best graphic presentations of CWT results are obtained in case of conformity of frequency spectra of signal $f(t)$ and wavelet $w(t)$.

For each scale s, function $Wf_s(\tau)$ is similar to the cross-correlation function of signals $w_s(t)$ and $f(t)$, and describes them both as the similarity measure of two signal form and their positional relationship to each other on the time axis.

It is known that values of the cross-correlation function are maximal in case of function coincidence [**Ошибка! Источник ссылки не найден.**]. In this case, values $Wf_s(\tau)$ are maximal for such τ as functions $f(t)$ and $w_s(t-\tau)$ are equal: $f(t)=w_s(t-\tau)$. It is obvious that besides shift τ, the equality of two functions in each point t is required for fulfillment of this condition.

The wavelet function used in CWT should meet a number of required conditions [**Ошибка! Источник ссылки не найден.**]:

1. limitation (localization) in time: $w(t)>0$, at $t\to\infty$;
2. sectional continuity of function $w(t)$;
3. integrability with zero equality.

The example of the mother wavelet-function satisfying all given conditions is Morlet wavelet (Figure 1).

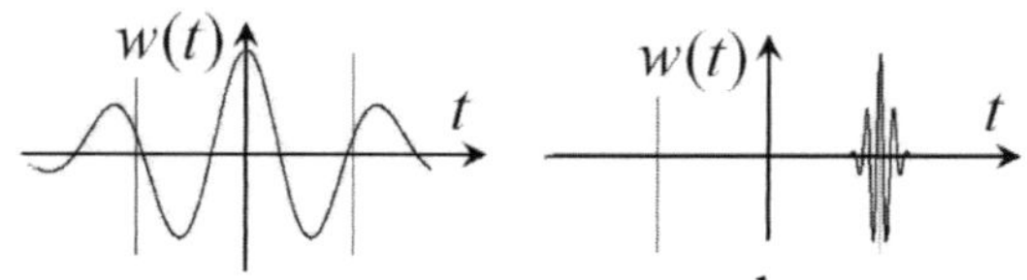

Figure 1. Wavelets of Morlet family: a) a mother Morlet wavelet obtained by scaling s=0,2; b) a mother Morlet wavelet obtained by a shift by τ =5 and scaling s=2.

4. Mother wavelet construction

Owing to the conditions of time localization imposed to the basic wavelet function, function $Wf_s(\tau)$ reaches a maximum value when wavelet $w_s(t-\tau)$ of scale s coincides more precisely with the local section of signal $f(t)$ [**Ошибка! Источник ссылки не найден.**]. Functions of a wavelet and a signal should be equal for exact coincidence at the local time interval.

We suggested using the function of the mother wavelet formed of the local section of the analyzed musical signal in the given paper. For a more restricted task, identification of musical instrument notes, the mother wavelet may be formed of the local section of the basic note signal function of this musical instrument $n_0(t)$ (Figure 2).

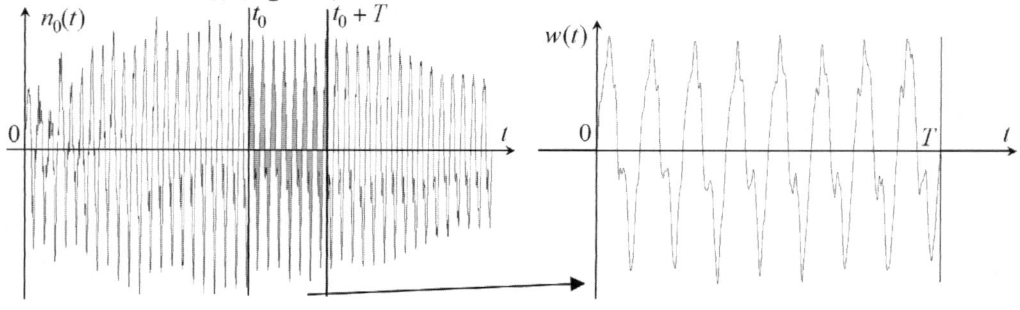

Figure 2. Function $n_0(t)$ and wavelet $w(t)$ formed on the basic function fragment.

For the formed wavelet, conditions 1, 2, 3 should be fulfilled as follows.

1. Limitation in time

$w(t)=0,\ t \notin [0,T]$;

$w(t)= n_0(t+t_0)$, at $t \in [0,T]$.

Here, t_0 characterizes time moment from which the values of wavelet function $w(t)$ are equal to values of the function of signal $n_0(t)$, and value T equals to the fragment duration of signal $n_0(t)$, coinciding with wavelet $w(t)$. Values t_0 and T are selected to satisfy the rest conditions required for wavelet functions [Ошибка! Источник ссылки не найден.].

2. Sectional continuity

The function of basic note $n_0(t)$ is continuous on the whole interval of existence as it describes oscillations of a physical body with finite mass in time and cannot have any breaks.

To support sectional continuity of the basic wavelet function, the condition of zero equality of initial and finite values of function $n_0(t)$ on interval $[t_0, T+t_0]$ should be fulfilled:

$n_0(t_0)=0$ and $n_0(T+t_0)=0$.

3. Integrability with zero equality

One of the properties of musical instruments is absence of harmonic components with frequency lower than frequency of the note pitch. Zero harmonic is absent in musical instrument signals as well [Ошибка! Источник ссылки не найден.]. This property allows supporting zero mean for $n_0(t)$ on interval $[t_0, T+t_0]$ at the integer number of periods in function $n_0(t)$ on a given interval [**Ошибка! Источник ссылки не найден.**].

Therefore, to form the wavelet possessing the highest selectivity to signal $n_0(t)$, it is necessary to use a periodic section of signal $n_0(t)$ with zero initial $n_0(t_0)$ and zero end $n_0(T+t_0)$ moments such as

$$\int_{t_0}^{T+t_0} n_0(t)dt = 0 .$$

Identification of musical notes comprises the determination of frequencies of notes pitch themselves, start time, and duration of their sounding. One of the properties of wavelet function, determining the main time-frequency selectivity of CWT, is wavelet localization in time and frequency regions [Ошибка! Источник ссылки не найден.]. During CWT, each wavelet of one family, obtained from one basic wavelet, forms a time-frequency window of the limited size. The window area for one family wavelet on the plane shift-scale is always constant [Ошибка! Источник ссылки не найден.] (Figure 3).

After a number of experiments, it was found out that changing the wavelet itself makes window geometric characteristics change [**Ошибка! Источник ссылки не найден.**]. So, increasing a number of periods in the basic wavelet (and, therefore, in all wavelets of the family), the window is extended along the time axis, narrowing relative to the scale axis and, on the contrary, decreasing a number of periods in the basic wavelet, the window is extended along scale axis s, narrowing along the time axis (Figure 4).

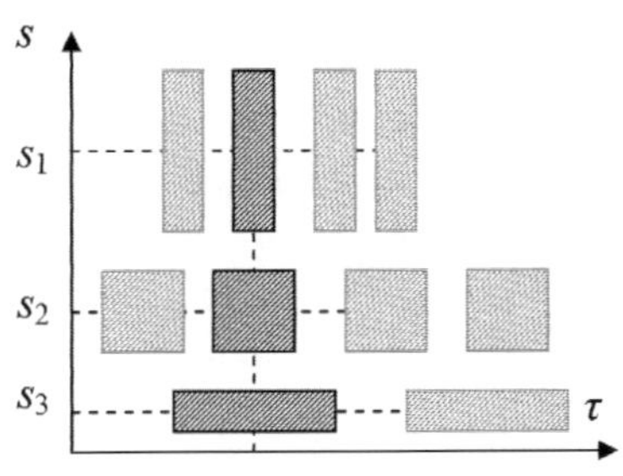

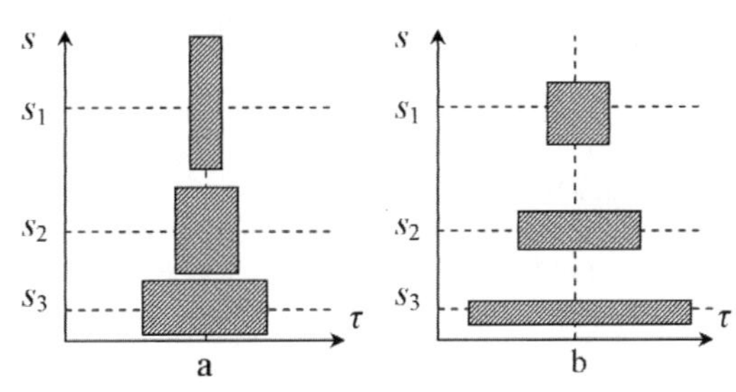

IOP Conf. Series: Journal of Physics: Conf. Series **803** (2017) 012039 doi:10.1088/1742-6596/803/1/012039

Figure 3. The window of wavelet time-frequency localization at different values of parameters of shift τ and scale s.

Figure 4. The window of time-frequency localization of wavelets with a number of periods equal to a) N=2 and b) N=8.

For the tasks of elementary components (notes) detection in a musical signal, two conditions should be followed for the wavelet family:

1. frequency resolution (the window height along scale axis Δs) should identify different frequencies of two adjacent notes;
2. time resolution (the window width along time axis $\Delta\tau$) should allow identifying all notes of minimum possible duration.

1. Frequency resolution

The experiment was carried out to estimate the resolution of artificial mother wavelets. The aim of the experiment was to determine a number of periods in the wavelet that allows identifying frequency scales m_i of all notes being in the signal simultaneously (at CWT wavelet scales s relative to the basic one are equivalent to m_i).

Chord «C major» of the first octave was used in the experiment. Frequencies of note pitches of this chord corresponds to harmonic signals with frequencies 261,6; 329,6; and 392 Hz [**Ошибка! Источник ссылки не найден.**]. The signal duration is chosen to be equal to 0,2 s:

$$f(t) = \sin(2\pi 261{,}6t) + \sin(2\pi 329{,}6t) + \sin(2\pi 392t).$$

Mother wavelets $fw_i(t)$ were constructed from the harmonic signal to study the test signal. The amount of harmonic signal periods in wavelets was 1, 2, 4, 8, and 16 periods, respectively (Figure 5).

CWT was carried out with all wavelet families $w_i(t)$ for test signal $f(t)$. For each transform, the results of graphical interpretations were interpreted in three-dimensional models [**Ошибка! Источник ссылки не найден.**]. 3-D models of CWT results for the families of wavelets $w_i(t)$ are presented in Figure 6. The ordinate axis of each illustration represents an axis of the wavelet s scale, the abscissa axis is the axis of time shift τ; τ_0, τ_1 – time of signal $f(t)$ beginning and ending. The magnitude of CWT results is presented by the gray scale where darker sections correspond to a higher magnitude of CWT results [**Ошибка! Источник ссылки не найден.**] (Figure 6).

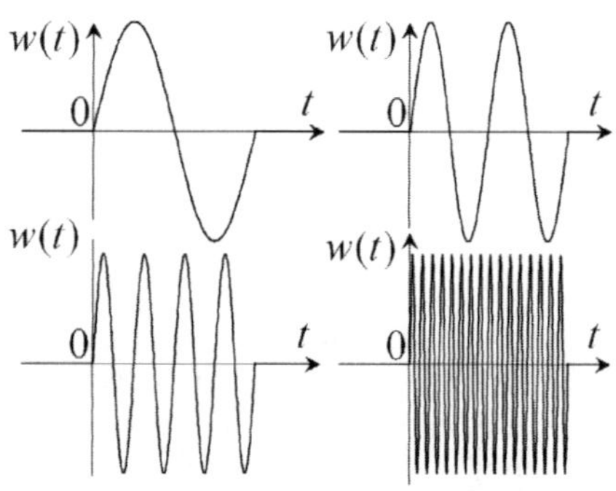

Figure 5. Mother wavelets with a number of periods of harmonic signal 1, 2, 4, and 16.

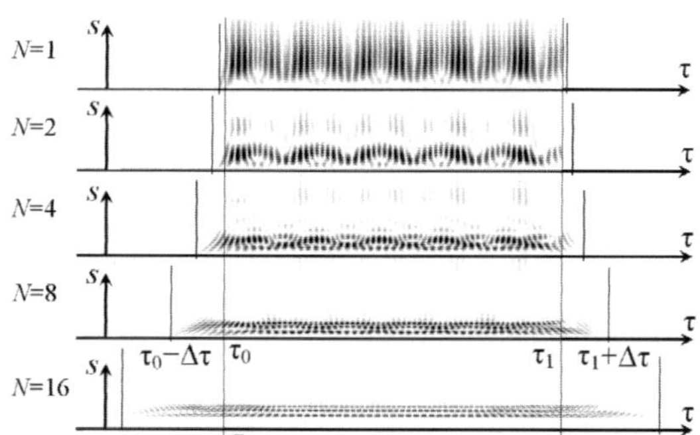

Figure 6. Graphical interpretations of CWT results for wavelets with different numbers of periods N of the harmonic signal.

Figure 6 shows that when the wavelet was used with one period, a region of uncertain results in the region of time shifts τ, with duration $\Delta\tau$ is rather low; therefore, the error of the signal time estimate is low. In this case, frequency resolution is so low that for the majority of scales s, CWT results have the same values indicating the presence of frequency components in the whole frequency range of research. However, there are three localized harmonics in test signal $f(t)$.

It is seen that for the wavelet with 16 periods over signal $f(t)$, on the interval from τ_0 to τ_1, there are high values only for CWT results for three scales s equivalent to frequencies 261,6; 329,6; and 392 Hz of signal $f(t)$. For the rest values of s, CWT results are almost equal to zero. Low time resolution produced uncertain results at the beginning and ending moments of the signal with magnitude intermediate values on intervals with duration $\Delta\tau$. It does not allow judging the changes of the test signal magnitude on this interval.

Thus, the wavelet family with one period of the sine signal during CWT gives high time resolution, but rather low frequency resolution. However, the wavelet family with sixteen periods gives high frequency resolution (all harmonics in the signal are identified definitely), but low time resolution.

2. Time resolution

Musical notation implies using notes and the rest symbols to mark out melody elements and gaps at which voices do not sound. The note duration as well as the rest duration are multiplied by duration t_1 of semibreve («whole note») – a note of maximum possible duration. The system of musical notation consists of alternating notes and the rest imposes strict constraints on the moments of the note sounding start, and the rest start. Moments of the note sounding start are sampled with sampling period t_d equal to duration of the shortest note. Both in classical and modern musical compositions, the shortest note by duration is the hemidemisemiquaver note with duration t_{64}=1/64·t_1 [**Ошибка! Источник ссылки не найден.**]. In practice, notes with duration t_{64} occur rather seldom due to the technical complexity of performance. In fact, duration t_{32}=1/32·t_1 (demisemiquaver note) may be considered as the shortest note. The fragment of a two-voice melody is shown in Figure 7. Each time, not more than two notes sound simultaneously. Note 2 is the shortest one. Notes 1, 3, and 4 are equal in duration and two times longer than note 2 and the rest. If t_d=t_{32}, then note 2 and the rest have duration t_{32}, and notes 1, 3, 4 have duration t_{16}.

At wavelet duration T=t_d, an envelope function of joint-correlation function $Wf_s(\tau)$ of the wavelet and the signal of one note coinciding with it by form (for concrete value s) regenerating into the autocorrelation function has the form of an equilateral triangle with maximum in the centre of note sounding [**Ошибка! Источник ссылки не найден.**] and width $2t_d$ (Figure 8, a). If all values of $Wf_s(t)$ are smaller than $Wf_{mist}(t)$, it would be ejected, and the note identification time is t_i. If $Wf_{mist}(t)$=0,5$Wf_{max}(t)$, then t_i=0,5t_d. This means that the note identification time equals to a half of its length.

If the wavelet duration is T=0,5t_d, then the envelope of correlation function $Wf_s(\tau)$ of the wavelet and one note signal, coincided with it in the form, has the form of an isosceles trapezium the width of the upper boundary of which is equal to 0,5t_d (Figure 8, b). If all values $Wf_s(t)$ are smaller than $Wf_{mist}(t)$, they are rejected, and then the note identification time is t_i. If $Wf_{mist}(t)$=0,5$Wf_{\max}(t)$, then t_i=0,5t_d. It means that note identification time equals to its length.

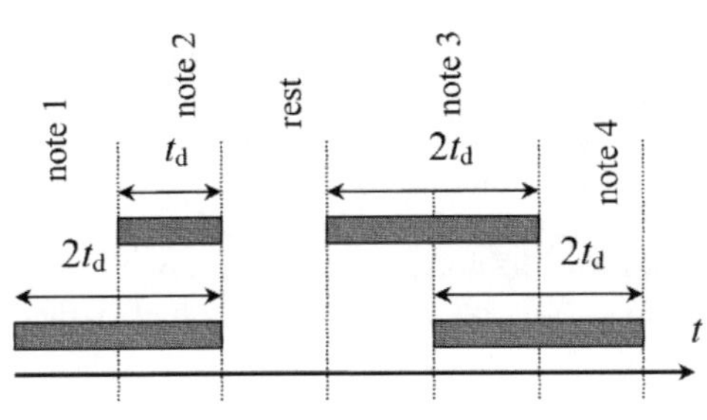

Figure 7. Time discretization of beginning and duration of notes and rests in a polyphonic melody.

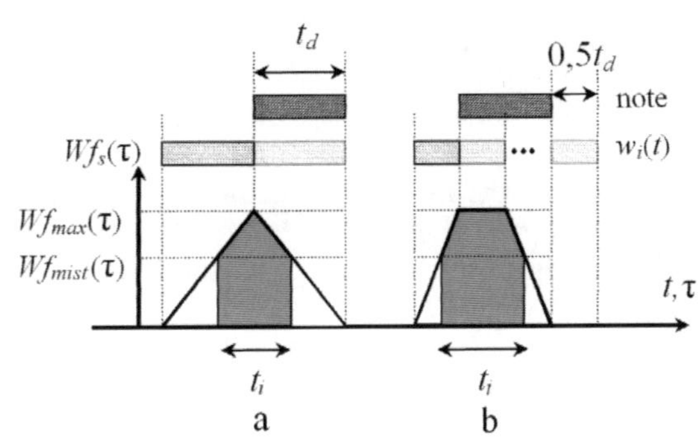

Figure 8. Envelopes of the correlation function of the note and wavelet $w(t)$ corresponding to the frequency of the scale note pitch.

Upon further decrease of wavelet width T and condition $Wf_{mist}(t)=0{,}5Wf_{\max}(t)$, note identification time t_i is constant and equal to t_d.

Conclusion: the wavelet of length T is not more than t_d, and should be used for time identification of the note with smallest length t_d.

Taking into account two conditions to duration T of wavelet in time:

1. $T \geq \dfrac{16}{\upsilon}$, where υ is the frequency of the identified note pitch;
2. $T \leq t_d$.

Let us calculate the boundary values of note pitch frequency υ, the reliable identification of which is possible both in time and frequency: $T = \dfrac{16}{\upsilon}$, $T = t_d$, therefore, $t_d = \dfrac{16}{\upsilon}$ or $\upsilon = \dfrac{16}{t_d}$. Dependence $\upsilon(t_d)$ determines minimal (boundary) frequency υ of the note pitch with duration t_d which is definitely identified by the wavelet with 16 periods.

Tempo of the majority compositions performance varies as a rule in the range of 60–180 beats per minute (BPM) that corresponds to the time of sounding for one semibreve (whole) note t_1=1,3...4,0 s. Therefore, note sounding with duration t_{32} at quick tempo (180 BPM) amounts to t_{32}=1,3/32=0,041 s. Therefore, the length of the wavelet capable of identifying the time interval of a note with the shortest duration at high tempo should amount to not more than $t_d=t_{32}$=0,041 s. The value of boundary frequency of the note is pitch identification equal to $\upsilon(t_d)$=16/0,041=390 Hz. This means that just identification of the beginning and ending time of note sounding is possible only for notes with pitch frequency higher than 390 Hz (Figure 9) (starting with note «G» of the Middle octave, the pitch frequency of which amounts to 392 Hz).

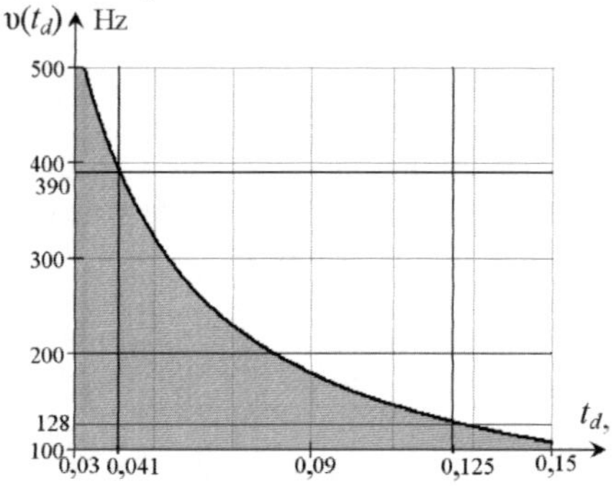

Figure 9. Dependence of boundary frequency of note pitch identification on its duration.

For slower tempo of composition performance or for tasks of detecting the note with the duration longer than t_{32}, the boundary frequency decreases. So, for example, for modern club dancing compositions, the reproducing tempo varies near the value of 120 BPM. And the shortest note is $td=t_{16}$=0,125 s. The value of boundary frequency of note pitch identification is $\upsilon(t_d)$=16/0,125=128 Hz. Identification of sounding time of note beginning and ending is possible only for notes higher than «C» of Bass octave with pitch frequency 130,8 Hz. The majority of musical compositions use notes at the range of the Middle, under Middle and above Middle octaves, which is higher than the range of Bass octave.

5. Conclusion

The suggested approach allows making mother wavelet families with defined selectivity. The wavelet developed from a fragment of a certain basic signal allows detecting the time-frequency behavior of the basic signal under study instead of separate time-frequency characteristics of the studied signal.

Mother wavelets for voices of various orchestra groups such as piano, organ, violin, bells, and trumpet were obtained experimentally. All wavelets contain 16 periods of a signal of a proper musical instrument with pitch frequency 55 Hz. While calculating CWT, the mother wavelet is scaled in such a manner that next wavelet $w_i(t)$ coincides by pitch frequency with note pitch frequency $n_i(t)$ of a musical instrument.

Construction of mother wavelet families, developed on the basic of musical instrument notes, showed the possibilities of detecting frequency and time parameters of certain instrument notes in one-voice and polyphonic tunes. Besides, it made it possible to identify a tune of a certain musical instrument on the background of sounding of another one in a number of experiments [**Ошибка! Источник ссылки не найден.**].

The authors suggested that the implementation of this technique in tasks, requiring detecting the fragments of certain families with the limited length in a signal against the background of other signals or disturbances, is to be researched further.

Acknowledgments
The reported study was funded by the Russian Foundation for Basic Research according to research project No. 16-37-00402 mol_a.

References

[1] Barthet M Kronland-Martinet R and Ystad S 2006 Consistency of Timbre Patterns in Expressive Music Performance Proc. 9th Int. Conf. on Digital Audio Effects DAFx-06 (Montreal, Canada) pp 19–24
[2] Devyatykh D Gerget O and Berestneva O 2014 Sleep Apnea Detection Based on Dynamic Neural Networks *Communications in Computer and Information Science* pp 556-67
[3] Elert G 2016 The Physics Hypertextbook (http://physics.info/music/)
[4] Fadeev A and Kochegurova E 2006 Preparation of the results of continuous wavelet-transform to automated treatment *Bulletin of the Tomsk Polytechnic University* **309** 32–35
[5] Hankinson A Burgoyne J A Vigliensoni G and Fujinaga I 2012 Creating a large-scale searchable digital collection from printed music materials *Proc. Advances in Music Information Research* (Lyon, France) pp 903–8
[6] Holopainen R 2012 Self-organised Sound with Autonomous Instruments: Aesthetics and experiments (University of Oslo) 404
[7] Janelle K Hammond and Kelly S 2011 Mathematics of Music *UW-L Journal of Undergraduate Research* **14** 11
[8] Kochegurova E and Gorokhova E 2015 Current derivative estimation of non-stationary processes based on metrical information *Lecture notes in artificial intelligence* **9330** (Springer) 512–19
[9] Kochegurova E and Fadeev A 2006 Wavelet analysis in the task of musical information identification *Proc. Vseros. Conf. Molodezh I Sovremennye Informatsionnye Tekhnologii* (Tomsk: TPU Press) pp 149–151
[10] 1. Makhijani R Shrawankar U and Thakare V M 2010 Opportunities and challenges in automatic speech recognition *Proc. Int. Conf. Biomedical Engineering and Assistive Technologies* (NIT Jalandhar, India)
[11] Mallat S 1998 *A Wavelet Tour of Signal Processing* (Academic Press Inc) p. 704
[12] McKay C and Fujinaga I 2010 Improving automatic music classification performance by extracting features from different types of data *Proc. ACM Int. Conf. on Multimedia Information Retrieval* 257–66
[13] Phinyomark A Limsakul C and Phukpattaranont P 2011 Application of Wavelet Analysis in EMG Feature Extraction for Pattern Classification *Measurement Sci. Rev.* **11** 45–52
[14] Sadowsky J 1996 Investigation of signal characteristics using the continuous wavelet transforms *Johns Hopkins APL Technical Digest* **17** 258–69

[15] Skirnevskiy I and Korovin A 2016 Optimal methods of segmentation of tomographic volume searching system: a preliminary review *Key Engineering Materials* **685** 857–62
[16] Wright D 2009 Mathematics and Music *Washington. Mathematical World* **28** 176

IOP Conf. Series: Journal of Physics: Conf. Series **803** (2017) 012040 doi:10.1088/1742-6596/803/1/012040

The leak location package for assessment of the time-frequency correlation method for leak location

V A Faerman[1], A G Cheremnov[2], V S Avramchuk[3], and D V Shepetovsky [4]

Tomsk Polytechnic University, 30, Lenina Ave., Tomsk, 634050, Russian Federation

E-mail: [1]vaf@tpu.ru, [2]8xandr@gmail.com, [3]avs@tpu.ru, [4]dshepet@tpu.ru

Abstract. The paper describes the simplest implementation of a software and hardware package for acoustic correlation leak location and results of its performance assessment for location of water leaks from a metallic pipe in laboratory conditions. A distinctive feature of this leak locator is the use of the software based on the time-frequency correlation analysis of signals, which was proposed in our previous papers. Comparative analysis results are given for the information content of classical and time-frequency cross-correlation functions as obtained during processing of experimental data. The results obtained justify comparatively higher efficiency of a time-frequency cross correlation method to solve the leak location task. Improved efficiency is determined by bandpass filtration embedded into the time-frequency cross-correlation function calculation.

1. Introduction

The housing and utilities sector constitutes an important part of any modern national economy. This importance is caused by heat and water supply being basic needs of any modern citizen. Due to an enormous number of customers municipal services rely on a complex infrastructure and employ numerous specialists. In particular, in Russia alone there are about 900 thousand km of underground water pipes currently in operation [1].

Efficient operation of the pipeline infrastructure is a complex and important task. Studies performed in Germany [2], USA [2] and Russia [1] show that loss of water during transportation may well reach 30% of the original volume. There are multiple factors causing this situation, including a high degree of wear in pipelines, practical impossibility of its simultaneous replacement and complexity of routine repair works due to prevalence of underground piping.

Timely detection and location of leaks are a precondition for fast damage control and mitigation. Multiple methods have been developed and are currently in use to perform this task, a prominent place among them is held by an acoustic correlation method, based on acquiring and correlation processing of acoustic emissions caused by leaking liquid.

The efficiency of this method when applied to location of leaks is largely determined by efficiency of digital signal processing algorithms used and their software implementations [3]. Thus, mathematical methods and algorithms implemented in the software have a significant influence on the accuracy and fidelity of leak location.

To assess effectiveness and efficiency of both hardware and software of correlation leak locators, there are several approaches, including processing of software-generated signals [3], use of the software and hardware virtual pipe [4], lab-based and field testing [5].

This paper considers the simplest implementation of the software and hardware leak seeker package developed to assess and study time-frequency correlation analysis of signals as applied to the leak location task. The principal goal of the laboratory research is to confirm the efficiency of the method, as previously detailed in a computational experiment [3]. Another objective of the work is to validate feasibility of a fully-functional software solution for leak locators implementing this method to find leak signals.

2. An acoustic correlation method of leak location

An acoustic correlation method of location was developed in the second half of the 20th century to solve the task of on-the-spot location of breaches in pipe walls to perform timely repairs [6, 7]. The method is based upon receipt of acoustic waves emitted when the liquid under pressure escapes the pipe by means of two acoustic transducers and the subsequent estimation of delay in the signals arriving through different measuring channels [7].

2.1. Principle of leak location

Principal sources of acoustic signals are turbulent fluctuations which accompany fluid outflow and cavitation (formation and collapse of gas bubbles) in leak localities due to a high local reduction of liquid pressure [8]. It worth noting that despite the fact that signals generated by the leak are weak with respect to the noise created by the continuous flow of liquid, correlation analysis makes it possible to extract information about location of the leak even when receivers are at a distance of hundreds of meters [9].

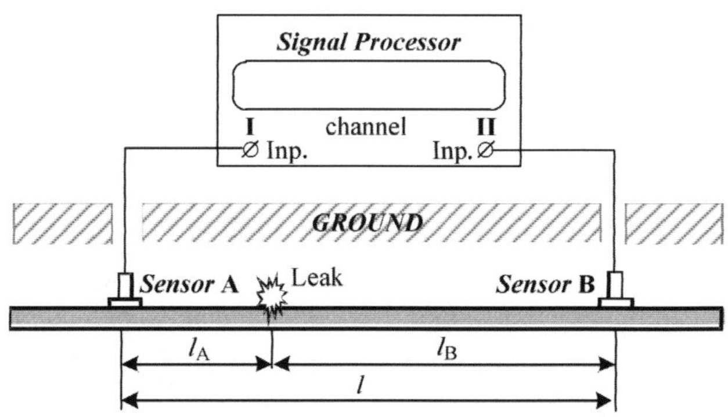

Figure 1. Leak location with an acoustic correlation method.

A pipeline section with equipment necessary for leak location with the acoustic correlation method is shown in Figure 1. The distance from the leak to any of the sensors (l_A and l_B) is found with formula

$$l_{A,B} = \frac{l \pm \tau_{leak} \cdot v}{2}, \quad (1)$$

where τ_{leak} - delay time of the leak acoustic signal, v - signal propagation speed along the pipeline axis, l – distance between the sensors.

2.2. Signal processing for leak location

Accuracy of leak location as per (1), depends on accuracy of τ_{leak} estimation and accuracy of v evaluation. It worth noting that for both τ_{leak} estimation [3,7,10], and v evaluation [11], there are different methods available. However, from a methodical point of view, τ_{leak} estimation is of more interest because the value of v is usually held as reproducible [4]. Modern leak location packages

IOP Conf. Series: Journal of Physics: Conf. Series **803** (2017) 012040 doi:10.1088/1742-6596/803/1/012040

reduce τ_{leak} estimation to a digital signal processing problem which is then solved by a specialized software.

Correlation analysis is a traditional mathematical tool for this task [7, 8]. In general, the analysis is calculation of the cross correlation functions (CCFs) of the studied signals and finding its peak value corresponding to the target value of delay.

At that, CCF calculation may be performed for the time domain [12], as well as for the frequency domain, as per Correlation Theorem [12, pp. 267-270]. However, the second way finds more practical application due to providing better speed when used with modern microprocessor equipment [12]. Thus, CCF computation is usually performed as per

$$r(\tau_j) = F^{-1}\left[F\left(y(t_i)\right) \cdot F^{*}\left(x(t_i)\right)\right], \tag{2}$$

where F is direct discrete Fourier transform (DFT) of digitized signal $x(t_i)$, F^* is a complex conjugate of direct DFT results $y(t_i)$, F^{-1} - inverse DFT.

3. A software and hardware package of the leak locator

To assess the software developed around time-frequency correlation analysis (FTCA) [6], a prototype leak locator was built. It is characterized by simple design, low cost and flexibility for experimenting. The latter is achieved by means of separation of data acquisition and data processing functionality and their implementation as separate units as per diagrams, shown in Figures 2 and 3.

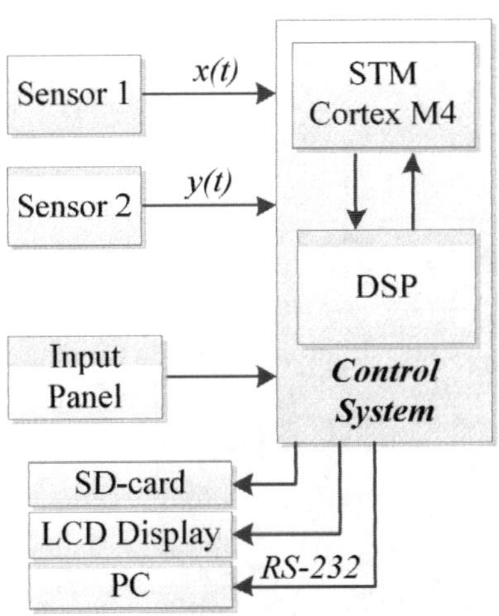

Figure 2. The functional diagram of the Data Acquisition Unit.

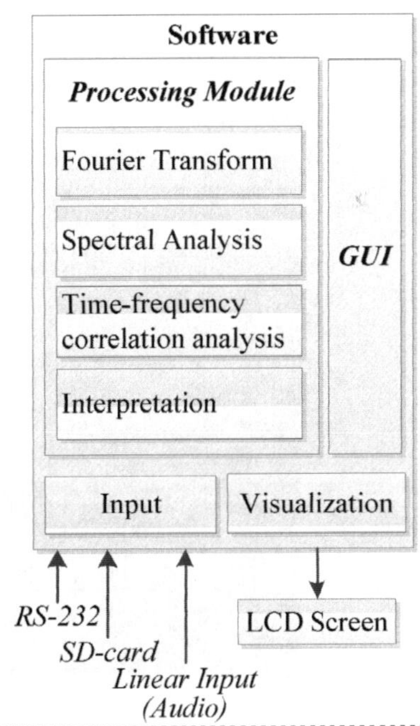

Figure 3. The functional diagram of the software of the Processing Unit.

As shown in the diagram above, a STM32 Cortex M4 microcontroller-based portable device provided with piezoelectric accelerometers DN-3 is used to acquire the signals. Layout of the portable device is shown in Figure 4.

Figure 4. The data Acquisition Unit (without sensors).

Data processing is provided by an external Processing Unit with the custom software preinstalled. The software is not limited to calculation of time-frequency CCFs, but allows one to use classical methods of spectral and correlation analysis. Necessary interface and visualization were designed and implemented in the software as well. This work used a Lenovo Yoga Tablet 2 tablet PC as the Processing Unit (shown in Figure 5).

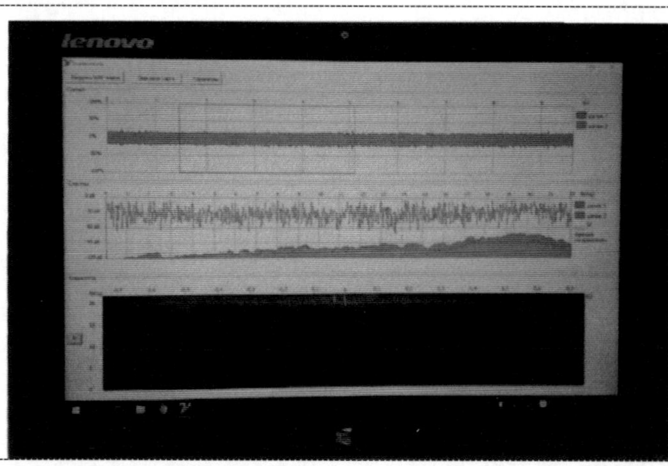

Figure 5. Lenovo Yoga Tablet 2 with the preinstalled software.

4. Leak locator assessment

4.1. Experiment description

A series of experiments to locate leaks on a 18 m steel pipe with 30 mm diameter and 2 mm wall thickness was performed with the prototype described above. Three holes were made into the pipe (2, 4 and 6 meters from one of the pipe ends) to model leaks. In each experiment, two holes out of three were covered with tight clamps completely stopping the fluid outflow. The fluid pressure was created with a water pump.

Recording of acoustic signals coming from sensors was performed with the device described above. Waveform Audio File Format was selected to store the signals and pass them to processing. Thus, raw data at the processing stage were represented by *.wav* files with the following parameters: channels – 2, bit depth – 8, sampling rate – 44100 per second.

4.2. Information content assessment method

As noted above, a sufficient condition for leak location would be discernibility of the correlation function peak which arises due to noises generated by the outflowing liquid. To assess information content of the CCF, i.e., discernibility of its peak, a quality coefficient may be used [13]

$$Q_{CCF} = \frac{N \cdot r(\tau_{leak})}{\left(\sum_{j=0}^{N-1} r(\tau_j)^2 \right)^{0.5}} \quad (3)$$

where N is number of CCF tic marks, $r(\tau_{leak})$ is the value corresponding to the leak signal. It is worth noting that coefficient $Q_{TF\ CCF}$ similar to (3) may be used for assessment of information content of the time-frequency CCF as well, as calculated with the algorithm given in [6].

4.3. Results and Discussion

Signal analysis used 5 seconds of audio recordings. The DFT window width was constant at 16384 tics. As per algorithm in [6], calculation of time-frequency CCF used the different number of non-overlapping frequency bands M: 20 (which corresponds to the width of a single frequency band of 1102 Hz) and 40 (width of 551 Hz). Results of the information content assessment are given in Table 1.

Table 1. Comparison of the information content in CCF and time-frequency CCF

Leak Position (m)	Q_{CCF}	$Q_{TF\ CCF}$ (M=20)	$Q_{TF\ CCF}$ (M=40)
2, 14	18.154	43.333	52.446
4, 12	12.133	24.576	27.078
6, 10	16.104	30.987	35.872
7, 9	16.322	34.336	42.756

From the results given in the table, we may draw a conclusion that filtration embedded into the time-frequency CCF algorithm provides better discernibility of the leak-caused correlation peak.

It is necessary to note that this information content is also influenced by the width of frequency bands, the highest information content corresponds to the case where the frequency band of the leak signal completely coincide with one of the time-frequency CCF frequency bands. The last fact is of special importance for development of formal criteria and algorithms aimed at automated detection and location of leaks.

5. Conclusion

The work describes implementation of the simplest software and hardware implementation for the acoustic correlation leak location and its application to the experimental location of pipeline leaks in laboratory conditions.

Distinctive features of the designed device are its simplicity and wide availability of components, as well as flexibility of the assembly necessary for laboratory studies of efficiency of algorithmic and software solutions used in the leak locator.

The leak locator was used for assessment of the previously proposed software solution for analysis of acoustic signals of the leaks based upon application of the original mathematical method of time-frequency CCF. The obtained results support applicability of the software and demonstrate high efficiency of the tested software in leak location tasks.

Advantages provided by time-frequency CA are deemed significant for location of leaks with the help of acoustic control means. Thus, the development of the software with functionality beyond a set of common signal analysis methods in frequency and time domains and including time-frequency correlation analysis appear an important task.

Acknowledgement

The reported study was funded by the Russian Foundation for Basic Research according to research project No. 16-37-00049 мол_а.

References

[1] Primin O G and Khramenkov S V 2012 The problems and ways of reducing water loses *Vodosnabzhenie i sanitarnaya tekhnika* **11** 31-37
[2] Martini A, Troncossi M and Rivola A 2015 Automatic leak detection in buried plastic pipes of water supply networks by means of vibration measurements *Shock Vib.* 165304
[3] Avramchuk V S and Goncharov V I 2013 Time-frequency correlation method for improving the accuracy in detecting leaks in pipelines *Adv. Mat. Res.* **650** 443-446
[4] Brennan M J, Kroll de Lima F, de Almeida F C L, Joseph P F and Paschoalini A T 2016 A virtual pipe rig for testing acoustic leak detection correlators: proof of concept *Appl. Acoust.* **102** 137-145
[5] Lapshin B M, Ovchinnikov A L and Chekalin A S 2011 Features of the search for leaks in pipelines of heat networks using the acoustic-emission method *Russ. J. Nondestruct+* **47(10)** 710-717
[6] Faerman V, Avramchuk V, Luneva E and Cheremnov A 2014 Prospects of frequency-time correlation analysis for detecting pipeline leaks by acoustic emission method *IOP C. Ser. Earth Env.* **21(1)** 12041
[7] Anway A R 1978 Method and apparatus for detecting and locating fluid leaks *US Patent* 4083229 A
[8] Fuller C R and Fahy F I 1982 Characteristics of wave propagation and energy distributions in cylindrical elastic shells filled with fluid *J. Sound Vib.* **81(4)** 501-518
[9] Fuchs H V and Riehle R 1991 Ten years of experience with leak detection by acoustic signal analysis *Appl. Acoust.* **33(1)** 1-19.
[10] Wen J, Li P and Wen Y 2007 A new method for unbiased time-delay estimation in noisy environment *Int. J. Adapt. Control* **21(7)** 623-634
[11] Li S, Wen Y, Li P, Yang L and Yang J 2014 Determination of acoustic speed for improving leak detection and location in gas pipelines *Rev. Sci. Instrum.* **85(2)** 024901
[12] Ifeachor E C and Jervis B W 2001 *Digital Signal Processing: A Practical Approach* (New Jersey: Prentice Hall) p 878
[13] Ionel R, Ionel S, Bauer P and Quint F 2014 Water leakage monitoring education: cross correlation study via spectral whitening *Proc. of Int. Conf. of IEEE Industrial Electronic Society* (24 February Dalas, USA) pp 2465-2471

Mathematical support for automated geometry analysis of lathe machining of oblique peakless round–nose tools

A V Filippov[1,2], S Yu Tarasov[1,2], O A Podgornyh[2], N N Shamarin[2], E O Filippova[2]

[1] Institute of Strength Physics and Materials Sciences SB RAS, Tomsk, Akademicheskii Lane 2/4, Russia
[2] Tomsk Polytechnic University, 30, Lenina Ave., Tomsk, 634050, Russia

E-mail: andrey.v.filippov@yandex.ru

Abstract. Automatization of engineering processes requires developing relevant mathematical support and a computer software. Analysis of metal cutting kinematics and tool geometry is a necessary key task at the preproduction stage. This paper is focused on developing a procedure for determining the geometry of oblique peakless round-nose tool lathe machining with the use of vector/matrix transformations. Such an approach allows integration into modern mathematical software packages in distinction to the traditional analytic description. Such an advantage is very promising for developing automated control of the preproduction process. A kinematic criterion for the applicable tool geometry has been developed from the results of this study. The effect of tool blade inclination and curvature on the geometry-dependent process parameters was evaluated.

1. Introduction

To develop the mechanical processing automation and condition monitoring, one needs a detailed understanding of both mechanics and kinematics of the process. The mistakes in kinematics description will result in the tool trajectory distortion, improper performance of the process system, premature tool/equipment failure, and degradation of both machining quality and accuracy [1-5]. For condition monitoring, various methods of non–destructive control such as acoustic emission, vibrometry, eddy current, inductive and etc can be applied [6-11]. In connection with it, the first thing to do is to establish true kinematics specific to the process described. To achieve this goal, one can use several approaches as follows. The first approach is to describe the process kinematics by analytic expressions, however, this traditional approach implies tediousness and high risk of error due to using complicated trigonometric functions. Therefore, it is not of common use for automated design and control applications. The second approach is to analyze the process kinematics using the 3D computer-aided design (CAD) modeling. It is proved to be good for analyzing complicated kinematics cases, such as a complicated form of a workpiece or/and a tool as well as a complex trajectory of the tool/workpiece relative motion. The use of this approach is limited due to its increased working time required for analysing a large number of the nodes modeling the contacting surfaces. The third, relatively new aproach is to describe the kinematics using the vector/matrix functions. It works well and fast to describe the process kinematics even for complex tool/workpiece trajectories and geometry. Another advantage of such a description is its potential for automated computations.

The peakless turning differs from the traditional one by more complex contacting conditions between the tool and the workpiece and, therefore, by complicated kinematics of the metal cutting process [12-18]. The peakless cutting tools having either a straight or round-nose blade are most commonly used in practice. The kinematics of cutting by the round-nose blade tools is not very well understood now. The basic versions of such tools possess the curved rake and clearance faces (Figure 1). The advantages of the peakless cutting tools in lathe operations are determined by their high wear resistance, high machining accuracy and low roughness combined with the high performance rate.

Summing up the above mentioned, the aim of this study is to describe the peakless turning by round-nose tools using vector/matrix transformations, to estimate the geometry changes and optimize this geometry.

2. Materials and methods

This study was carried out by developing a 3D CAD model of oblique turning using a peakless round-nose tool with the specified rake and clearance surfaces curvatures (Figure 1). Theoretical analysis of the process geometry and kinematics has been conducted using the Mathcad software package.

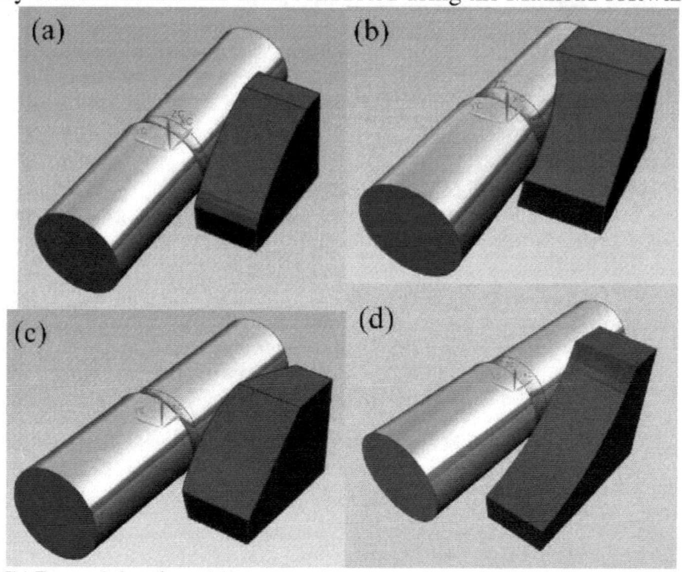

Figure 1. The 3D CAD models of peakless round-nose tool turning by rake (a, d) and clearance (b, c) faces

3. Results and discussions

To computerize the process, it is necessary to set the initial position, shape and geometry of the tool, i.e a tool-in-hand system. The orientation of the system axes is defined by unit vectors $\overrightarrow{x_u}$, $\overrightarrow{y_u}$,, $\overrightarrow{z_u}$. The x_u axis orientation coincides with that of the cutting blade.

$$\overrightarrow{x_u} = \begin{pmatrix} 1 \\ 0 \\ 0 \\ 0 \end{pmatrix}, \qquad \overrightarrow{y_u} = \begin{pmatrix} 0 \\ 1 \\ 0 \\ 0 \end{pmatrix}, \qquad \overrightarrow{z_u} = \begin{pmatrix} 0 \\ 0 \\ 1 \\ 0 \end{pmatrix}, \qquad \vec{s} = \begin{pmatrix} -1 \\ 0 \\ 0 \\ 0 \end{pmatrix}.$$

The cutting blade geometry is given by the rake ($\alpha_и$) and clearance ($\gamma_и$) angle rotation matrices as follows:

$$rot\alpha=\begin{pmatrix}1 & 0 & 0 & 0\\ 0 & \cos(\alpha_u) & -\sin(\alpha_u) & 0\\ 0 & -\sin(\alpha_u) & \cos(\alpha_u) & 0\\ 0 & 0 & 0 & 1\end{pmatrix},\quad rot\gamma=\begin{pmatrix}1 & 0 & 0 & 0\\ 0 & \cos(\gamma_u) & -\sin(\gamma_u) & 0\\ 0 & -\sin(\gamma_u) & \cos(\gamma_u) & 0\\ 0 & 0 & 0 & 1\end{pmatrix}$$

The blade curvature is given by the parametric equations below:

$$\begin{cases} x=R\cdot\cos(\chi);\\ y=R\cdot\sin(\chi).\end{cases}$$

Here, χ is the parametric angle determining a point position on the blade; R is the blade curvature radius.

The blade position in the kinematical coordinate system, i.e. when installing the tool in the corresponding equipment, is given by the rotation angle (ω) matrices, whereas the current point position on the blade with respect to the workpiece is determined by rotation angle ψ.

$$rot\omega=\begin{pmatrix}\cos(\omega) & 0 & \sin(\omega) & 0\\ 0 & 1 & 0 & 0\\ -\sin(\omega) & 0 & \cos(\omega) & 0\\ 0 & 0 & 0 & 1\end{pmatrix},\quad rotV=\begin{pmatrix}1 & 0 & 0 & 0\\ 0 & \cos(-\psi_c) & -\sin(-\psi_c) & 0\\ 0 & -\sin(-\psi_c) & \cos(-\psi_c) & 0\\ 0 & 0 & 0 & 1\end{pmatrix}.$$

As soon as the cutting speed vector direction is changed with the current point position and is determined by basic plane inclination angle ψ_c, it becomes necessary to find its value with allowance for angle χ. For this purpose, the relationship equations have been derived for these angles as shown below.

- Rake face curvature blade:

$$\begin{cases}\sin(\chi)=\pm\left(\dfrac{R\cdot\cos(\omega)+r\cdot\sin(\psi)}{R}-a\sin(\cos(\omega))\right),\\[2ex] \sin(\psi)=\dfrac{\sin(\chi)+a\sin(\cos(\omega))-R\cdot\cos(\omega)}{R}.\end{cases}$$

- Clearance face curvature blade:

$$\begin{cases}\sin(\chi)=\pm\dfrac{r\cdot\sin(\psi)}{R\cdot\sin(\omega)},\\[2ex] \sin(\psi)=\dfrac{R\cdot\sin(\chi)\cdot\sin(\omega)}{r}.\end{cases}$$

The sign of the χ - angle depends on that of the ω-angle, so that for «+ω» and «-ω» it will be «-χ» and – «+χ», respectively. The sign of the blade curvature radius is determined by the blade curvature type as follows: for convex and concave blades it will be «+R» and «-R», respectively.

The tool shape and geometry in each point are determined by vectors $\vec{a}$, $\overrightarrow{n\alpha}$ and $\overrightarrow{n\gamma}$, so that

$$\vec{a}=rot\omega\cdot rotR\cdot\overrightarrow{x_u},$$

where rotR is to set the coordinate system rotation along the cutting blade length as follows.

For the rake face curvature blade:

$$rotR=\begin{pmatrix}\cos(\chi) & 0 & \sin(\chi) & 0\\ 0 & 1 & 0 & 0\\ -\sin(\chi) & 0 & \cos(\chi) & 0\\ 0 & 0 & 0 & 1\end{pmatrix}.$$

For the clearance face curvature blade:

$$rotR = \begin{pmatrix} \cos(\chi) & -\sin(\chi) & 0 & 0 \\ \sin(\chi) & \cos(\chi) & 0 & 0 \\ 0 & 0 & 0 & 0 \\ 0 & 0 & 0 & 1 \end{pmatrix}.$$

The normal curvature rake and clearance face vectors for the curvature blade are defined as follows: :

- The normal curvature rake vector for a blade curved with respect to the rake face:

$$\overrightarrow{n\gamma} = rot\omega \cdot rot\gamma \cdot rotR \cdot \overrightarrow{z_u}.$$

- The normal curvature clearance vector for a blade curved with respect to the rake face:

$$\overrightarrow{n\alpha} = rot\omega \cdot rot\alpha \cdot \overrightarrow{y_u}.$$

- The normal curvature rake vector for a blade curved with respect to the clearance face:

$$\overrightarrow{n\gamma} = rot\omega \cdot rot\gamma \cdot \overrightarrow{z_u}.$$

- The normal curvature clearance vector for a blade curved with respect to clearance:

$$\overrightarrow{n\alpha} = rot\omega \cdot rot\alpha \cdot rotR \cdot \overrightarrow{y_u}.$$

Static planes are:

- Working plane Ps is the plane containing both stroke speed ($\vec{V}$) and feed ($\vec{s}$) directions;
- Reference plane Pv is the plane drawn through the current point perpendicular to the current stroke speed direction in this point.
- Tool cutting edge plane Pτ is the plane tangential to the cutting edge and perpendicular to the reference plane in the current point.
- Sectional plane Pn is the plane perpendicular to the line of intersection of reference and tool cutting edge planes.

$$\overrightarrow{nv} = \vec{V} = rotV \cdot \overrightarrow{z_u}, \quad \overrightarrow{ns} = \frac{\vec{s} \times \vec{V}}{\left|\vec{s} \times \vec{V}\right|}, \quad \overrightarrow{n\tau} = \frac{\vec{a} \times \vec{V}}{\left|\vec{a} \times \vec{V}\right|}, \quad \overrightarrow{nn} = \frac{\vec{V} \times \overrightarrow{nn}}{\left|\vec{V} \times \overrightarrow{nn}\right|}.$$

Static tool angles, such as cutting edge inclination angle λ, approach angle ψ, clearance angle α, and rake angle γ are introduced by equations (1) as follows:

$$\lambda_c = \overrightarrow{nt} \cdot \frac{\vec{a} \times \overrightarrow{nn}}{\left|\vec{a} \times \overrightarrow{nn}\right|}, \varphi_c = \overrightarrow{nv} \cdot \left(\frac{(\overrightarrow{ns} \times \vec{V}) \times \overrightarrow{nn}}{\left|(\overrightarrow{ns} \times \vec{V}) \times \overrightarrow{nn}\right|} \right), \alpha_c = \overrightarrow{nn} \cdot \left(\frac{\overrightarrow{nv} \times (\overrightarrow{n\alpha} \times \overrightarrow{nn})}{\left|\overrightarrow{nv} \times (\overrightarrow{n\alpha} \times \overrightarrow{nn})\right|} \right), \gamma_c = \overrightarrow{nn} \cdot \left(\frac{(\overrightarrow{n\alpha} \times \overrightarrow{nn}) \times \overrightarrow{nt}}{\left|(\overrightarrow{n\alpha} \times \overrightarrow{nn}) \times \overrightarrow{nt}\right|} \right) \quad (1)$$

A transition from the static to kinematic coordinate system is by substituting a static speed vector for the kinematic one in the above shown formulas:

$$\overrightarrow{V_p} = rotV \cdot \begin{pmatrix} 0 \\ 0 \\ V \\ 0 \end{pmatrix} + \begin{pmatrix} -s_M \\ 0 \\ 0 \\ 0 \end{pmatrix}$$

where $s_M = s_n \cdot n$, s_n is the feed, mm/rev; n is the spindle rotation rate, rpm (rotation per minutes).

By the example of cutting tools with convex rake (Figure 2) and clearance (Figure 3) faces, it was shown that the static angles changed so that coloured areas corresponded to the geometry parameter ranges providing the positive clearance angle values. The latter condition is necessary for workability of the tool in lathe machining.

IOP Conf. Series: Journal of Physics: Conf. Series **803** (2017) 012041 doi:10.1088/1742-6596/803/1/012041

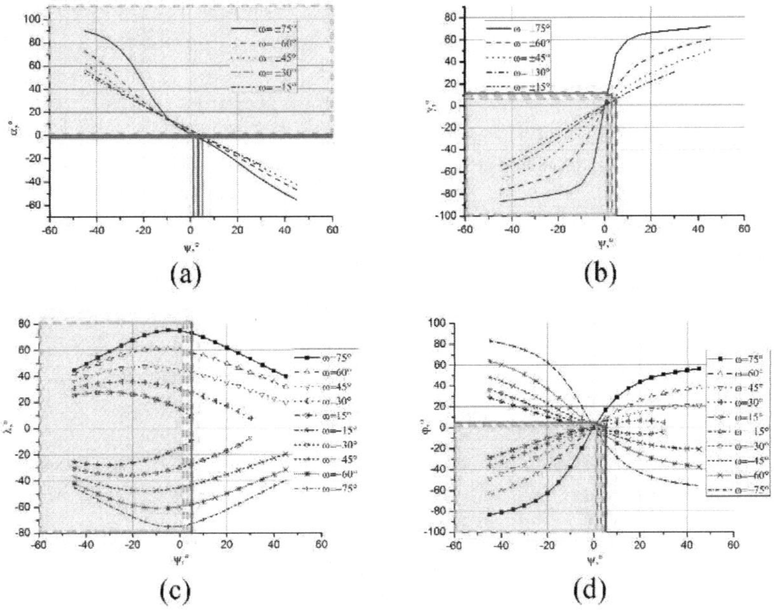

Figure 2. The peakless tool geometry for concave (a-d) and convex (e-h) rake surfaces: clearance angle (a, e), rake angle (b, f), edge inclination angle (c, g), approach angle (d, h)

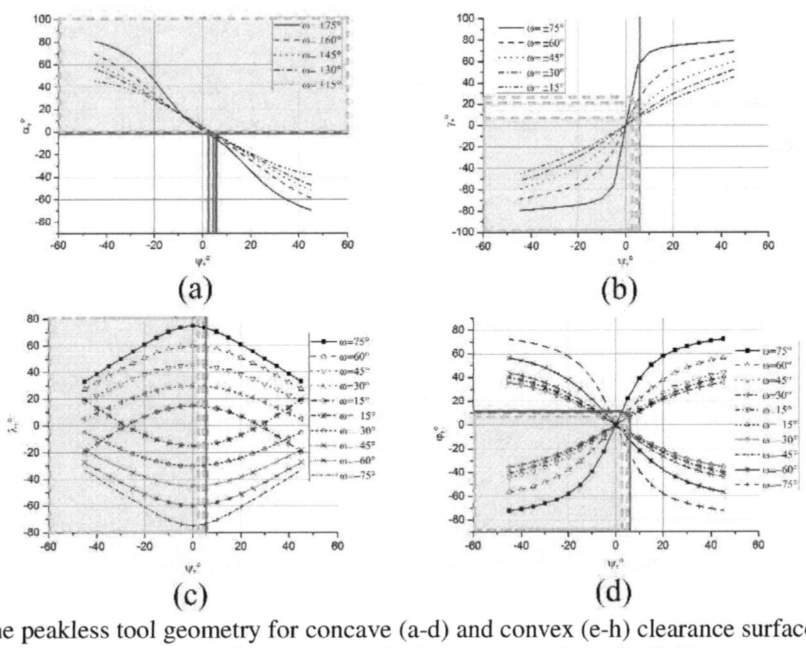

Figure 3. The peakless tool geometry for concave (a-d) and convex (e-h) clearance surfaces: clearance angle (a, e), rake angle (b, f), edge inclination angle (c, g), approach angle (d, h)

4. Conclusion

The cutting tool geometry parameter dependencies have been obtained in the course of this work and then verified by those obtained from 3D CAD models. The reconstruction of sections was carried out using a 15° step of the reference plane inclination for all used calculations of blade inclination angles.

The results of verification showed that the computational error is not higher than 0.5°, thus, confirming the applicability of the vector/matrix transformation approach to accurate and adequate description of the oblique peakless tool turning kinematics.

The main kinematic criterion limiting the peakless tool geometry is the necessity of using only positive clearance angle values.

The ranges of φ_c, λ_c, γ_c angles vary as dependent on the blade curvature and inclination. The higher ω-angle values serve to increase both the ranges of φ_c and γ_c. The blade curvature influences mainly the static edge inclination angle. The applicablity range of tool geometry parameters is greater for the round-nose clearance face tools.

It was shown by the analysis of both static and kinematic angles that the difference between them was not higher 3° for the same type of tool geometry behavior. Hence, it is reasonable to take into account only static angles in studying the geometry of lathe machining of the oblique peakless round-nose tool.

5. Acknowledgments

The work was carried out in the framework of the Fundamental Research Program of the State Academies of Sciences for 2013–2020 and within the framework of the Program for improving the competitive ability of NR TPU.

References

[1] Gubaidulina R H, Petrushin S I and Galeeva A A 2014 *Apl. Mech. and Mat.* **682** 613-616
[2] Nekrasov R Yu, Starikov A I and Lasukov A A 2015 *IOP Conf. Ser.: Mat. Sc. and Eng.* **91 (1)** 012042
[3] Chuykov R S, Mokhovikov A A and Chuykov S S 2014 *Apl. Mech. and Mat.* **682** 510-514
[4] Il'Yaschenko D P, Chinakhov D A, Danilov V I, Schlyakhova G V and Gotovshchik Yu M 2015 *IOP Conf. Ser.: Mat. Sc. and Eng.* **91 (1)** 012006
[5] Kovalevskaya Z G, Klimenov V A and Zaitsev K V 2014 *Apl. Mech. and Mat.* **682** 459-463
[6] RubtsovV E, Kolubaev E A, Kolubaev A V and Popov V L 2013 *Tech. Phys. Let.* **39(2)** 223–225
[7] Kolubaev E A, Kolubaev A V and Sizova O V 2010 *Tech. Phys. Let.* **36(8)** 762–765
[8] Dobrynin S A, Kolubaev E A, Smolin A Y, Dmitriev A I and Psakhie S G 2010 *Tech. Phys. Let.* **36(7)** 606–609
[9] Kolubaev A V, Ivanov Yu F, Sizova O V, Kolubaev E A, Aleshina E A and Gromov V E 2008 *Tech. Phys.* **53(2)** 204–210
[10] Kolubaev E A, Lobach M I, Goncharenko I M and Kolubaeva Yu A 2013 *Inorg. Mat.: Appl. Res.* **4 (2)** 178-182
[11] Kolubaev A V, Kolubaev E A, Vagin I N and Sizova O V 2005 *Tech. Phys. Let.* **31(10)** 813-816
[12] Filippov A V 2015 *Rus. Eng. Res.* **35 (5)** 385-388
[13] Filippov A V 2015 *Rus. Eng. Res.* **35 (5)** 381-384
[14] Raphael G and Stone B J. 1990 *CIRP Annals – Manuf. Tech.* **39 (1)** 425–428
[15] Stone B J, Bonikowski E J, Chapple D J and De Barr A E 1980 *CIRP Annals – Manuf. Tech.* **29 (1)** 275–280
[16] Nee A Y C and Venkatesh V C 1985 *CIRP Annals – Manuf. Tech.* **34 (1)** 121–124
[17] Grzesik W 1991 *Int. J. Mach. Tools Manuf.* **31 (3)** 329-344
[18] Monka P, Monkova K, Balara M, Hloch S, Rehor J, Andrej A and Somsak M 2015 *Int. J. Adv. Manuf. Tech.* **85 (9-12)** 2325-2343

The methodology of semantic analysis for extracting physical effects

M A Fomenkova[1], V A Kamaev[1], D M Korobkin[1], S A Fomenkov[1]

[1] Volgograd State Technical University, 28, Lenina Ave., Volgograd, 400005, Russia

E-mail: dkorobkin80@mail.ru

Abstract. The paper represents new methodology of semantic analysis for physical effects extracting. This methodology is based on the Tuzov ontology that formally describes the Russian language. In this paper, semantic patterns were described to extract structural physical information in the form of physical effects. A new algorithm of text analysis was described.

1. Introduction

Nowadays, the semantic analysis is one of the most topical and growing areas in computer science.
In this work, the semantic analysis is considered as a tool for extracting structured physical information in the form of physical effects (PE). Today the task of identifying PE descriptions in the science texts to supplement the database of PE is the information basis for the new technical solutions development. It is an important and actual task.
Today, there is the system of identifying PE descriptions. It was realized using the semantic analyzer "Semantix" [1]. The system and the semantic analyzer that are described in this work were realized using the Tuzov ontology.

2. Program systems and tools for semantic analysis

Analysis of the effectiveness of software systems for semantic analysis is shown in Table 1, Table 2.

Table 1. The comparison table of efficiency of software systems

Program system	An algorithm of target entities identifying	The system's flexibility (the ability to customize settings, adding new entities)	License	Accuracy extracting (1 - 75% 2 - 65 - 75%, 3 - 65% less)	Completeness (1- more than 60%, 2 – less than 60%)
Extracting facts from the text files, RCO [2]	Patterns search	The parameters are hard-coded	Close	1	2
Attensity Text Analytics [3]	Neural network technology	The parameters are hard-coded	Close	2	1
NetOwl Extractor [4]	Neural network technology	The ability to add new entities	Close	1	1
IOFFE [1]	Patterns	The ability to add new	Open	3	2

	search	patterns

Table 2. Semantic analysis systems

Software	Russian support	License	Base technology	Flexibility (focus on the subject area, the ability to customize, code availability)
Stanford nlp [5]	No	Open	Machine Learning Technologies	No
Malt parser [6]	Yes	Open	Machine Learning Technologies	No
Link grammar parser [7]	No	Open	Relations grammar	No
AGFL [8]	Yes	Open	AGFL-grammar	No
Tomita parser [9]	Yes	Open	CF-grammar and key words dictionaries	No

As is seen from the tables, most of the systems are flexible enough for adjustment to the subject area. It is necessary to create a new approach to work with the physical science text.

3. Semantic analyses for extracting structured physical knowledge in the form of physical effects of the Tuzov ontology

Ontology of the Russian language [10] is a formal description of the Russian language proposed by V.A. Tuzov (semantic roles are defined on the basis of semantic classes and morphological information). The base of computer semantics uses the functional model of the language:

1) the language is an algebraic system $\{f_1, f_2, \ldots f_n, M\}$, where f_i - basic functions of a language and M - the structure of a language, which is a set of basic concepts $m_1, \ldots m_r$ and their hierarchy;

2) each language sentence can be represented as the superposition of basic functions f_i, and every word of the language is expressed by these functions. The exception - basic concepts m_j, belonging to M;

3) grammar is inextricably linked with the language semantics, which basis is the semantic dictionary that describes more than one hundred thousand lexical units (words and phrases), and each word is described as a semantic formula consisting of basic functions.

Table 3. Some basic functions

Function	**Description**
Caus (x,y)	x is the reason y
Loc(x,y)	x is in y

The dictionary entry contains a header word and its interpretation in the semantic language:

EXPOSURE $ 15142 (Caus_o (AGENT: SOMETHING $ 1 ~ Gen, Lab (OBJECT:! Acc, LOCATION: Prep)))

3.1. Physical Effect

There is a formal description of a physical effect [11]. It consists of the following structure (A, B_1, B_2, C): A - an input stream of matter, energy or signals; B_1 - the initial state of a physical object B; B_2 - the final state of the physical object B; C - the flow of matter, energy or signals;

At the department, CAD, of Volgograd State Technical University, the physical effects fund was established. This fund has more than 1,200 PE descriptions.

3.2. The model of structured physical information representation in natural language texts

The model of structured physical information representation in natural language texts was created to retrieve descriptions of physical effects [12]:

$$M_{PE} = \langle C, D, B, R_C, R_B \rangle, \qquad (1)$$

where C - the set of predicates (relations), to describe the PEs in the text, $c_i \in C$;

D - semantic roles and case arguments in predicates $D_i \subset D$ - a list of roles / arguments of cases agreed with predicate c_i; $d_j \in D_i$;

B – a number of elements to describe PE (A, B, C), $B_k \in B$,

$$\forall c_i \in C \ \exists d_j \in D_i \ [d_j \xrightarrow{def} B_k]$$

where $B_k \in$ {input (A), output (C), object (B)}; def - the operator that is associated with the role / case of the d_j argument at the c_i predicate;

R_C – the relationship on C×D, pair $(c_i,d_j) \in R_C$ uniquely identifies the item of the PE description, consistent with the role of predicate / case d_j

R_B - the relationship at Rc×B, pair $((c_i,d_j),B_k) \in R_B$ defines a set of concepts corresponding to the element of PE description b_k, $b_k \in B_k$.

3.3. The algorithm of semantic analysis for extracting the semantic roles

The algorithm of semantic analysis for extracting the semantic roles of Agent, Object, Place arguments is shown in Figure 1.

As a result of domain analysis, we can see the presence of semantic ambiguity. Semantic ambiguity arises because sometimes it is possible to match one semantic role of the Tuzov ontology (role "Agent", "Object", "Place") and several elements of the physical effect description ("Input", "Output", "Object" of physical effect).

To remove this kind of semantic ambiguity, one software module was developed. It is required for matching the existing descriptions of physical effects and semantic roles described in the templates on the basis of the Tuzov ontology.

The approach is to compare the semantic roles, elements of the physical effects descriptions and specific words, which may show the ambiguity. It was realized using the field (PE Description in the natural text) in the database of physical effects [13].

<Semantic role> - <Item of PE description>.

On the basis of statistics of this bunch, a template for the PE extracting is formed for each predicate.

The algorithm for constructing correspondences of semantic roles in the ontology and elements of the PE descriptions for the subject area of the predicates is shown in Figure 2. Thus, the overall physical effects extraction algorithm is shown in Figure 3.

4. Results

The main performance indicators are the accuracy and completeness of extraction.

Accuracy is characterized by the number of correctly retrieved elements from the total number of elements of PE description.

$$P = N_k / N_n \qquad (2),$$

P - extraction accuracy of the PE, N_k number of correctly retrieved elements, N_n – the number of elements found in the text.

Completeness of the PE elements extraction indicates the amount of elements relative to the total number of PE elements in the text descriptions.

$$R = N_n / N \qquad (3),$$

R - completeness of extraction, N_n – the number of PE elements found in the text. N – a total number of elements of the PE elements in the text.

F-measure is calculated as 4:

$$F = \frac{(\beta^2 + 1)PR}{\beta^2 P + R} \qquad (4),$$

$$\beta^2 = \frac{1-\alpha}{\alpha} \qquad (5),$$

where $\alpha = 0.3$

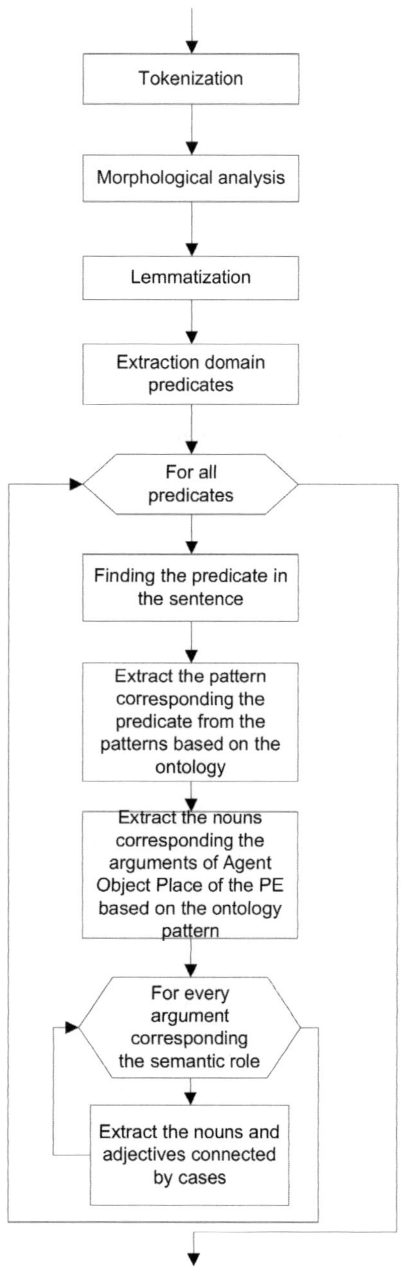

Figure 1. The algorithm of semantic analysis for extracting the semantic roles of Agent, Object, Place arguments.

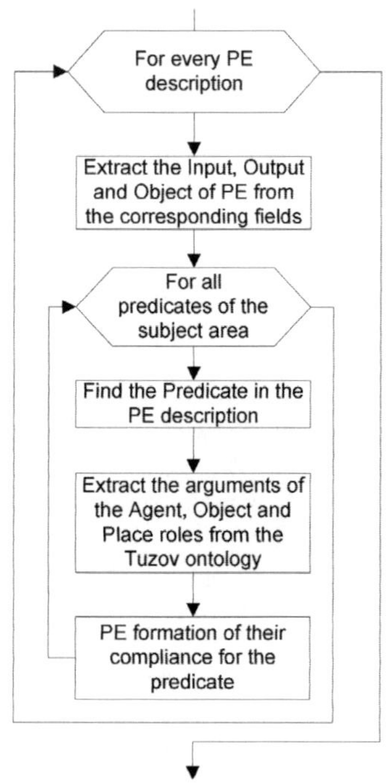

Figure 2. The algorithm for constructing correspondences of semantic roles in the ontology and elements of the PE descriptions for the subject area of the predicates.

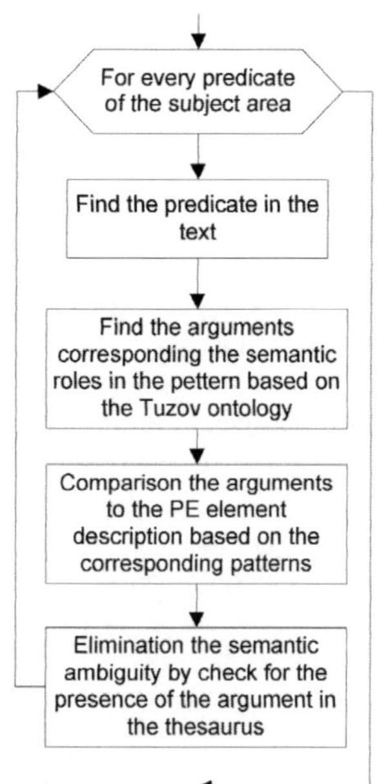

Figure 3. The extraction algorithm of overall physical effects.

Tests were carried out on the basis of the database of physical effects developed at the department of VSTU. 100 physical effects were selected. Tests were conducted on the base of physical effects descriptions in the database and compared with the fields "Input", "Output" and "Object" of the physical effects. Tests were also conducted by the example of 31 patent documents (field "description").

The results were compared with the results of program IOFFE [1] on the basis of the semantic analyzer "Semantix".

The results of the effectiveness comparing are shown in Tables 4 and 5.

Table 4. Analysis of efficiency using the PE database

	Accuracy (%)	Completeness (%)	F-measure
Our system	68	59	61.5
IOFFE	57	53	54.14

Table 5. Analysis of the effectiveness using the patent array

	Accuracy (%)	Completeness (%)	F-measure
Our system	53	46	47.9
IOFFE	46	41	42.38

Efficiency Analysis showed that the developed system improves the efficiency to 4% for accuracy and completeness – to 7%.

Sample. Patent description: "The photoelectric conversion element may be a photodiode having a p-n junction or a pin junction, a phototransistor, or the like. When the incident light hits the semiconductor junction of the cell, this light leads to the appearance of the photoelectric effect, in which electric charges arise." [14]

PE Input: "Light, any other electromagnetic radiation (energy - eV)".

PE Output: "The electric charge (electron emission), (J)".

PE Object: "Photoconductive material (photoconductor)".

The results of the program show the results of the physical effect elements extraction:

PE Input: "Light". PE Output: "electric charge". PE Object: "phototransistor".

5. Conclusion

The method described in this article allowed increasing efficiency of the PE elements extracting. The semantic analyzer based on the Tuzov ontology was created to increase the accuracy and completeness of the method. The approach was tested on the PE database and the patent array.

6. Acknowledgments

This research was financially supported by the Russian Fund of Basic Research (grants No. 15-07-09142 A, No. 16-07-00534 A) and the Russian Ministry of Education in scope of the base part (project 2586 of task N 2014/16).

References

[1] Korobkin D M, Fomenkov S A, Kamaev V A, Fomenkova M A 2016 Multi-agent model of ontology-based extraction of physical effects descriptions from natural language text *Information Technologies in Science, Management, Social Sphere and Medicine' (ITSMSSM 2016)* (Atlantis Press) pp. 498-501

[2] Pleshko V V and Ermakov A У 2009 Semantic interpretation in text analysis computer systems *Information Technologies* **155 (7)**

[3] Taylor J 2011 First Look – Attensity 5.5 *Decision, management, solutions.*

[4] Krupka G R, Hausman K.I 1998 Description of the NetOwl TM Extractor System as used for MUC-7 (1998) *Proceedings of the MUC-7.*

[5] Manning C D, Raghavan P, Schütze H 2008 *Introduction to information retrieval* (Cambridge University Press)

[6] Nivre J, Hall J and Nilsson J 2006 MaltParser: A Data-Driven Parser-Generator for Dependency Parsing. *In Proceedings of the fifth international conference on Language Resources and Evaluation (LREC2006)* (Genoa, Italy) pp 2216-2219

[7] Tukeyev U A, Melby A K, Zhumanov Zh M 2011 Models and algorithms of translation of the Kazakh language sentences into English language with use of link grammar and the statistical approach *Proc. of IV Congress of the Turkic World Math. Society* (Baku) p 474

[8] Azarova I 2002 The matching of AGFL subcategories to Russian lexical and grammatical groupings *Proceedings of the Second AGFL Workshop on Syntactic Description and Processing of Natural Language.*

[9] Ogogrodnik P B, Serebryannaya L V 2014 Text analysis with Tomita parser *Proceeding of The International Conference, BSUIR, Minsk, 29th October 2014* pp 230-231

[10] Tuzov V A 1998 Computer Linguistics. *St. Petersburg, St. Petersburg State University Press*

[11] Korobkin D, Fomenkov S, Kolesnikov S., Lobeyko V, Golovanchikov A 2015 Modification of Physical Effect Model for the Synthesis of the Physical Operation Principles of Technical System *Creativity in Intelligent Technologies and Data Science. CIT&DS 2015*: First Conference (Springer International Publishing) pp 368-378

[12] Korobkin D M, Fomenkov S A, Kolesnikov S G 2014 Method of Ontology-Based Extraction of Physical Effect Description from Russian Text *Knowledge-Based Software Engineering : Proceedings of 11th Joint Conference, JCKBSE 2014* (Springer International Publishing) pp 321-330

[13] Fomenkov S A, Kolesnikov S G, Korobkin D M, Kamaev V A, Orlova Y.A 2014 The Information Filling of the Database by Physical Effects *Journal of Engineering and Applied Sciences* **9 (10-12)** 422-426

[14] Patent US 20100051095 A1 "Hybrid Photovoltaic Cell Using Amorphous Silicon Germanium Absorbers With Wide Bandgap Dopant Layers and an Up-Converter"

Method of monaural localization of the acoustic source direction from the standpoint of the active perception theory

V E Gai[1], I V Polyakov[1], M S Krasheninnikov[1], A A Koshurina[1], R A Dorofeev[1]

[1]Nizhny Novgorod State Technical University n.a. R.E. Alekseev, Minin St., 24, Nizhny Novgorod, 603950

E-mail: iamuser@inbox.ru

Abstract. Currently, the scientific and educational center of the "Transport" of NNSTU performs work on the creation of the universal rescue vehicle. This vehicle is a robot, and intended to reduce the number of human victims in accidents on offshore oil platforms. An actual problem is the development of a method for determining the location of a person overboard in low visibility conditions, when a traditional vision is not efficient. One of the most important sensory robot systems is the acoustic sensor system, because it is omnidirectional and does not require finding of an acoustic source in visibility scope. Features of the acoustic sensor robot system can complement the capabilities of the video sensor in the solution of the problem of localization of a person or some event in the environment. This paper describes the method of determination of the direction of the acoustic source using just one microphone. The proposed method is based on the active perception theory.

1. Topicality

Hearing allows us to perceive sounds of various types (music, speech, etc.). The development of the robot, interacting with someone is impossible without an implemented acoustic sensor system. The sensor system of this type is required for the robot, first, to determine the direction of the acoustic source. Usually, in robotic systems, the problem of determination of the direction of the acoustic source is achieved by using two microphones. Taking into consideration the time difference between the arrival of the signal at the microphones, the direction of the acoustic source is estimated. The accuracy of determination of the direction of the acoustic source is limited by the size of the array of microphones. If the array size is small (microphones are located close to each other), it is difficult to determine the direction of the acoustic source accurately. But it is hard to use the arrays of large microphones for small robots. Consequently, the problem of determination of the direction of the acoustic source using just one microphone is quite important. The present work is devoted to solving the problem of monaural localization of the acoustic source direction.

There are several solutions to the problem of monaural localization of the acoustic source. The method, described in [1], uses the transfer functions of the head created by using a dummy and statistical speech models, based on the Gaussian mixture model for a particular speaker, to create a set of Gaussian mixture models for speech from various directions. Then, the models obtained at the testing stage are used to determine the direction of the source of speech.

In [2], the acoustic source localization is performed using a single microphone or an artificial ear. Acoustic signal $y(t)$, registered by the microphone is represented as a set of some acoustic sources $s(t)$, additive white Gaussian noise $w(t)$ and transfer function $h_\theta(t)$, depending on the direction of the acoustic source:

$$y(t) = h_\theta(t) * s(t) + w(t).$$

The calculation of signal decomposition $y(t)$ into the most likely $s(t)$ and $h_\theta(t)$ is performed in the frequency domain. For it, the hidden Markov models are generated for different audio signals $s(t)$. Simulation $h_\theta(t)$ for different angles is performed using Gaussian noise recorded from different directions. When using the method, the assessment of angle θ is performed using the maximum likelihood method.

2. Localization in the direction of the acoustic source from the standpoint of the active perception theory

2.1. Introduction into the theory of active perception

The system of determination of the acoustic source direction can be represented as a pattern recognition system. It is known that the detection system comprises three processing stages: pre-treatment, calculation and decision making [3]. To implement the stages of pre-processing and calculation of the acoustic signal features, it is proposed to use the theory of active perception [4].

Pretreatment means the performance of the Q-transformation, which consists in the application of the addition operation of the original signal segments:

$$g(t) = \sum_{k=(t-1)\cdot L+1}^{t\cdot L} f(k), t = \overline{1, N},$$

where L – the number of readings, within the segment, N – the number of signal segments, g – the result of application of the Q-transformation to signal f, $f(k)$ – k –reading of signal f, $g(t)$ – t-reading of te signal g.

Formation of the feature description of the original signal is the application of the set of Walsh filters of the Hartmut system to signal g:

$$\mu(k, c(t)) = \sum_{i=0}^{M-1} F_k(i) \cdot g(((t-1) \cdot M + 1) : (t \cdot M)),$$

where $\mu(k, c(t))$ is the result of application of a sets of Walsh filters of the Hartmut system to signal g, $k = \overline{0, M-1}$, $t = \overline{0, |c|-1}$, $c = \{1, P, 2 \cdot P, 3 \cdot P, ..., N - T \cdot P\}$ is a set of the offset values for signal g, $|c|$ – is a cardinal number of set c, P - is an offset value for signal g ($1 \le P \le M$), M – is the number of the used filters. Thus, the feature description of the signal represents the size of matrix $M \times |c|$ and each line of the feature description is a result of the U-transformation the signal segment.

The consequent application of the Q-transformation and filter systems to the signal, implement U-transformation, which is basic in the theory of active perception.

U-transformation has the lowest possible computational complexity, since for its realization only simple operations are used – addition and subtraction. Standard transformations require the implementation of convolution, and the level of weighting coefficients – the operation of arithmetic multiplication.

The theory of active perception is not limited to the formation of the spectral representation of the signal. The structure of the theory includes the *Group algebra* section on the analysis of dependencies between the spectral expansion coefficients. Detected dependences allow us to use them at the stages of decision-making and understanding of the analyzed signal. Let each filter $F_i \in \{F_i\} \equiv \boldsymbol{F}$ correspond to coordinate-defined binary operator $V_i \in \{V_i\} \equiv V$; Then components $\mu_i \neq 0$ of vector μ is permissible to put in line with operator Vi or $\overline{V_i}$ depending on the component sign. As a result, vector μ is associated with a subset of operators $\{V_i\}$, with a design similar to the filter, but having a different meaning of the vector elements ($+1 \leftrightarrow 1$; $-1 \leftrightarrow 0$). By specifying set-theoretic operations of multiplication and addition on set $\{V_i\}$, we have the algebra of signal description of one-dimensional

IOP Conf. Series: Journal of Physics: Conf. Series **803** (2017) 012043 doi:10.1088/1742-6596/803/1/012043

Boolean functions. Taking into account the reversals, all in all, there are 15 operators that can be used in the formation of feature descriptions, as the V_0 operator receives only the direct value.

On the set of operators, the algebra group is formed (the stage of synthesis) of the analyzed signal:

1) The family of algebraic structures (called the complete groups) $\{P_{ni}\}$ kind of $P_{ni} = \{V_i, V_j, V_k\}$ with the power of 35;

2) The family of algebraic structures (called the closed groups) $\{P_{si}\}$ kind of $P_{si} = \{V_i, V_j, V_k, V_r\}$ with the power of 105, where each group is formed of a pair of complete groups connected in a certain way.

Among the complete groups, we can allocate the complete groups for the operations of addition and the operation of multiplication, among the closed groups - closed groups and closed sets.

Two groups (complete or closed) are called incompatible if they include operators with the same numbers, but with different signs.

With closed and complete groups, we perform the spectral correlation analysis. Complete groups reveal the correlation between the operators. Closed group reveal the correlation between the complete groups. If the set of operators is an alphabet, then the set of groups is more complex grammatical descriptions of an observed signal: a complete group is a word, a closed group is a phrase.

Using the spectral representation of signal μ, formed a set of operators describing the signal, and then the set of complete and closed groups:

$$V = GV[\mu],\ P_{na} = GP_{na}[\mu, V],\ P_{nm} = GP_{nm}[\mu, V],\ P_s = GP_s[\mu, V, P_{na}, P_{nm}],\ P_c = GP_c[\mu, V, P_{na}, P_{nm}],$$

where GV – is an operator for calculation of feature descriptions V using spectral representation of the signal on the basis of operators, GP_{na} (GP_{nm}) – on the basis of complete groups for the operations of addition P_{na} (multiplication, P_{nm}), GP_c (GP_s) – on the basis of closed groups P_s (closed sets P_c).

3.2. Proposed approach: the implementation of stages of the recognition system

Using feature descriptions described above, we can get integrated feature descriptions in the form of histograms of the frequency of the operator occurrence, complete and closed groups (two-dimensional, three-dimensional)

$$h_V = H\,[V, \Gamma],\ h_{na} = H\,[P_{na}, \Gamma],\ h_{nm} = H\,[P_{nm}, \Gamma],\ h_s = H\,[P_s, \Gamma],\ h_c = H\,[P_c, \Gamma],$$

where h_V – histogram of operators, h_{na} (h_{nm}) – histogram of complete groups for operations of addition (multiplication), h_s (h_c) – histogram of closed sets (closed groups), Γ – dimensions of histogram: $1d$ – one-dimensional histogram, $2d$ – two-dimensional histogram, $3d$ – three-dimensional histogram. In the two-dimensional histogram, we consider the possible occurrence of pairs of groups in the description of one of the signal segments, in the three-dimensional – triples are considered.

Two-dimensional histograms, which are the matrices, have the following properties:

1) Zero main diagonal;

2) Symmetry relating to the main diagonal.

Thus, the number of significant elements is a two-dimensional histogram $(H_{dim} \cdot H_{dim} - H_{dim}) / 2$, where H_{dim} – number of columns (rows) of the matrix.

Taking into consideration the incompatibility of some complete and closed groups, we may reduce the dimension of the proposed feature systems. The compression considering the incompatibility is only possible for the histograms with dimension greater than or equal to two. Thus, taking into account the given arguments, dimensions of the feature systems based on the histogram are shown in table 1. In brackets, we see the compression ratio, which is calculated as the ratio of the number of features before compression to the number of features after compression.

By increasing the dimension of the proposed systems of features, the number of features is growing as an exponential function. In this regard, it was decided to limit ourselves to two-dimensional histograms of groups and three-dimensional histogram of operators.

Table 1. The dimension of the feature systems before and after the compression

System of features	Before compression			After compression		
Dimension	$1d$	$2d$	$3d$	$1d$	$2d$	$3d$

h_V	1×30	30×30	$30 \times 30 \times 30$	1×435 (2,06)
h_{na} (h_{nm})	1×140	140×140	$140 \times 140 \times 140$	1×5050 (3,88)
h_s (h_c)	1×840	840×840	$840 \times 840 \times 840$	1×106030 (6,65)

4. Simulation experiment

4.1. Installation for carrying out experiments

The experiment was performed in the laboratory. The installation for the experiment includes a recording device (a microphone connected to a sound card or a smart phone), a speaker and a device for turning the microphone and a reflector for a predetermined angle (see figure 1.a). As a sound recording device, we use an omnidirectional microphone. A speaker is a plate with a non-uniform surface. As a playback device, a broadband speaker is used. Figure 1.b shows a prototyping device consisting of a simplified model of the external human ear as the most successful natural catcher of speech sounds. The model includes a small elliptical screen and a channel with the length of about 3 cm up to the microphone.

4.2. Description of the experiment

The simulation experiment aims to study the accuracy of the direction localization of the acoustic source, depending on the layout, signal duration, as well as signs of the system. In the experiment, the analyzed range of angles is from 0 ... 180 degrees, at a pitch of 15 degrees. The process of teaching of the system is to register audio recording with duration of L_1 seconds from each direction (see figure 2). As a result, we get 13 records. Each record is divided into parts, each part is two seconds long. Then, for each of these parts a feature description is made, which is stored in the database. Thus, by L_1 seconds of a training signal, $(L_1 / 2) \cdot 13$ records are formed in the data base.

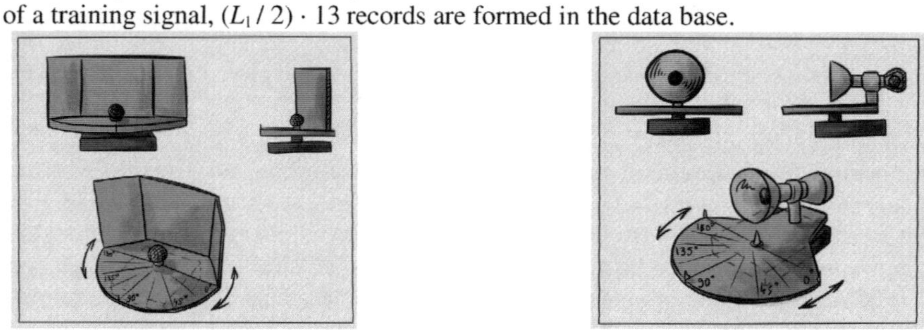

a) Installation model No. 1 b) Installation model No. 2

Figure 1. The installation used for audio recording

The process of determination of the acoustic source direction is as follows:

1) acoustic signal h with duration of L_2 seconds is registered;

2) the recorded signal is divided into segments; each segment is 2 seconds long;

3) for each segment a feature description is formed (see paragraph 3.2);

4) based on the classifier, an assumptive direction of the angle is determined for each segment;

5) the final decision about the direction to the acoustic source for signal, h is generally accepted as the most frequently occurring angle value obtained for the separate signal segments.

When performing a computational experiment, we will determine the accuracy of the direction of the acoustic source, depending on the features of the system, the installation model, the length of training signal (L_1), and the duration of test signal (L_2). The solution of the problem of classification is performed based on the method of support vector machine (SVM) and k-nearest neighbor (KNN).

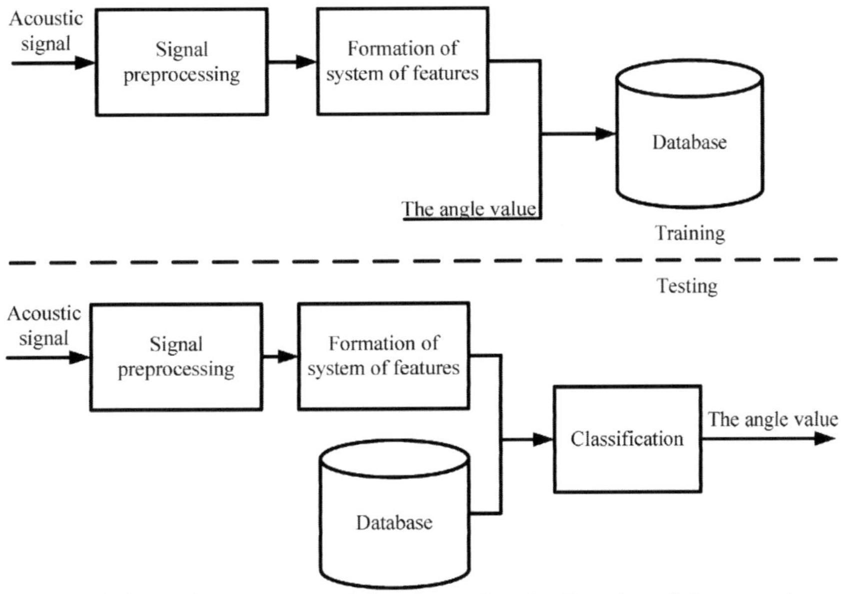

Figure 2. An information system model to determine the direction of the acoustic source

4.3. Notable results

Table 3 shows the error in determining the direction of the acoustic source under different conditions [2]. The experiments were conducted with different types of acoustics; the recording duration of the test was 5 minutes. The recording truck included four types of acoustic-reflecting walls (A, B, C, D) with varying degrees of complexity of the form (C – the most difficult).

Table 3. An average error in degrees of localization of the of the acoustic source direction

Model type of the recording system / Acoustic Type	A	B	C	D
Broadband noise	42.6	8.8	4.3	22.3
Speech	67.8	19.3	7.7	21.35
Dog's barking	55.7	14.2	18.3	60.28
The waterfall noise	42.1	11.8	9.3	42.7
Pure tone	88.7	89.1	86.4	82.6

4.4. Results

Tables 4 and 5 show the evaluation results of the average error in determining the direction of a sound source by using different signs systems classifiers and L_1 and L_2 values.

Table 4. The average error in determination of the direction (classifier KNN)

L_1	L_2	System of features						
		h_V, 2d	h_V, 3d	h_{na}, 1d	h_{na}, 2d	h_{nm}, 1d	h_{nm}, 2d	h_c, 1d
40	20	64.61	61.15	51.92	54.80	60.57	60.57	65.19
40	10	60.00	63.75	63.75	68.94	55.67	64.61	62.88
40	5	61.00	62.01	62.30	61.00	62.88	62.01	60.28
40	2	63.75	51.25	59.76	63.23	57.17	62.82	60.63

Table 5. The average error in determination of the direction (classifier SVM)

L_1	L_2	System of features

		h_V, 2*d*	h_V, 3*d*	h_{na}, 1*d*	h_{na}, 2*d*	h_{nm}, 1*d*	h_{nm}, 2*d*	h_c, 1*d*
40	20	2.30	2.57	2.88	2.15	2.57	2.19	5.30
40	10	12.98	4.59	10.67	3.01	6.63	0.20	8.07
40	5	22.50	10.52	16.87	10.52	12.98	6.40	8.94
40	2	34.78	27.98	25.90	24.92	26.19	22.26	25.73

Conclusions on the results of experiments:

1) the accuracy of the results obtained is compatible with known works [2], but in the proposed implementation the duration of the test signals is significantly reduced (20 seconds instead of 5 minutes);

2) using an SVM classifier provides better accuracy in determining the direction of the sound source, compared with the classifier KNN (tables 4 and 5);

3) using an SVM classifier provides the highly accurate determination of the sound source direction (according to the value of the average error – in the table 5);

4) reduction of the test signal duration leads to an increase of the average error in determination of the sound source direction (see table 5).

The developed method can be used in a universal rescue vehicle [5].

5. Conclusion

This paper describes the method of the monaural acoustic source localization. The solution is performed from the standpoint of the active perception theory. The investigations of the proposed method showed the possibility of reducing the time of decision-making at the expense of reducing the duration of the analyzed signal, as compared to known approaches. The software implementation of the method was used in the mobile robot control system. The performed work is a stepping-stone in the development of new methods and means of search and rescue of persons in distress at sea.

6. Acknowledgments

This work was carried out at the Nizhny Novgorod State Technical University named after R.E. Alekseev, with financial support from the government in the face of the Russian Ministry of Education (the unique identifier of the project: RFMEFI57714X0105.

References

[1] Fuchs A K, Feldbauer C, Stark M 2011 *Proc. Int. Conf. Speech Communication Association* (27-31 August 2011, Florence, Italy) p 2521

[2] Saxena A, Ng A Y 2009 *Proc. Int. Conf. on Robotics and Automation* p 1737

[3] Perez-Meana H 2007 *Advances in Audio and Speech Signal Processing: Technologies and Applications* (Igi Global) p 374

[4] Utrobin V A 2004 Physical interpretation of the elements of image algebra, J. *Advances in Physical Sciences* **47** 1017-1032

[5] Krasheninnikov M, Kulashov A, Shapkin V, Koshurina A 2013 The concept and methodology of creating the universal life-saver with rotary-screw mover *J. Lecture Notes in Electrical Engineering* **195** 477-490.

The analysis of the root quality factors of a power unloading system

S A Gayvoronskiy[1], **T Ezangina**[1], **I Khozhaev**[1], **Lan Gunbo**[1]

[1]Department of Automatics and Computer Systems, National Research Tomsk Polytechnic University, Lenin Avenue 30, Tomsk, Russia

E-mail: eza-tanya@yandex.ru

Abstract. The article focuses on the development of the technique for the analysis of the robust quality of the interval automatic control systems with the affine uncertainty of their characteristic polynomial coefficients. This technique is based on the edge route of the polyhedron of the system with interval parameters and its mapping on the root plane. Theoretical results are used for the analysis of the root robust quality factors of the power unloading system.

1. Introduction

The automatization of modern manufacturing processes with mechanical and electrically-powered equipment requires the consideration of the control objects features, namely, the variation of their parameters during operation. These objects can have integral parameters which vary within certain limits on the beforehand unknown laws [1, 2]. Systems with such kind of objects are called the interval automatic control systems (IACS).

The designer of IACS has to solve the problem associated with the analysis of the system stability at any variations of the object parameters within known ranges. If the interval system of automatic control is stable, we can state that it has robust stability. Kharitonov's theorem is used for the analysis of the robust stability of the interval automatic control systems [3]. This theorem states that for stability of the interval automatic control systems with a polynomial, it is sufficient when four polynomials formed from the extreme values of coefficients, alternating in pairs (two lower and two upper values), are steady.

The estimation of the robust stability can be root indices: the degree of robust stability α and the degree of robust oscillation $\mu = tg\varphi$. These indices can be found on the basis of the arrangement of poles localization $\Gamma([\alpha])$ of the interval automatic control systems (Figure 1).

It is well known that in order to define the values of robust stability and an oscillation degree at interval uncertainty of the polynomial coefficients, it is sufficient to find its roots in all 2^m vertices of the polyhedron of the interval coefficients, where m is the number of interval coefficients. After that, we choose the desired values [4].

IOP Conf. Series: Journal of Physics: Conf. Series **803** (2017) 012044 doi:10.1088/1742-6596/803/1/012044

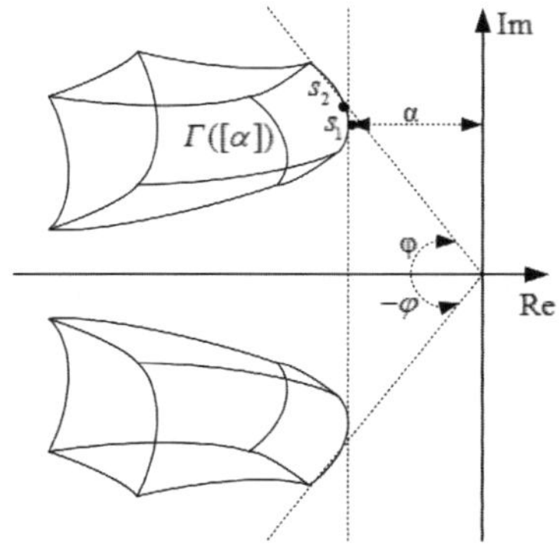

Figure 1. The domain of poles localization of the interval automatic control system

2. Problem statement

If the real parameters of the interval automatic control system are linearly included in the polynomial coefficients, there is a special method based on the edge theorem [5]-[9] to perform the quantitative analysis of the robust stability of the interval automatic control system more precisely. According to this theorem, the interval automatic control system has robust stability on all edges of the polyhedron of the interval parameters of the interval automatic control system. By mapping these edges on the root plane, we can define quantitative assessment of robust quality factors by the values of the interval parameters corresponding to the worst modes of the interval automatic control system operation. However, it is known that the boundary of the poles localization domain of the interval automatic control system is defined not by all edges, but only by exterior ones, which specify the boundary edge route [9]. Therefore, it is important to solve the problem associated with the route construction.

3. The definition of root quality factors of the interval automatic control system on the basis of the edge route of the polyhedron of the interval parameters

Let us consider the procedure of the boundary edge route construction. It consists of the following stages.

1) The calculation of the characteristic polynomial of the IACS of the following image:

$$\sum_{i=1}^{m} T_i \cdot A_i(s) + B(s) = 0, \tag{1}$$

where $T_{i\min} \le T_i \le T_{i\max}$, $T_{i\min} = \underline{T_i}$, $T_{i\max} = \overline{T_i}$.

2) The definition of vertices coordinates V_q ($q \in \overline{1,2^m}$ – the number of the vertex) of the polyhedron of interval parameters $P_T = \{T_i \,|\, \underline{T_i} \le T_i \le \overline{T_i}, i = \overline{1,m}\}$.

3) The choice of any complex root node U_q from the poles of the transfer functions of edge branches

$$W_i^{Vq}(\Delta T_i, s) = \frac{\Delta T_i \cdot A_i(s)}{Q^{Vq}(s) + B(s)}, \tag{2}$$

where $Q^{Vq}(s) = \sum_{i=1}^{m} T_i^{V_q} \cdot A_i(s)$; $T_i^{V_q}$ – is the value of the interval parameter of the IACS in the V_q vertex.

4) The definition of the outlet angle of the edge branches by all interval parameters from the U_q node found in paragraph 3. This definition is based on (2).

These angles can be found from the phase equation [10]. Let a characteristic polynomial be of n power and polynomial A_i be of r power. We can write down outlet angles Θ_{Ti}^{Vj} of the edge branch from node U_q by the increase of T_i:

$$\Theta_{T_i}^{V_q} = 180° - \sum_{k=1}^{n-1} \Theta_k + \sum_{l=1}^{r} \Theta_l \ , \tag{3}$$

and the reduction of T_i:

$$\Theta_{T_i}^{V_q} = -\sum_{k=1}^{n-1} \Theta_k + \sum_{l=1}^{r} \Theta_l \ , \tag{4}$$

where Θ_k and Θ_l are angles defined by the vectors from node U_q to the k pole and l zero of function (2). If $\Theta_{T_i}^{V_q} \notin [0°;180°]$, then the resultant angle is $\Theta_{T_i} = \Theta_{T_i}^{V_q} - 360°$.

5) The conditional test is

$$\Theta_{\max}^{V_q} - \Theta_{\min}^{V_q} < 180° \ , \tag{5}$$

where $\Theta_{\max}^{V_q}$, $\Theta_{\min}^{V_q}$ - relatively maximal and minimal outlet angles. If the statement of the problem is (5), then U_q is on the boundary of the poles localization.

6) Arranging $\Theta_{T_i}^{V_q}$, $i = \overline{1,m}$ in an increasing order and in accordance with this arrangement to write down the sequence of edges $R_{T_i}^{V_q}$, $i = \overline{1,m}$, forming the direct edge route.

7) Finding faces G_{ij}, edge images $R^q{}_i$ and $R^q{}_j$ which can intersect.

It is necessary to get conditions of these intersections. Let us write down the equations of face plane reflection G_{ij}:

$$T_i A_i(s) + T_j A_j(s) + \sum_k T_k^q A_k(s) + B(s) = 0 \ . \tag{6}$$

If in (6), we specify $s = s_r = \alpha + j\beta$, $r \in \overline{1,n}$ and distinguish real and imaginary components we will get the combined equations consisting of two linear equations which combine s_r with two variables T_i and T_j:

$$\begin{cases} \mathrm{Re}\, A_i(\alpha,\beta)\cdot \mathrm{Im}\, A_j(\alpha,\beta) - \mathrm{Re}\, A_j(\alpha,\beta)\cdot \mathrm{Im}\, A_i(\alpha,\beta) = 0; \\ \mathrm{Re}\, A_j(\alpha,\beta)\cdot \mathrm{Im}\left[\sum_k T_k^q \cdot A_k(\alpha,\beta) + B(\alpha,\beta)\right] - \mathrm{Im}\, A_j(\alpha,\beta)\cdot \mathrm{Re}\left[\sum_k T_k^q \cdot A_k(\alpha,\beta) + B(\alpha,\beta)\right] = 0, \end{cases} \tag{7}$$

If the system of the image as in (7) has solutions for all combinations of the interval parameters by $\beta \neq 0$ then boundary S_r consists of intersecting edge branches $R^q{}_i$ and $R^q{}_j$. If the images of two successive edges $R^q{}_i$ and $R^q{}_j$ of face G_{ij} intersect, it is required to add two opposite edges for this face in the direct edge route.

8) Mapping of the obtained boundary edge route on the root plane and the definition of root robust quality factors of the system on its basis.

4. The quality analysis of the power unloading system with interval parameters

Power unloading systems (PUS) should be capable of ensuring smooth and accurate shifting of the cargo within the working area by direct manual operator's action over the cargo applying small directed stress, which steers the cargo along the path and assigns the velocity of cargo shifting. PUS is a two mass system with a flexible connection. The rope is used for connecting the cargo and an electric drive.

The analysis of dynamic properties of the properties of power unloading systems (PUS) is a very urgent problem since the availability of the rope in the system can cause the oscillation in this system. The system is required to be a high-performance one and be able to shift the cargo of any mass and length of the rope from the known ranges.

The analysis of the transfer function of the PUS on disturbance under the steady-state condition with different transfer functions of the controller has shown that it is reasonable to use the PI-controller, which ensures astaticism of the first order. Its transfer function is as follows: $W_c(s) = (k_1 + k_2 s)/s$. The given controller has two setting parameters k_1, k_2, determining the quality of the transient processes of the PUS. As a result of mathematical description of PUS [12], the block diagram (Figure 2) is constructed.

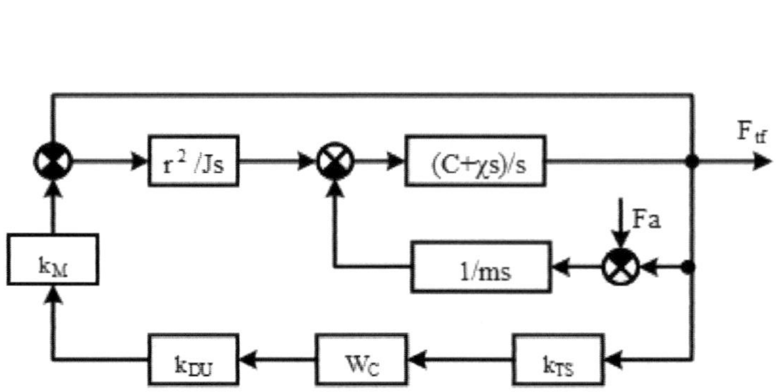

Figure 2. The block diagram of the PUS

k_M is the motor gain coefficient; k_{DU} is the transfer coefficient of the amplifying device; k_{TS} is the tension sensor transfer coefficient; $J = 0.7$ kgm^2 is the moment of inertia of the electric drive of PUS; $C=C_{sr}/l$ is the coefficient of the cable rigidity, $C_{sr} = 1.8 \cdot 10^4\ N/m$ is the specific coefficient of the cable rigidity; $\chi=\chi_{sd}/l$ is the coefficient of the rope resilience losses, $\chi_{sd} = 0.8 \cdot 10^4\ N \cdot s/m$ is the specific coefficient of the rope damping; $r=0.15\,m$ is the block radius; $k_1=1, k_2=0.03$ are parameterizations of the PI controller, $m=[40;200]\,kg$ is the cargo mass; $l=[90;130]\,m$ is the rope length; $k = k_{DU} \cdot k_M \cdot k_{TS} = [3;35]$ is the gain coefficient of the electric part of the electric drive. The input signal of the PUS is external action force F_a which drives the section of cargo in the weightlessness environment. The output signal of the PUS is the tighting force of cable F_t.

Thus, we will get the transfer function of the closed PUS, which input signal is the operator's stress and the output signal is the variation of the rope tensile stress:

$$W(s) = \frac{J\chi_{y\partial} \cdot s^2 + C_{y\partial} J \cdot s}{J[l][m] \cdot s^3 + (J+[m]r^2(1+k_2 k)) \cdot \chi_{y\partial} s^2 + (C_{y\partial}(J+[m]r^2) + k[m]r^2(\chi_{y\partial} k_1 + k_2 C_{y\partial})) \cdot s + [m]C_{y\partial} k k_1 r^2}.$$

A characteristic polynomial of PUS with affine uncertainty is as follows: $[T_1] \cdot A_1(s) + \frac{1}{[T_2]} \cdot A_2(s) + [T_3] \cdot A_3(s) + B(s) = 0$, where $[T_1]=[l]$; $[T_2]=[m]$; $[T_3]=[k]$; $A_1(s)=Js^3$; $A_2(s) = J(\chi_{y\partial} s + C_{y\partial})s$; $A_3(s) = r^2(\chi_{y\partial} k_2 s^2 + (\chi_{y\partial} k_1 + C_{y\partial} k_2)s + C_{y\partial} k_1)$; $B(s) = r^2 s(\chi_{y\partial} s + C_{y\partial})$. It is necessary to construct the boundary edge route of the polyhedron of SUP interval parameters on the basis of the developed algorithm of robust stability and find the values of root robust quality factors on its reflection.

The parametric polyhedron is formed by three interval parameters and includes eight vertices: $V_1(\underline{T_1},\underline{T_2},\underline{T_3})$, $V_2(\underline{T_1},\overline{T_2},\underline{T_3})$, $V_3(\underline{T_1},\overline{T_2},\overline{T_3})$, $V_4(\overline{T_1},\underline{T_2},\underline{T_3})$, $V_5(\overline{T_1},\overline{T_2},\underline{T_3})$, $V_6(\overline{T_1},\overline{T_2},\overline{T_3})$, $V_7(\overline{T_1},\underline{T_2},\overline{T_3})$, $V_8(\underline{T_1},\underline{T_2},\overline{T_3})$. The poles of SUP $[-2.04-j3.36 \;\; -2.04+j3.36 \;\; -1.24]$ are defined in the first vertex of the polyhedron. Three sets of zeroes are also defined in the first vertex of the polyhedron (roots $A_1(s)=0$ have values $[0 \;\; 0 \;\; 0]$, roots $A_2(s)=0$ have values $[-2.25 \;\; 0]$ and roots $A_3(s)=0$ have values $[-33.3 \;\; -2.25]$). Outlet angles $\Theta_{T_2}^{V1} = 13.48^\circ$, $\Theta_{T_1}^{V1} = -8.19^\circ$, $\Theta_{T_3}^{V1} = 78.26^\circ$ for root node $U_1 = -2.04+j3.36$ were also found. By these angle parameters, boundary condition U_1 is fulfilled, therefore V_1 belongs to the edge route. Since $\Theta_{T_1}^{V1} < \Theta_{T_2}^{V1} < \Theta_{T_3}^{V1}$, we can write down the consequence of the interval parameters variation from vertex V_1: $T_1 \to T_2 \to T_3 \to T_1 \to T_2 \to T_3$. This sequence corresponds to the direct edge route illustrated in Figure 3.

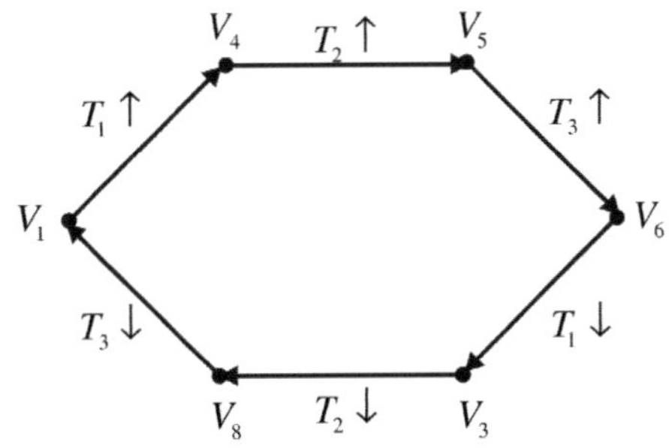

Figure 3. The direct edge route

Let us check whether the boundary edge route has intersecting edges. Three faces G_{31}, G_{21} and G_{32}, converge in vertex V_1, where the indices correspond to the indices of the interval parameters. Let us work out combined equations for each face (7). Since there is no solution for combined equations G_{31}, G_{21} and G_{32}, it means that there is no intersection for the edge. As a result, the boundary edge route preserves the image illustrated in Figure 3. Its mapping on the root plane is shown in Figure 4

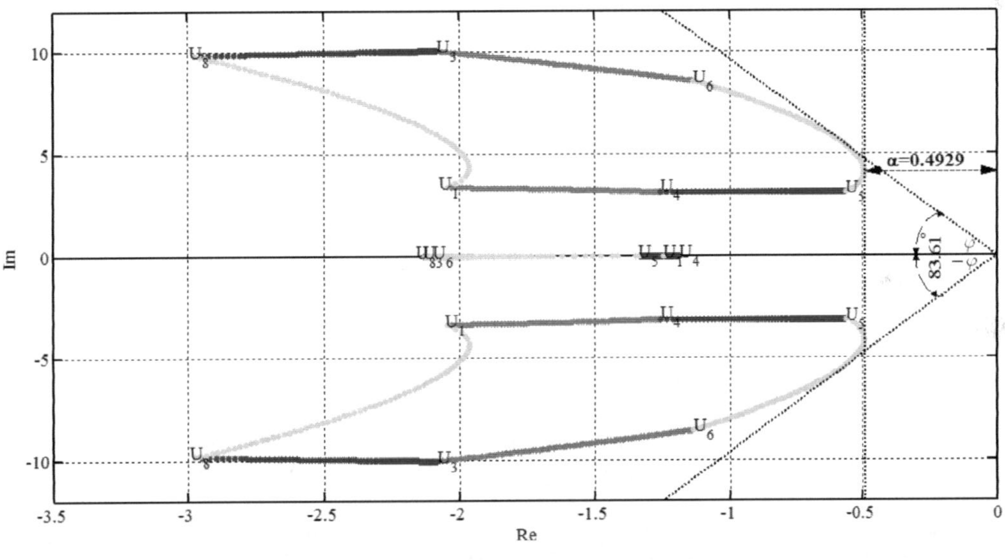

Figure 4. The interval root locus

As we can see from Figure 4, the image of the edge route defines the degree of robust stability $\alpha = 0.49$ and the degree of robust oscillation μ=9.6, which corresponds to the sector with angle $\varphi = \pm 84^{\circ}$. It is to be noted that these quality factors are defined by the image of one edge T_3.

Let us construct the transient processes of the power unloading system by different values of parameters T_1, T_2, T_3 on the edge route (Figure 5). From Figure 5, we can define their maximal decay time $t_{max} = 7.15\,\text{s}$ and the value of overcontrol $\sigma_{max} = 46\%$. The maximal decay time corresponds to the desired degree of robust stability α.

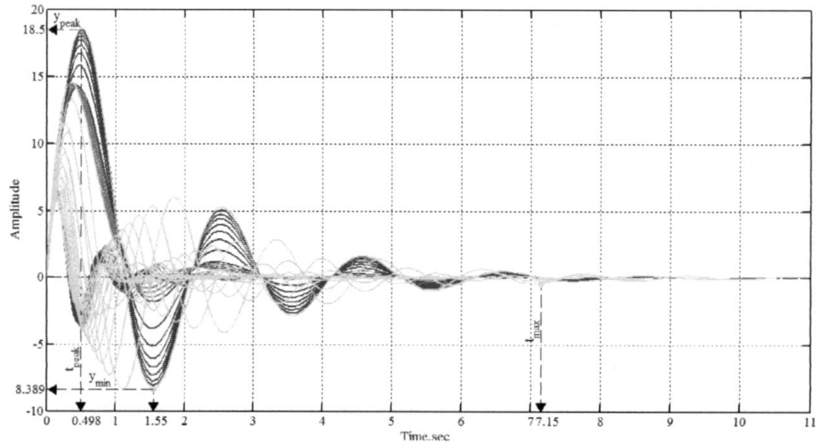

Figure 5. Transient processes on the boundary edge route

5. Conclusion

The algorithm of the boundary edged route of the polyhedron of the interval parameters of the system was developed in the given work. The algorithm was developed on the basis of the edge theorem. It was established that in order to define the rooted robust quality factors, it is sufficient to reflect only the edges which form the boundary edged route on the root plane. The developed algorithm was tested on a real power unloading system with three interval parameters.

Acknowledgments

The reported study was partially supported by RFBR, research project № 15-01-99447

Reference

[1] Gusev Yu.M., Efanov V.N., Krymskiy V.G., Rutkovskiy V.Yu. 1991 *Technical Cybernetics* **1** 3–23

[2] Zhmud V, Semibalamut V, Vostrikov 2015 Feedback systems with pseudo local loops *Proc. of Int. Conf. on Testing and Measurement: Techniques and Application* 411-417

[3] Kharitonov, V. L. 1978 Asymptotic stability of an equilibrium position of a family of systems of linear differential equations *Differential Equations* **14(11)** 2086–2088

[4] Barmish B. R., Tempo R. 1990 The robust root locus *Automatica* **26(2)** 283-292

[5] Rimsky G.V., Nesenchuk A.A. 1996 Root locus methods for robust control systems quality and stability investigations *in Proc. of IFAC 13th Triennial World Congress* San Francisco USA 469-474

[6] Soh Y.C., Foo Y.K. 1989 Generalized edge theorem *Systems and Control Letters* **12(3)** 219-224

[7] Neimark Y.I. 1992 Robust stability and D-partition *Automation and Remote Control*, **53(7)** 957–965

[8] Keel L.H., Bhattacharyya S.P. 1999 *Automatica* **35** 1717-1724

[9] Vadutov O.S., Gayvoronskiy S. A. 2003 Application of edge routing to stability analysis of interval polynomials *Journal of Computer and Systems Sciences International*, **42(6)** 833-838

[10] Uderman E. G. 1972 *Root Locus Method in Control Theory*. (M: Science) 24-27

[11] Zhmud V, Sablina G, Dimitrov L, Trubin V 2016 The use of bypass channel for feedback control of oscillatory object well-known as difficult one for control *SIBCON-2016* 7491767

[12] Ezangina T. A., Gayvoronskiy S. A. 2014 The synthesis of the robust Stabilization system of cable tension for the test Bench of weightlessness simulation *Advanced Materials Research* **1016** 394-399

Maximization of the robust stability degree of interval systems by means of a linear controller in the presence of limits

S A Gayvoronskiy[1], T Ezangina[1], I Khozhaev[1], Lan Gunbo[1]

[1]Department of Automatics and Computer Systems, National Research Tomsk Polytechnic University, Lenin Avenue 30, Tomsk, Russia

E-mail: eza-tanya@yandex.ru

Abstract. The authors of this article developed the technique of combined parametric synthesis of a linear controller on the basis of the coefficient method and the method of mathematical programming capable of ensuring the maximization of the degree of robust stability of a control system. The article also presents the numerical illustration of the PI controller synthesis of the position stabilization system of an underwater object.

1. Introduction

The issue of intensification of the parameterization of a controller by unfavorable system modes of operation is urgent by designing robust controllers for control systems with interval parameters. One of the methods to solve this problem can be the acquisition of such robust stability margin which would be able to ensure the maximal operation speed in the system by most unfavorable parameters of the control object. The research of domestic and foreign scientists [1-10] is of interest for the solution of the above-specified problem. The approach [3] is mainly employed in these works, where standardized polynomials on a certain parameter are used. By applying the coefficient methods for the solution of this problem on the basis of the interval coefficient stability and oscillation indices in [11], the sufficient conditions for the parameterization of the controller are obtained, which ensures the maximal lower estimate of the degree of robust stability. Since such approach allows defining only a quasi maximal degree of the robust system stability, it is desirable to maximize the stability till the maximal degree increases the operation speed of the integral control system. To cope with this problem, we suggest using the theory of mathematical programming, as it was done for the steady-state system in [4]. Thus, it is necessary to solve the problem associated with the definition of the maximal degree of stability of the linear system and parameters of its controller on the basis of the interval characteristic polynomial if there are restrictions on a Q factor and system oscillation.

2. Robust expansion of the mathematical programming technique

Let the interval characteristic polynomial be specified as $D(s,\vec{k})=\sum_{i=0}^{n}[d_i(\vec{k})]s^i$, where $[d_i(\vec{k})]=[\underline{d_i(\vec{k})};\overline{d_i(\vec{k})}]$, $\vec{k}$ – the vector of parameterization of the controller . It is required to carry out robust expansion for employing the method of mathematical programming to solve the assigned task. It is suggested using the vertex interval characteristic polynomial:

$$D^q(s,\vec{k}) = \sum_{i=0}^{n} d_i^q(\vec{k}) s^i \ , \tag{1}$$

which corresponds to the worst mode of the interval control system operation. When $s = \alpha + j\beta$, the vertex interval characteristic polynomial has the following image:

$$D^q\left(\vec{k},\alpha,\beta\right) = \operatorname{Re} D^q\left(\vec{k},\alpha,\beta\right) + \operatorname{Im} D^q\left(\vec{k},\alpha,\beta\right). \tag{2}$$

Based on (2), let us work out combined equations:

$$\begin{cases} \operatorname{Re} D^q\left(\vec{k},\alpha,\beta\right) = 0; \\ \operatorname{Im} D^q\left(\vec{k},\alpha,\beta\right) = 0; \\ \dfrac{\partial \operatorname{Re} D^q\left(\vec{k},\alpha,\beta\right)}{\partial \alpha} = 0; \\ \dfrac{\partial \operatorname{Im} D^q\left(\vec{k},\alpha,\beta\right)}{\partial \alpha} = 0; \\ \ldots \\ \dfrac{\partial^c \operatorname{Re} D^q\left(\vec{k},\alpha,\beta\right)}{\partial \alpha^c} = 0; \\ \dfrac{\partial^c \operatorname{Im} D^q\left(\vec{k},\alpha,\beta\right)}{\partial \alpha^c} = 0. \end{cases} \tag{3}$$

The solution of the given system is the parameterization of the $\vec{k}$ controller and the value of the maximal degree of α stability in the worst mode of the interval control system. The number of system equations is defined by the number of the decision variable.

3. The synthesis technique of a robust controller

Employing the coefficient method [11] of the controller, the synthesis of a quasi maximal degree of $\eta_{\max}$ robust stability, the definition procedure of the corresponding vertex of the polyhedron of interval coefficients of the interval characteristic polynomial and combined equations (3), the technique allowing maximizing the controller parameters of a quasi maximal degree of $\eta_{\max}$ robust stability was developed. The technique includes the following stages:

To find the parameterization of the $\vec{k}$ controller, ensuring the quasi maximal degree of robust stability $\eta_{\max}$.

To construct the vertex route of the polyhedron of interval coefficients of the interval characteristic polynomial [13] based on the found parameterizations of the controllers.

To represent the found vertices on the rooted plane and to define vertex V^*, which image is the nearest to the imaginary axes.

To find vertex V^*, it is required to solve nonlinear equations (1) and define the limited values of robust quality factors and the parameterizations of controller $\vec{k}$ which ensure these quality factors.

4. Numerical example

Let us carry out the parametric synthesis of the robust PI controller $W_C(s) = (k_1 + k_2 s)/s$ for the position stabilization system of a descent underwater vehicle (DUV) on the basis of the developed technique. The mathematical model in the form of the block diagram illustrated in fig. 1 is created on the basis of the differential equations describing the dynamics of the certain components of the stabilization system of a DUV. The system has the following constant parameters [1]: k_{m2}=0,3 (Nm/A) the transfer constant of a shock-absorbing hoist (SAH) drive on torque; k_{e2}=1 (Vs/rad) is a back EMF coefficient of the motor of SAH ; J_2=0,5 (kgm^2) is the moment of inertia of SAH; R_2=0,1 (m) is a drum radius of SAH and interval parameters, $\chi = [35;200]$ (Ns/m) is an equivalent value of the damping factor of the cable; $C = [800;4.25\times10^3]$ (N/m) is an equivalent value of stiffness of the cable; $m_{DUV} = [4.47\times10^3;4.65\times10^3]$ (kg) is mass of the DUV.

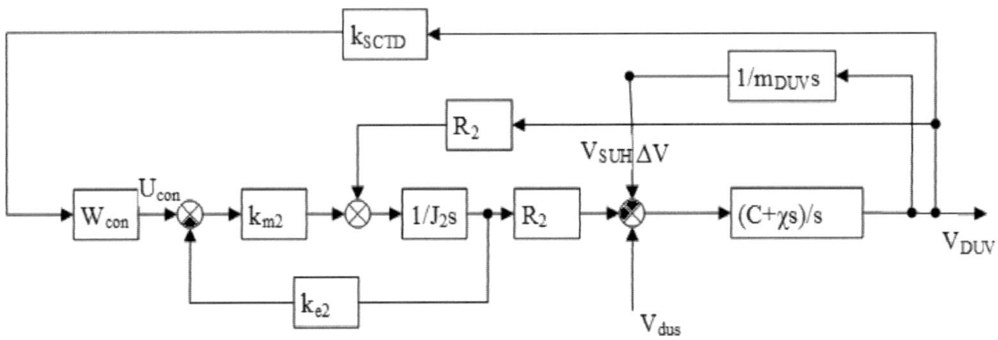

Figure 1. The structure diagram of the stabilization system of a DUV

The transfer function on the extraneous signal is obtained on the basis of the system structure: $W_{зам}(s) = F_н / V_{ЗАМС} = [b_3]s^3 + [b_2]s^2 + [b_1]s / [a_3]s^3 + [a_2]s^2 + [a_1]s^1 + [a_0]$, where $[b_1] = [1072800;\ 5928750]$; $[b_2] = [762135;4371000]$; $[b_3] = [31290;279000]$. $[a_0] = [240 + 214560k_1;1275 + 1185750k_1]$; $[a_1] = [35930.5 + 214560k_2 + 9387k_1;198565 + 1185750k_2 + 83700k_1]$; $[a_2] = [2912.5 + 9387k_2;15405 + 83700k_2]$; $[a_3] = [894;930]$.

Let us specify at the first stage the synthesis of the tolerable index of oscillation $\delta_Д = 4.8\times10^{-2}$ from the expression in [11]. We will get relation $k_1(k_2) = \frac{(15405 + 83700k_2)^2 - 8863941.6 + 52931880k_2}{3736368}$.

Then, it is required from the expression in [11] to find dependence k_2 on η_{max}:

$$k_2(\eta_{max}) = \sqrt[3]{-\frac{Q1(\eta_{max})}{2} + \sqrt{-\frac{Dis(\eta_{max})}{108}}} + \sqrt[3]{-\frac{Q1(\eta_{max})}{2} + \sqrt{-\frac{Dis(\eta_{max})}{108}}},\quad Dis(\eta_{max}) = -4e(\eta_{max})^3 - 27Q1(\eta_{max})^2,$$

$$e(\eta_{max}) = \frac{\gamma(\eta_{max})^3}{3\nu(\eta_{max})^2} - \frac{\upsilon(\eta_{max})}{\nu(\eta_{max})},\quad \nu(\eta_{max}) = -\frac{\lambda}{\delta_Д \underline{a_2}}\underline{b_1}^3,\quad Q1(\eta_{max}) = \frac{(2\gamma(\eta_{max}))^3}{27\nu(\eta_{max})^3} - \frac{\upsilon(\eta_{max})\gamma(\eta_{max})}{3\nu(\eta_{max})} + \frac{\vartheta(\eta_{max})}{\nu(\eta_{max})},$$

$$\gamma(\eta_{max}) = \frac{1}{\delta_Д \overline{b_1}}\underline{b_1^2}\,\overline{b_0} - \lambda(\overline{b_1}(\underline{b_0} - \eta_{max}\overline{b_1}) + \frac{\underline{b_1^2}}{\delta_Д \underline{a_2}}(2\underline{a_1} + \frac{1}{\underline{b_1}}\underline{b_1}\overline{a_1}) - \underline{b_1}\overline{b_0}),$$

$$\upsilon(\eta_{max}) = \frac{\overline{b_0}}{\underline{b_1}}(\frac{2\underline{a_1 b_1}}{\delta_Д} - \underline{b_0 a_2}) - \lambda(\frac{\underline{b_1 a_1}}{\delta_Д \underline{a_2}}(\underline{a_1} - \frac{2\underline{b_1}\overline{a_1}}{\underline{b_1}}) + \overline{b_1}(\underline{a_0} - 2\eta_{max}\overline{a_1}) + \overline{a_1}(\underline{b_0} - \frac{\underline{b_1}\overline{b_0}}{\underline{b_1}})),$$

$$\vartheta(\eta_{max}) = \frac{\overline{b_0}}{\underline{b_1}}(\frac{\underline{a_1}^2}{\delta_Д} - \overline{a_2 a_0})) - \lambda(\overline{a_1}(\underline{a_0} - \eta_{max}\overline{a_1})) - \frac{\underline{b_1}\overline{a_1}}{\underline{b_1}}(\frac{\underline{a_1}^2}{\delta_Д a_2} - \overline{a_0})).$$

After that, on the basis of dependences $k_1(k_2(\eta_{max}))$ and $k_2(\eta_{max})$ we will form a set of inequalities [11]

$$\begin{cases} \underline{F_g(k_1,k_2,\eta_{max})} = \underline{d_1}(k_1\ (k_4(\eta_{max})),k_4(\eta_{max})) - \overline{d_2(k_4(\eta_{max}))}(n-m-1)\eta_{max} \geq 0; \\ \underline{F_0(k_1,k_2,\eta_{max})} = \underline{d_0}(k_1\ (k_2(\eta_{max}))) - \overline{d_1(k_1\ (k_2(\eta_{max})),k_4(\eta_{max}))}\eta_{max} + \underline{2d_2(k_2(\eta_{max}))}\dfrac{\eta_{max}^{\ 2}}{3} \geq 0; \\ \underline{\delta_1(k_1,k_2,\eta_{max})} = \dfrac{\overline{d_1^2(k_1\ (k_2(\eta_{max})),k_2(\eta_{max}))}}{\underline{d_0(k_1\ (k_2(\eta_{max})))d_2(k_2(\eta_{max}))}} \geq \delta_Д. \end{cases}$$

Solving this system, we will find $\eta_{max} = 0.056$ and define $k_2(\eta_{max}) = 0.078$, $k_1\ (k_2(\eta_{max})) = 0.079$.

By these parameterizations of the controller, let us find the vertex coordinates in which the system has its minimal degree of stability. Let us construct the marginal edge route of the polyhedron of the interval characteristic polynomial coefficients of the position stabilization system of a descent underwater object and reflect it on its rooted plane. To construct this route, we will use the algorithm developed for the interval characteristic polynomial with the interval uncertainty of coefficients [13]. The obtained route has an image shown in Figure 2.

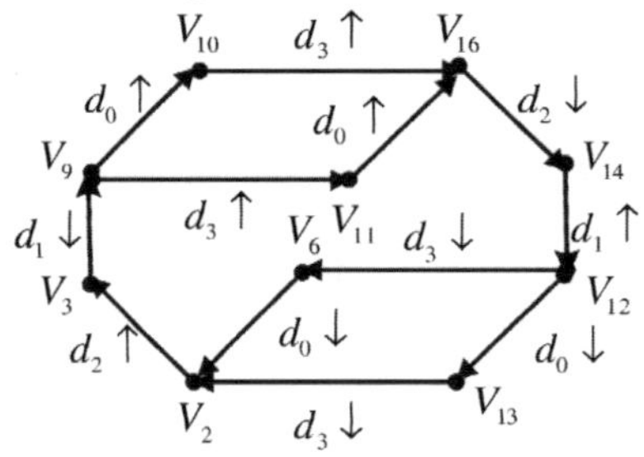

Figure 2. The vertex-edge route

Since all the edge branches by the reflection of the route are convex relative to imaginary axes, in order to define α it is sufficient to reflect only the vertices of the route on the rooted plane. Figure 3 illustrates only those vertices which define the mapped parameter.

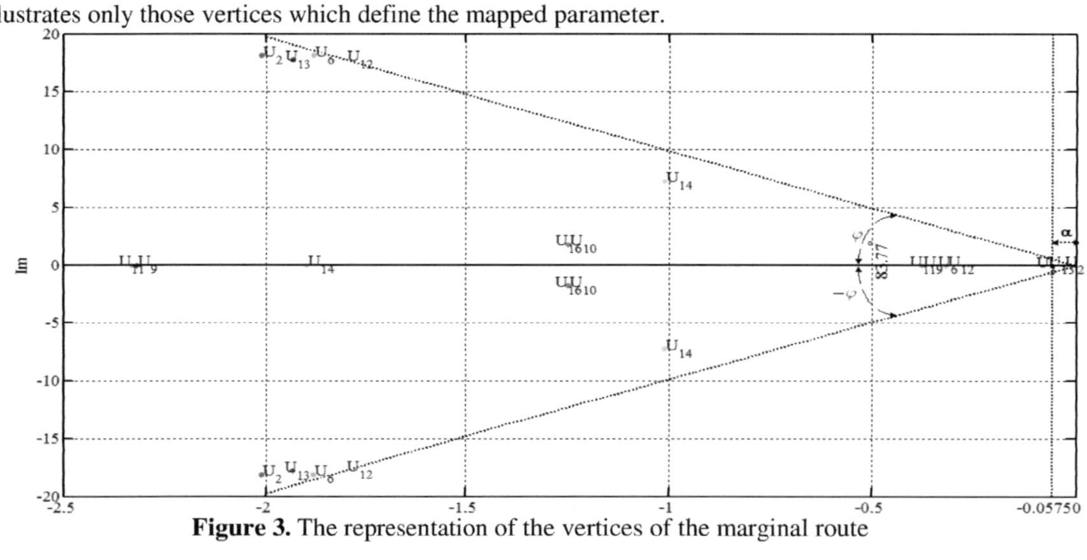

Figure 3. The representation of the vertices of the marginal route

As a result, we will get $\alpha = 0.057$ in vertex $V_2(\underline{d_0};\overline{d_1};\underline{d_2};\underline{d_3})$. At the third stage of the synthesis, let us substitute vertex coordinates V_2 in (2) and get system (3)

$$\begin{cases} \operatorname{Re} D^2(k_1,k_2,\alpha,\beta)=0; \\ \operatorname{Im} D^2(k_1,k_2,\alpha,\beta)=0; \\ \dfrac{\partial \operatorname{Re} D^2(k_1,k_2,\alpha,\beta)}{\partial\alpha}=0; \\ \dfrac{\partial \operatorname{Im} D^2(k_1,k_2,\alpha,\beta)}{\partial\alpha}=0. \end{cases}$$

The solution of the given system is the parameterization of PI controller $k_1 = 0.011$, $k_2 = 0.16$ and the maximal robust degree of stability $\alpha = 0.07$ of the position stabilization system of a descent underwater object.

5. Conclusion

To increase the operation speed of the integral control system, the technique of a combined parametric synthesis of a robust controller was developed. This technique allows maximizing the quasi maximal robust stability on the basis of two-stage parametric synthesis employing the coefficient methods and the method of mathematical programming. The technique was tested on a numerical illustration.

Acknowledgments

The reported study was partially supported by RFBR, research project № 15-01-99447

Reference

[1] Wang L. 1995 Robust strong stabilizability of interval plants: it suffices to check two t vertices *System and control letters* **26** 133-136

[2] Volkov A.N., Zagashvili Y. V. 1997 The method of synthesis of automatic control systems with maximum stability under constraints *Izvestiya RAS, Ser. Computer and Systems* **3** 12-19

[3] Kim D.P. 2003 Synthesis regulator maximum degree of stability *Drive Technology* **4** 52-57

[4] Tatarinov A.V, Tsirlin A.M 1995 Mathematical programming problems containing complex variables, and the maximum degree of stability of linear dynamical systems *Bulletin RAS. Ser. Computer and Systems* **1** 28-33

[5] Keel L.H., Bhattacharyya S.P. 1999 Robust stability and performance with fixed-order controllers. Automatica **35** 1717-1724

[6] Zhmud V, Semibalamut V, Vostrikov 2015 Feedback systems with pseudo local loops *Proc. of Int. Conf. on Testing and Measurement: Techniques and Application* 411-417

[7] Hollot C.V. 1990 Extreme point results for robust stabilization of interval plants with first–order compensator *Proceedings of American Control Conference San Diego, CA, May 1990* 2533-2538

[8] Voevoda A A 1994 Design of linear control systems for plants with interval parameters using the motion separation method *Izvestiya Akademii Nauk: Tekhnicheskaia Kibernetika* **5** 233-241

[9] Nam B.W, Hong B.W., Kim Y.S. 2013 Effects of Passive and Active Heave Compensators on Deepwater Lifting Operation International Journal of Offshore and Polar Engineering **23(1)** 33-37

[10] Nesenchuk A.A. 2002 Parametric synthesis of qualitative robust control systems using root locus fields *Proceedings of the 15th Triennial World Congress of The International Federation of Automatic Control (IFAC)* Barcelona, Spain 21–26 July 2002 387-387

[11] Ezangina T. A., Gayvoronskiy S. A. 2014 The Control of the Descent Submersible in the Confused Sea Environment *Applied Mechanics and Materials*, **536-537** 1200-1206

[12] Zhmud V, Sablina G, Dimitrov L, Trubin V 2016 The use of bypass channel for feedback control of oscillatory object well-known as difficult one for control *SIBCON-2016* 7491767

[13] Vadutov O.S., Gayvoronskiy S. A. 2003 Application of edge routing to stability analysis of interval polynomials *Journal of Computer and Systems Sciences International* **42(6)** 833-838

Bionic models for identification of biological systems

O M Gerget

Tomsk Polytechnic University, 30, Lenina ave., Tomsk, 634050, Russia

E-mail: gerget@tpu.ru

Abstract. This article proposes a clinical decision support system that processes biomedical data. For this purpose a bionic model has been designed based on neural networks, genetic algorithms and immune systems. The developed system has been tested on data from pregnant women. The paper focuses on the approach to enable selection of control actions that can minimize the risk of adverse outcome. The control actions (hyperparameters of a new type) are further used as an additional input signal. Its values are defined by a hyperparameter optimization method. A software developed with Python is briefly described.

1. Introduction

Improvement of intellectual support for decision making aimed at minimizing adverse outcomes is commonly discussed nowadays [1-4]. Clinical decision support systems are developed to aid functional diagnostics.

Data analysis and machine learning can be efficiently used to improve diagnostic systems. In particular, solutions based on deep learning, Gradient Boosting and Random Forest [5, 6] perform strongly. However, they can rarely be observed in practice and are mainly used for research purposes.

This project is aimed at developing a clinical decision support system that can be used to identify functional statuses of biosystems in the context of energy-information processes. The following tasks have been set to reach the overall objective:

1. To characterize behaviors of biological systems in clearly defined functional states (diagnostics).
2. To define control actions that can minimize the risk of system transition into adverse conditions.

Let us consider the following algorithm for the proposed clinical decision support system:

1. To create an operational image of an investigated biosystem based on functional characteristics of the system at some time t_i.
2. To calculate integral parameters I^*_{adapt} for time t_i. To construct cumulative curves.
3. To forecast changes in functional characteristics over time (across different forecast horizons).
4. To identify the forecasted functional state. To calculate forecasted values I^*_{adapt} and functional reserves.
5. To select optimal control actions at $t_{i+1}, t_{i+2}, \dots$.
6. To display possible control actions to enable selection of the optimal action.

2. Materials and methods

For the needs of this study, a bionic model is a mathematical model and a related software that reflect functioning and organization of a biosystem.

Presentation of a diagnosed subject (biosystem) in the form of a bionic model that is based on neural networks (NS), genetic algorithms (GA) and immune systems (IS) can facilitate identification of developing systems. Thus, each diagnosed subject (each functional state of a human) is described in a bionic model <NS, GA, IS, A> where A stands for tuning algorithms.

Combination of NS, GA and IS in bionic models allows an exchange of information and control actions among systems, which enables improved operability and interpretability of obtained values.

2.1. *Energy-informational approach for integral evaluation of functional states*

An approach [7] where the Fullback information measure is considered as a measure of preferred behavior of a biological object is used to evaluate a functional state of a biosystem in the developed clinical decision support system. For that matter, the problem of calculating the informational integral parameter to characterize a functional state of a human can be summarized by the following equation:

$$I = \sum_{j=1}^{n} P_0(\Delta x_j) \ln \frac{P_0(\Delta x_j)}{P_1(\Delta x_j)} = \sum_{j=1}^{n} P_0(\Delta x_j) \ln(P_0(\Delta x_j) - P_1(\Delta x_j)),$$

where n is a number of parameters;

$P_0(\Delta x_j)$ is the probability that characterises a preferable state of the object, i.e. the situation when deviation of j from its normal value is 0.

A normal value is a state when all variables stay within their standard range (for homogeneous groups of diagnosed subjects).

$P_1(\Delta x_j)$ is the probability that x_j stays within its standard range:

$$P_1(\Delta x_j) = P(|x_j - x_0| < \delta) = 2F\left(\frac{\delta}{\sigma}\right) - 1,$$

where x_0 is the average value of x_j; δ is some assigned deviation value;
σ is standard deviation of x_j; F is the Laplace's function.

$P_0(\Delta x_j)=1$, because the preferred state is set at the 0 deviation. Thus, the integral parameter of an object's functional state can be calculated as (1):

$$I_{adapt} = \frac{1}{n} \sum_{j=1}^{n} \ln \frac{1}{P_1(\Delta x_j)}, \quad (1)$$

where n is a number of parameters;
$P_1(\Delta x_j)$ is a probability of deviation of x_j from its preferred state.

Threshold values were obtained for I_{adapt}, on the basis of the levels used in biocybernetics (see Table 1).

Table 1. Correlation of biological object's states to threshold values.

Threshold values	Gradation level	Object's state
$I_n = 0.507$	$\|x - x_0\| < \pm 0.5\sigma$	Satisfactory adaptation
$I_{ten} = 2.01$	$\|x - x_0\| < \pm 1.5\sigma$	Tension of functional systems
$I_{st} = 3.09$	$\|x - x_0\| < \pm 2.0\sigma$	Supertension of functional systems
$I_f = 4.39$	$\|x - x_0\| < \pm 2.5\sigma$	Failure of adaptation

Note: Integral parameter of satisfactory adaptation (I_n), integral parameter of tensed functional system (I_{ten}), integral parameter of supertensed functional system (I_{st}), integral parameter of failed adaptation (I_f).

2.2. *Bionic models in forecasting functional states*

In the designed clinical decision support system, neural networks are responsible for forecasting changes in the parameters that characterize functional states over time and for selecting appropriate control actions. The choice of neural networks is determined by

- High efficiency of forecasting;
- Possibility to implement a committee of models to enable optimal forecasting;
- Possibility to use recursive neural networks for variable structure vectors.

Let us consider some object, which functional state is characterised with parameters x_i and actions u that affect the object's functional state. The functional state is defined by the energy-informational approach.

Let us assume that the value of the i-th parameter at some time t+1 can be calculated by the following equation:

$$x_i(t+1) = f(x_i(t), x_i(t-1), \ldots, u_j) .$$

Then it is the problem of time series forecasting. To solve this problem, a neural network is created for each parameter and trained on the basis of available data sampling.

Forecasting algorithms are based on recurrent neural networks [8, 9] that enable precise identification of behavior. Sequence-to-sequence learning is the main advantage of the aproach with a random length vector at the entry and a random length vector at the output.

2.3. *Selection of control actions and minimization of the risk of system transition into an adverse functional state*

Any functional state of a diagnosed subject at any specific time is characterized by integral parameter, I_{adapt}, which is a function of parameters. Besides, values of control actions $U = \{u_1, u_2, \ldots\}$ are assigned for each diagnosed subject. Let us assume that desirable values $I_{\kappa p}(t), t \in [t_0, t_k]$ are assigned for a preset control horizon at any specific time. The loss function [10] is calculated as a difference between the desired value of the functional state and the current state at some time $C_t = I_n - I_{adapt}$. Then, the optimization algorithm can take the following form:

$$J(x;u) = \frac{1}{k} \sum_{i=t_0}^{t_k} (I^*_{adapt_i} - I_{n_i})^2 \rightarrow min,$$

where $J(x;u)$ is an objective function; $I^*_{adapt_i}$ is a forecasted value of the integral parameter.

The use of the genetic algorithm takes the form of generating optional control actions on the basis of parameters obtained by the proposed energy-informational approach. The generated control actions should be aimed at minimizing the risk of adverse outcome.

The genetic algorith is operated according to the following procedure:

1. The initial population of n chromosomes is generated.
2. Each chromosome is evaluated.
3. A pair of parent chromosomes is selected by one of selection methods.
4. A crossover of two parents is carried out with pc probability to generate two offspring chromosomes.
5. The offspring chromosomes mutate with pm probability.
6. Steps 3-5 are repeated until a new generation with n chromosomes is produced.
7. Steps 2-6 are repeated until a termination criterion is reached.

IOP Conf. Series: Journal of Physics: Conf. Series **803** (2017) 012046 doi:10.1088/1742-6596/803/1/012046

2.4. Identification of functional states of biological systems and detection of inter-state transition based on implementation of an immune system

Immune systems are integrated into the bionic model to identify functional states based on the informational model. The use of the immune algorithm is reasonable, because an addition of new state parameters results in the change of the algorithm only (addition of cells in the immune algorithm); it does not affect formal characterization, neither requires additional training or retraining.

A negative selection algorithm is used as a base algorithm [11, 12]. The number of systems that use the characteristics of immune systems corresponds to the number of states. Each immune system responds to its state and neglects other states. Thus, we not only understand the current state of the object, but also detect where it can transition. Overlapping characteristics in several states can be considered as the level of immunity.

A control action (type of intervention) is created at the output of the immune algorithm and sent to the genetic algorithm.

3. Software development

Currently, a software is being developed for the clinical decision support system based on a multidimensional grid. Each cell of the grid contains values that reflect efficiency of solutions obtained for specific hyperparameters of the model and of the used numerical technique. The number of hyperparameters depends on the dimension of the grid.

In this study hyperparameters are sets of a priori assigned parameters of a model. Hyperparameters remain unchanged during training. In a neural-network-based model, hyperparameters can include the number of layers and neurons, the type of activation functions and their factors, as well as interconnections between the neurons.

Hyperparameters of the learning algorithm include a training pace; a type of parameter initialization; an objective function; a regulation mechanism; an approach to divide training data into training, test and validation sets; and the type of data normalization.

The designed software enables the use of several hyperparameters as input signals. Such hyperparameters are called control actions. To forecast some value $x(t)$, it is defined on the basis of previous values $(x(t-1), x(t-2), \ldots, x(t-p))$ and control action $u(t)$. In the medical context, $u(t)$ can be interptreted as a therapeutic intervention.

Software optimization of hyperparameters is based on the genetic algorithm, where the fitting function is determined by the difference in integral parameters.

The software of the clinical decision support system is developed in Python and consists of the main module and secondary modules (pybrain, numpy, pandas and scikit_learn) that can be connected while running the code of the main module. These modules are complete forecasting models including neural network models and training algorithms. The main module is responsible for importing two dataframes with initial data and hyperparameters that are used as arguments in the instance constructors of the Model and Teacher classes. It also fills the grid cells.

4. Results

The clinical decision support system based on the bionic model was tested on data from pregnant women during all three trimesters of pregnancy. The following therapeutic measures were considered as control actions $u(t)$ [13]: 1) therapeutic exercises, respiratory gymnastics, music therapy, aqua gymnastics; 2) therapeutic exercises, respiratory gymnastics, pharmacotherapy; 3) aqua gymnastics, respiratory gymnastics.

Integral parameters [12] were calculated and cumulative curves were constructed (Figure 1) at the first stage of the investigation to characterize functional states of each diagnosed subject.

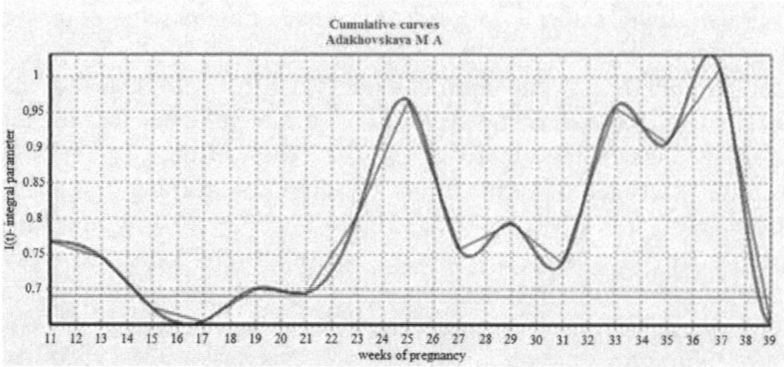

Figure 1. A sample cumulative curve calculated by (1) where blood substances were used as x_i parameters.

At the second stage, forecasted values of the obtained parameters, i.e. blood parameters of the pregnant women, and a forecasted value of $I^*_{adapt_i}$ were calculated by neural networks.

The genetic approach was used to minimize the risk of deviation of the obtained integral parameter [14] from its desired value (healthy performance) and select therapeutic intervention at the final stage. The resulting values are given in Table 2.

Table 2. Selected control actions to minimize the risk of adverse outcome

Trimester	Values of loss function C_t calculated for women with somatic diseases		
	Therapeutic measures. Group 1	Therapeutic measures. Group 2	Therapeutic measures. Group 3
1	0.4	0.7	0.1
2	0.2	0.7	0.3
3	0.1	0.5	0.4

The values in Table 2, calculated for a specific woman, impose therapeutic measures of groups 3-1-1 in appropriate trimesters of the pregnancy.

5. Conclusion

Neural networks, genetic and immune algorithms can be used to create bionic models that can be trained with joint gradient and bionic algorithms. In this paper, the influence of independent variables on dependent parameters is studied with due regard to control actions.

Efficient determination of parameters and hyperparameters in the model and the training algorithm allows minimization of the risk of deviation between actual and desired integral parameters. This enables selection of optimal control actions.

6. Acknowledgments

This work was (in part) supported by the Russian Governmental Programme "Nauka", No. 4.1660.2016.

References

[1] Shcherbakov M, Kamaev V and Shcherbakova N 2013 *Proceedings of the 7th IFAC Conference on Manufacturing Modelling, Management, and Control* **46(9)** 1027-1032

[2] Yurchenko M , Kochegurova E, Fadeev A and Piletskaya A 2015 *Proceedings of the 2014 International Conference on Engineering Technologies, Engineering Education and*

Engineering Management 847-851

[3] Massel L and Massel A 2012 *Bulletin of Tomsk Polytechnic University* **321(5)** 135-140

[4] Kochegurova E and Gorokhova E 2015 *Lecture Notes in Computer Science* **9330** 512-519

[5] http://scikit-learn.org/stable/modules/generated/sklearn.ensemble.RandomForestClassifier.html

[6] Carmesin H 1994 *Physics Letters* **A 188(1)** 27-31

[7] Fokin V, Pekker Ya, Berestneva O and Gerget O 2012 *Bulletin of Tomsk Polytechnic University* **321(5)** 120-124

[8] Wang H and Song G 2014 *Neurocomputing* **134** 289-295

[9] Gerget O, Marukhina O and Cherkashina Yu 2016 *Key Engineering* **685** 957-961

[10] Scherbakov M, Brebels A, Scherbakova N, Tyukov A, Yanovsky T and Kamaev V 2013 *World Applied Sciences Journal* **24(24)** 171-176

[11] Dasgupta D 2006 *Iskusstvennuye immunnye sistemy i ih primeneniye* (Artificial immume systems and their applications, Moscow: Fizmatlit) Russian

[12] D'haeseleer P, Forrest S and Helman P 1996 *Proceedings 1996 IEEE Symposium on Security and Privacy* 110-119

[13] Krivonogova T, Evtushenko I, Trenkaeva N, Gladkih L and Gerget O 2008 *Kompleksniy podhod k ozdorovleniuy beremennyh zhenschin* (Complex approach to health improvement of pregnant women, Tomsk) p. 183 Russian

[14] Gerget O and Kochegurov A 2012 *Bulletin of Tomsk Polytechnic University* **135(5)** 117-120

Simulation of economic agents interaction in a trade chain

I A Gimanova[1], A S Dulesov[1], N V Litvin[1]

[1]Khakass State University named after N.F. Katanov, 92, Lenina Ave., Abakan, 655009, Russia

E-mail: gimanowa@gmail.com

Abstract. The mathematical model of economic agents interaction is offered in the work.It allowsconsidering the change of price and sales volumesin dynamics according to the process of purchase and sale in the single-product market of the trade and intermediary network. The description of data-flow processes is based on the use of the continuous dynamic market model. The application of ordinary differential equations during the simulation allows one to define areas of coefficients - characteristics of agents - and to investigate their interaction in a chain on stability.

1. Introduction

The research of the regularities that occur in trade systems attracts interest of all participants: producers, intermediaries, consumers, and the new companies appearing in the market as well. The head of any line-up of a network has to consider these processes to predict the consequences of the administrative resolutions.

A set of conditions and factors influences the dynamic processes that occur in the system. That is why, there is the need of deep and comprehensive study of problems of agents interaction for intermediary systems in the market conditions.

The trade system means a non-steady dynamic system, represented in the form of a set of chains of economic agents where characteristics are changed in the course of time.

The review of scientific works showed a variety of approaches on how to research trade systems [1-3].When developing this model of economic agents interaction, the data-flow processes proceeding in intermediary chains of the system were considered [4, 5].

2. The model of economic agents interaction

It is given: There are restrictions in the chain of economic agents consisting of the producer, intermediaries and the end user. They are connected with:

- the money given for purchase of goods that depends on the predicted demand for goods according to season fluctuations and other influences of environment;
- the inability to sale and/or to purchase a particular quantity of goods that depends on a type of goods, conditions of its storage, the type and volume of a warehouse.

Economic agents own data about the situation in the market and can influence the demand of $D(t)$ and the supply of $S(t)$ of these goods in the market.

It is required: to develop the model of economic agents interaction of a trade chain.

IOP Conf. Series: Journal of Physics: Conf. Series **803** (2017) 012047 doi:10.1088/1742-6596/803/1/012047

Task solution: The trade and intermediary chain is represented in the form of a multiagent system where the interaction of agents occurs among themselves, the consumer market and the market of resources (fig. 1).

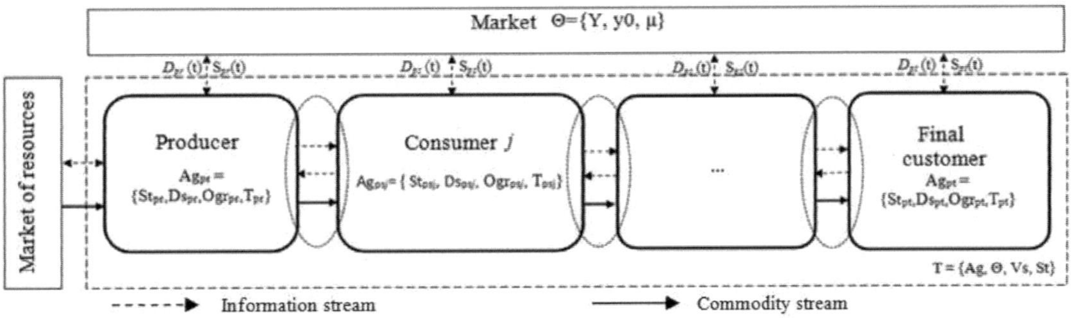

Figure 1. The trade and intermediary chain.

The consumer market is described by:

$$\Theta = \{Y(t), y0, \mu\},$$

where Y(t) – a set of market conditions, y_0 – a reference state of the market, μ: Y(t) ×Dsag → Y(t) – a function of market behavior, Dsag(t) = $\{Dsag_i(t) \mid i = 1..n\}$ – a set of actions of chain agents. Interactions between agents occur in the local micromarkets on the basis of mechanisms of information exchange and making the decision of the purchase-sale satisfying the buyer and the seller.

The trade and intermediary chain is described as

$$T = \{Ag(t), \Theta, Vs(t), St(t)\},$$

where Ag= $\{ Ag_i \mid i = 1.\, n \}$ – a great number of agents, Vs(t) – a set of the relations between agents, St(t) = $\{ St_1(t) \times St_2(t) \times ...\times St_n(t) \}$ – a set of conditions of chain, $St_i(t)$ – a condition of the *i*-agent.

Each *i*-agent is described as follows:

$$Ag_i(t) = \{St_i(t), Dsag_i(t), Ogr_i(t), T_i\},$$

where $Dsag_i(t)$ – actions of the *i*-agent, T_i: $St_k \times Ds_i \times Ogr_i \rightarrow St_{k+1}$ – a function of transitions from a condition of *k* to a condition of *k*+1, $Ogr_i(t)$ – a set of restrictions of the *i*-agent.

3. Mathematical methods of definition of chain characteristics

It is given: Producers, intermediaries and consumers of the market of one product form uniform related retail chain stores which can be presented in the form of connected terminating focused graph G=⟨V, E, H⟩, where V – aset of tops of the count, E – aset of arches, H of display H: X→ V×V.

Tops of $i_k \in V$, i=1..*n*, k=0..3 are interpreted as economic agents of retail chain stores of the *k*-level, arch (i_k, j_{k+1}) as the micromarkets where the prices of $pi_k j_{k+1}(t) \geq 0$, according to whichan exchange of production of the volume of $Qi_k j_{k+1}(t) \geq 0$ between various tops of the count is established.

It is required: to develop mathematical methods of definition of economic agent characteristics in their interaction.

Task solution: In figure 2, the chain of retail chain stores consisting of two agents of *k*– and *k*+1 – levels where $i_k = i$ – the seller, $j_{k+1} = j$ – the buyer presented to $Q_{i,j}(t)$ – volumes of the realized goods going on arch (i_k, j_{k+1}) , $p_i(t)$ – the goods price for *i*-m the market, $F_{i,j}(t) \geq 0$ – the financial means coming from the *j*-buyer to the *i*-seller.

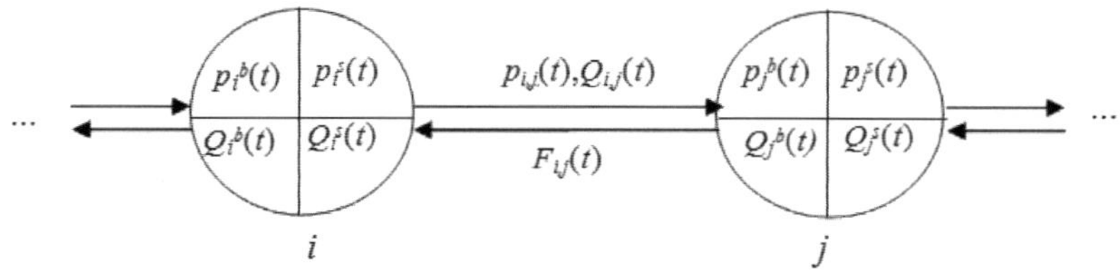

Figure 2. The line-up of retail chain stores.

Let us use the continuous dynamic market model in the shape of

$$\frac{dp(t)}{dt} = a_{DS}(D(p) - S(p)),$$

where $D(p)$ – a demand function, S(p) – a supply function, a_{DS} - a coefficient.

After replacement of $\Delta Q(p) = D(p) - S(p)$ and $a = 1/a_{DS}$, we obtain

$$\Delta Q(p) = a\frac{dp(t)}{dt}.$$

Each i-agent – an intermediary at any level – can act as a buyer and as a seller, where p_i^b(t) – the price of goods of the intermediary agent as a buyer, as, p_i^s(t) – the price of goods of the intermediary agent as a seller.

Main characteristics of subjects of the trade system of the single-product market (end user, buyer, seller):

- a coefficient of goods conductivity of the i-agent:

$$R_i(t) = Q_i(t) / p_i(t);$$

- a coefficient of the price change of the agent depends on the change of volumes of consumption of goods in time $[t_i, t_j]$

$$a_1(t) = \frac{\Delta p(t)}{\Delta Q(t)} = \frac{p(t_j) - p(t_i)}{Q(t_i) - Q(t_j)};$$

- a coefficient of volume change of goods of the agent from the change in price $[t_j, t_k]$.

$$a_2(t) = \frac{\Delta p(t)}{\Delta Q(t)} = \frac{p(t_k) - p(t_j)}{Q(t_j) - Q(t_k)}.$$

Goods movement occurs from the producer (seller) to the consumer (buyer) under the influence of forces of supply and demand.

Such economic agents in a trade chain as the end user, the buyer, the seller are considered. Their trial functions of the price and sales volume are presented in table 1.

Table 1. Ratios between the price and sales volume.

	Price function	Volume function
Seller	$p(t) = a_2(t)\frac{dQ(t)}{dt}$	$Q(t) = \frac{1}{a_2(t)}\int_{-\infty}^{t} p(t)dt$
Buyer	$p(t) = \frac{1}{a_1(t)}\int_{-\infty}^{t} Q(t)dt$	$Q(t) = a_1(t)\frac{dp(t)}{dt}$
Final customer	$p(t) = Q(t) \setminus R(t)$	$Q(t) = R(t) \cdot p(t)$

4. Results and Discussion

The mathematical model of advance of commodity streams is formed for the allocated chains from the system based on preservation regularities of the material stream in a network knot, cost preservation of goods in a network contour and characteristics of each agent according to table 1.

The scheme of a current circuit «the seller – the final customer» is represented in figure 3, where communication of the active seller and the final buyer through the micromarket is simulated.

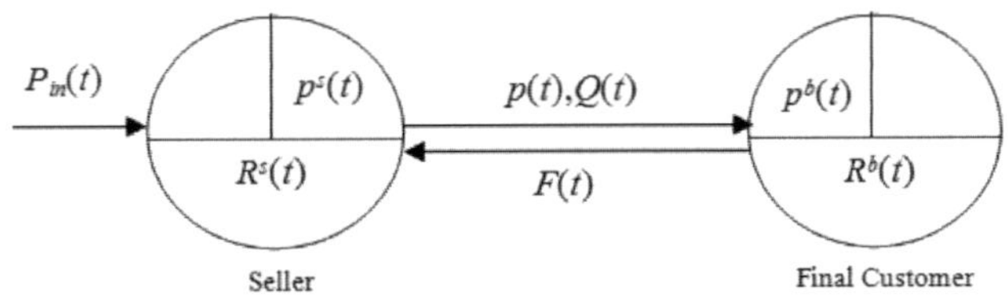

Figure 3. The connection scheme of a chain «the seller – the final customer».

This connection is described in the form of the system of ordinary differential equations of the first order with constant coefficients with starting conditions $Q(0) = Q_0$, $p(0) = p_0$,

$$\begin{cases} Q(t)/a_2(t) + Q'(t)/R(t) = Q_0(t) \\ p(t)/a_2(t) + p'(t)/R(t) = p_0(t) \end{cases}.$$

The simulation function is the following:

$$\begin{cases} Q(t) = Q_y + (Q_0 - Q_y)e^{-t/E_S} \\ p(t) = p_y + (p_0 - p_y)e^{-t/E_S} \end{cases}, \quad E_S = \frac{Q_0 - Q_y}{p_0 - p_y} \cdot \frac{P_y}{Q_y} = a_2 \frac{p_y}{Q_y}.$$

In periods of time $[\Delta t_{1,6}, \Delta t_{1,13}]$, the system simulation with step Δt=1 is carried out a week with coefficients values

$$a_2 = \frac{Q_0 - Q_y}{p_0 - p_y} = 79.902 > 0, \quad E_S = \frac{Q_0 - Q_y}{p_0 - p_y} \cdot \frac{P_y}{Q_y} = 0.949 > 0.$$

The transient period made T = 7 weeks (Table 3).

Table 2. The simulation at step Δt=1 week.

i	$\Delta t_{1,i}$, week	Price, p_i	Sales volume, Q_i	Model price, $p_{i,mod}$	Ratio error, $\delta(p_{i,})$, %	Model volume, $Q_{i,mod}$	Ratio error, $\delta(Q_{i,})$, %
0	6	19.00	1602.55	19.00		1602.55	
1	7	19.36	1718.75	19.326	0.176	1628.579	5.25
2	8	19.50	1617.80	19.439	0.313	1637.649	1.23
3	9	19.50	1560.28	19.479	0.108	1640.81	5.16
4	10	19.50	1597.55	19.493	0.036	1641.91	2.78
5	11	19.50	1452.55	19.497	0.015	1642.295	11.37
6	12	19.50	1395.41	19.499	0.005	1642.43	13.07
7	13	19.50	1544.46	19.50	0	1642.48	6.35

To estimate the accuracy of the mathematical model, i.e. the hit rate of values of exit pupils of an actual object and values of the same parameters calculated by means of the model, the following relative accuracy has been used:

$$\delta = \max_{i=1..7} \delta_i = \max_{i=1..7}\{\frac{|Q_{i,\text{mod}} - Q_i|}{Q_i}, \frac{|p_{i,\text{mod}} - p_i|}{p_i}\}, \Rightarrow \delta = 13.07\ \%.$$

During the change in price of the seller in the course of purchase and sale, we can see the movement to a stable state (fig. 4). Values $Q_0<Q_y$ and $p_0<p_y$, that is why, the price and volume functions increase up to the established values. The seller and buyer interaction will lead to the equilibrium price and volume, i.e. agentsinteraction is steady.

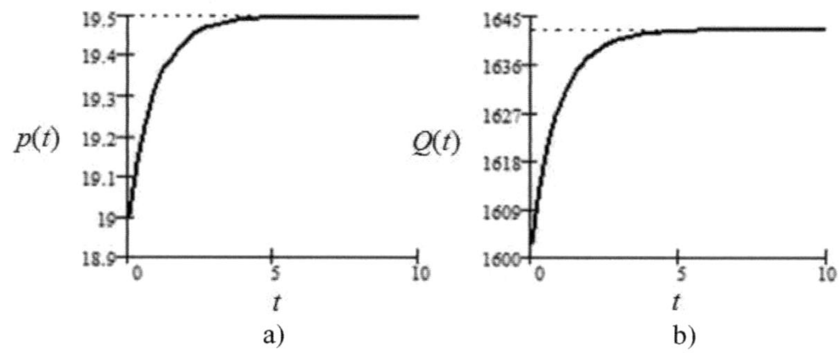

Figure 4. Changes of functions: a) prices of $p(t)$; b) volume of $Q(t)$.

5. Conclusion

The developed model of economic agents interaction in a trade and intermediary chain taking into account their dynamics allows one to prove decisions made on the choice of administrative decisions by agents when pricing and sales volumes, andredistribution management of commodity streamsas well.

The developed mathematical methods of definition of economic agents characteristics when interacting with an intermediary chain allow tracing the change in price and sales volumes of goods in dynamics, to reveal stability boundaries of interrelation of agents in a network and to estimate potential opportunities in advance of goods.

6. Acknowledgments

The reported study was funded by the RFBR, according to research project No. 16-46-190398.

References

[1] Oliveira J B, Lima R S, Montevechi J A B 2016 *Simulation Modelling Practice and Theory* **62** 166-191
[2] Groves W, Collins J, Gini M, Ketter W 2014 *Decision Support Systems* **57** 274-284
[3] Denkena B, Henjes J, Henning H 2011 *CIRP Journal of Manufacturing Science and Technology* **4(1)** 9-14
[4] Dulesov A S, Kurynova I A 2011 *Modern problems of science and education* **5**
[5] Dulesov A S, Kurynova I A 2012 *Modern problems of science and education* **4**

A hardware-software complex for modelling and research of near navigation based on pseudolites

A B Gladyshev, D D Dmitriev, E A Veysov, V N Tyapkin

Siberian Federal University, 79 Svobodny st., 660041 Krasnoyarsk, Russia

E-mail: a-glonass@yandex.ru

Abstract. The paper considers a hardware-software complex for research of characteristics of accuracy and noise immunity of a near navigation system based on pseudolites. The complex is implemented on the basis of the "National Instruments" hardware platform and "LabView" coding environment. It provides a simulated navigation field, the analysis of the received signals, the determination of the errors of measurement of navigation parameters for pseudolites signals, comparing the measured error with the characteristics of a standard GNSS receiver.

1. Introduction

Currently, a near navigation system is widely adopted technologies based on the use of pseudolites [1–3]. In such systems, the measurement of navigation parameters is carried out basing on signals pseudolites. Their structure is similar to that of signals from navigation satellites GLONASS / GPS. The main advantages of near navigation systems based pseudolites are an opportunity to improve the geometric factor, the formation of high-power signals. The result is the accuracy of the navigation determinations system and improved noise immunity.

Despite all the advantages of navigation systems based on pseudolites, there are a number of unsolved problems [4–8]:

1. An increased pseudosatellite signal level requires an enhanced dynamic range of the navigation receiver.
2. Multipath pseudolites navigation signal error increases with code and phase measurements of navigation parameters.
3. Great synchronization errors of time scales accuracy of pseudolites lead to deterioration in the accuracy of navigation determinations.
4. Weak electromagnetic compatibility of signals pseudolites and navigation satellites.

Solving these problems requires further theoretical and experimental studies. This leads to the creation of the hardware-software complex - an experimental prototype of the near navigation system based on pseudolites.

2. Requirements for the hardware-software complex

Basic requirements for the hardware-software complex are formed based on the navigation structure of the system, in which the main element is a grouping of pseudolites. The structure of the hardware-software complex should include pseudolites prototypes for forming the navigation field with the specified parameters. Thus, it should provide [7–9]:

- determination of the accuracy characteristics of navigation parameters measurement methods;
- analysis of the noise immunity of the receiver pseudolites signals;
- the ability of flexible changes of parameters of pseudolites signals;
- the ability to control the geometric factor of pseudolites grouping.
- flexible management of all components of the hardware-software complex.

3. The structure of the hardware- software complex

As a result of analysis of possible options of the construction of hardware systems for the study of such systems [10–12], a variant of creation of the hardware-software complex, based on hardware platform National Instruments and software LabVIEW, was suggested. In addition, it should provide for the possibility of using test equipment from other manufacturers, and a specialized software.
Based on these requirements, the block diagram of the hardware-software complex was developed (Figure 1).

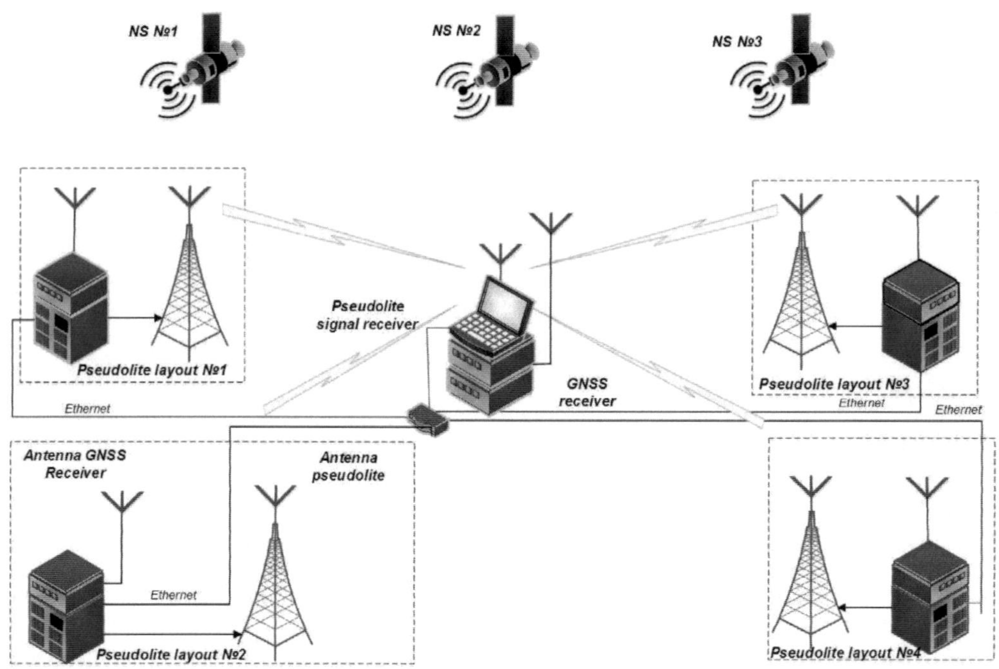

Figure 1. The structure of the hardware-software complex for modeling and researching the systems of near navigation on the basis of pseudolites

Its structure consists of the following parts:
- four pseudolites layouts;
- receivers of pseudolites signals and satellite navigation signals, connected to a receiving antenna;
- a portable signal analyzer;
- personal computers for processing of experimental results.

The main task of the hardware-software complex is the formation of an artificial navigation field in a given region of space. This task is performed by four pseudolites layouts that emit navigation signals of the required structure. Transmitting antennae of pseudolites have accurate geodetic reference to the terrain. The exact coordinates of the pseudolite antenna are included in its navigation message and are used in the solution of the navigation task in the receiver.

4. The operating principle of components of the hardware-software complex

The basis of the pseudolite layout is a hardware platform "PXI" of National Instruments Company, which operates under the control of software LabVIEW. A pseudolite layout consists of the following components (see Figure 2):

- Vector generator of signals NI PX1e-5673E;
- RF attenuator NI PXI-5695;
- Computer controller NI PXIe-8880;
- Chassis NI PXIe-1085;
- GNSS receiver (such as MRK-101).

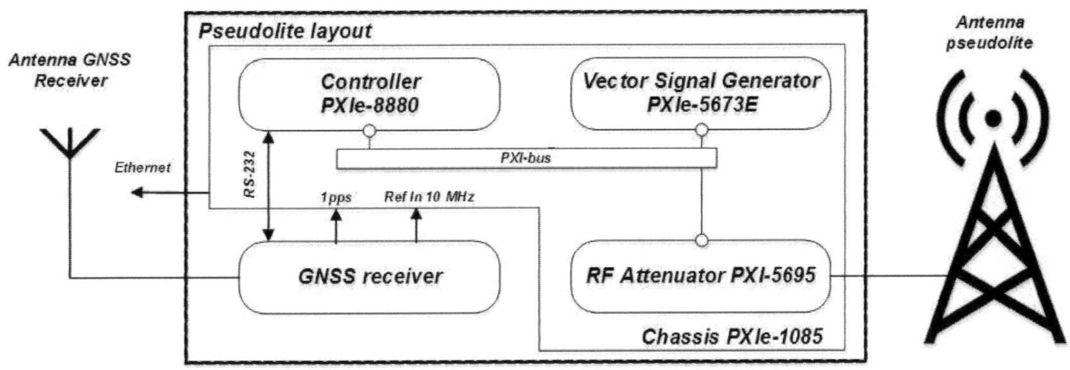

Figure 2. A block diagram of the pseudolite layout

The navigation signals are generated by vector generator NI PXIe-5673. Modern technologies used in the vector generator, allow arranging the full direct digital synthesis in the formation of radio signals. Vector Generator NI PXIe-5673 consists of three functional modules: a signal generator of the arbitrary waveforms "NI PXIe-5450", HF generator "NI PXI-5652" and a vector modulator "NI PXIe-5611" (Figure 3).

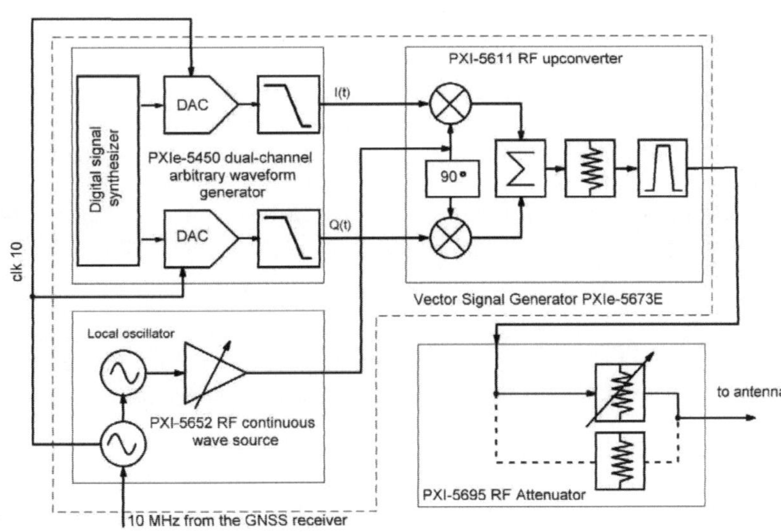

Figure 3. A functional scheme of the pseudolite layout based on vector generator PXIe-5673

The digital signal synthesizer, located in arbitrary waveform generators NI PXIe-5450, is formed by a digital code in phase I (t) and quadrature Q (t) components of the modulating sequence of the navigation signal. This code is generated by the algorithm specified in special software module (subVI) LabVIEW. By changing the algorithm, it is possible to create a variety of the navigation signal structures and explore their effectiveness.

The digital code of the modulating sequence of the navigation signal is converted into an analog signal DAC. Phase and quadrature components of the analog signal from the DAC output are filtered by a low-pass filter (LPF) and arrive at the output of the generator of the arbitrary waveform signal. Thus, the output of NI PXIe-5450 formed modulating sequence corresponding to the structure of the navigation signal.

In vector generator NI PXIe-5673 of signals, direct transfer of signals from the baseband to an RF band is applied. The inputs of vector modulator NI PXIe-5611 receives phase and quadrature components of the modulation signal sequence and navigation RF signals at the carrier frequency with HF generator NI PXI-5652.

In vector modulator NI PXIe-5611 there is a transfer of in-phase and quadrature components of the navigation signal to the frequency of pseudolite radiation. In addition, the formation of an integrated navigation signal, its amplification and filtering within the necessary bandwidth occur. As a result, the output of the vector modulator (it is the output of the vector generator) enters the high-frequency navigation signal in the band specified frequency code channel.

To solve the problem of the navigation provision of users, similarly to GNSS, the time scales of pseudolites are to be synchronized and their signals - coherent. The peculiarity of the formation of a highly coherent navigation pseudolite signal is to synchronize all functional units of the vector generator by means of the highly stable reference signal of 10 MHz from GNSS receiver MRK-101.

Permissible relative error of the reference signal in frequency in the mode of GNSS receiver tracking after the signals of navigation satellites is less than $\pm 3 \cdot 10^{-10}$. The generator of pseudolite time scale synchronizes all the functional components of the vector generator: RF generator oscillator NI PXI-5652, a digital signal synthesis scheme, the DAC and other nodes requiring synchronization. If necessary, synchronization vector generator signals can be realized by the highly stable time and frequency standard.

To extend the range of power variation of the output signal, two-channel programmable RF attenuator NI PXI-5695V was introduced in the pseudolite scheme. When maximum power of the output signal of the vector generator is equal to 10 mW, HF attenuator allows changing the power of the signal at the output of the pseudolite in the range of 1 mW to 10 mW. This wide range of changes in signal strength will allow investigating the effect of the signal power level of the reliability of navigation definitions under the impact of noise and assessing the sensitivity of the receiving devices. If we take the values of the gains of the receiving and transmitting antennas to be equal to 3 dB, we can determine that the signal level with a power of 1 mW to 10 mW will provide an opportunity to carry out experimental research at a distance of 500 m to 50 km from the pseudosatellite layout, when the signal level at the receiver input is161 dBW.

The hardware-software complex allows carrying out experimental studies with different types of radiated signals. Since vector signal generator NI PXIe-5673 can generate signals in the range from 85 MHz to 6.6 GHz, and the LabVIEW software environment allows generating any code signal structure, then, at the pseudosatellite layout output, it is possible to obtain signals both with frequency and code division.

Values of signal frequencies emitted by pseudolites layouts are selected in the operating range of the standard frequency of the GNSS receiver. However, to eliminate the mutual influence of the GLONASS signals and the signals from the pseudolites layout, the signal frequencies of the latter are formed in the frequency band free from standard satellite signals. Figure 4 shows the frequency bands for signals pseudolites layouts in the L1 band. Frequency bands for their performance in the range of L1 are equal to 1593…1598 MHz and 1605…1610 MHz, the width of each band is 5 MHz. In this frequency band, it is possible to form up to 16 frequency channels. Selection and distribution of frequencies for band L2 can be produced in a similar manner. In this case, the range of signal emission of the pseudolites layouts will be within 1238…1243 MHz and 1248…1253 MHz.

The configuration of location of pseudolites layouts and their number in the near navigation system under development depend on the terrain and the size of the area where the navigation provision is required. Using the hardware-software complex will provide practical experimental data of dependence of to the geometric factor on the number of pseudolites layouts and their relative position. This will allow calculating the potential accuracy of coordinates measurement, the accuracy of determining the spatial position of the consumers of navigation information, as well as the performance of the noise immunity of consumers of the developed near navigation system.

To reduce the error associated with multipath signals of navigation in difficult terrain conditions, it is advisable to use a directional antenna with the directional diagram of a special shape as a transmitting antenna in order to create the required navigation field.

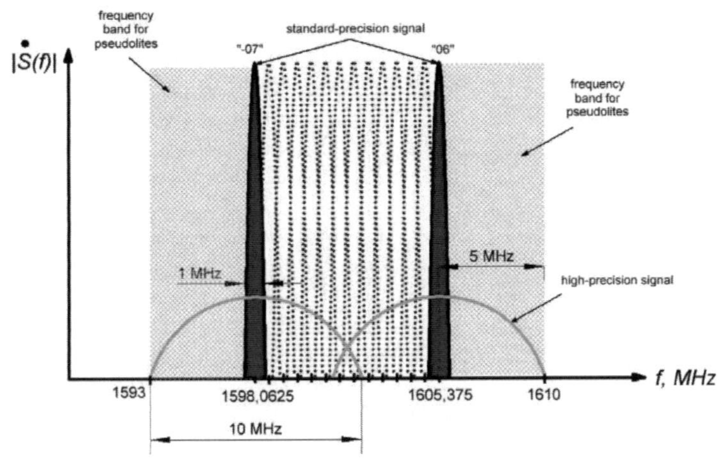

Figure 4. Distribution of frequency signals for the pseudolites layouts for the L1 band

5. The methodology of the research with the help of the hardware-software complex

Two measuring receivers are placed on the ground: angle GNSS receiver (of MRC-101 type) and a finished signal receiver of pseudolites connected to the same antenna. Pseudolites frequency signals are in the operating range of the GNSS frequencies. If possible, it is advisable to carry out topographic binding of the pseudolite antenna to the points of the state geodetic network.

Received and processed navigation information from two receivers is supplied via communication lines for data processing to the PC. Then, using special software, the processing of the experimental data, calculation of measurements errors under different external conditions: changing of the geometric factor and power signals of pseudolites layouts, application of navigational signals of various types, under the influence of interference.

To measure the level of the received navigation signal, a signal analyzer is connected to the input of the reception path through the directional coupler.

The means of the hardware and software complex make it possible to evaluate the measurement error of navigation parameters for pseudolites signals by comparing the coordinates obtained from the GNSS receiver and from other sources under the impact of various external factors.

6. Conclusion

The proposed version of the hardware-software complex provides large-scale opportunities for modeling and researching the systems of near navigation on the basis of pseudolites. This complex allows us to estimate the validity of selection of the navigation signal parameters, optimum geometrical arrangement of pseudolites, efficiency of the developed receiver processing algorithms at the stages of designing. A special software will allow us to automate the process of measuring the accuracy of the navigation options, to explore the noise-immunity of the developed near navigation system.

7. Acknowledgments

The study was performed owing to the grant of the Russian Science Foundation (project №16-19-10089).

References

[1] Kim C., So H., Lee T., Kee C. 2014 A pseudolite-based positioning system for legacy GNSS receivers. *Sensors (Switzerland).* **14 (4)** 6104-6123

[2] Rzepecka Z., Rapiński J., Cellmer S., Chrzanowski A. 2014 Effect of additional distance measurements on satellite positioning. *Acta Geodynamica et Geomaterialia.* **11 (1)** 35-44

[3] Xu R., Chen W., Xu Y., Ji S. 2015 A new indoor positioning system architecture using gps signals. *Sensors (Switzerland).* **15 (5)** 10074-10087.

[4] Wan X., Zhan X. 2012 The research of multipath and linear error for pseudolites applications. *Applied Mechanics and Materials.* **130** 2890-2893

[5] Shen L., Li L., Zhang H., Bao T. 2015 A near-far effect mitigation method of pseudolites based on array signal. *Lecture Notes in Electrical Engineering.* **340** 647-656

[6] Alexandru R.-C., Elena-Simona L. 2015 Comparison of detection techniques for multipath propagation of pseudolite signals used in dense industrial environments. *Energy Procedia.* **100** 1294-1300

[7] Liu C., Lin P., Zhang J., Zhao X. 2015 An improved pseudolite location model of GPS/pseudolite integrated positioning. *Zhongguo Kongjian Kexue Jishu/Chinese Space Science and Technology.* **35 (1)** 75-83

[8] Sultana Q., Sarma A., Kumar A., Malik M. 2012 Enhancement of DOP in stratolite-based navigation systems. *IETE Journal of Research.* **58 (6)** 476-482

[9] Borsoev V.A., Galeev R.G., Grebennikov A.V., Kondratiev A.S. 2011 *Scientific Bulletin MSTU CA.* **164** 17-23

[10] Krasnenko S.S., Pichkalev A.V. 2011 *Reshetnev Readings: Materials of the XV International Scientific Conference.* **2** 184-185

[11] Krasnenko S.S., Pichkalev A.V., Nedorezov D.A., Lapin A.Yu., Nepomnyashchii O.V. 2014. *Vestnik SibSAU.* **1 (53)** 30-34

[12] Dmitriev D.D., Gladishev A.B., Tyapkin V.N., Fateev Yu.L. 2016 Hardware-Software Complex for Studying the Characteristics of GNSS Receiver. *International Siberian Conference on Control and Communications (SIBCON)*

Computer simulation of functioning of elements of security systems

A V Godovykh, B P Stepanov, A A Sheveleva

Tomsk Polytechnic University, 634050 Russia, Tomsk, Lenin Avenue, 30

e-mail: aas@yandex.ru

Abstract. The article is devoted to issues of development of the informational complex for simulation of functioning of the security system elements. The complex is described from the point of view of main objectives, a design concept and an interrelation of main elements. The proposed conception of the computer simulation provides an opportunity to simulate processes of security system work for training security staff during normal and emergency operation.

1. Introduction

Safe functioning of the industrial facility cannot be possible without creation of a security system. Usually, an automated systems execute the management of the security system. Such systems presuppose the presence of people [1, 2]. In this case, the security staff should perform competent and prompt actions, because it defines the capabilities of the system to resist threats.

Objectives of the security staff are: to observe the operation of facility, to react to alarm signals, to manage the response forces for capture of adversary. Training of the security staff is necessary to increase the protection level of an object [3]. However, such kind of trainings on the basis of the real security system are impossible. Depending on the field of activity of the staff, different training methods can be implemented. It can be consideration of technical and legal documentation, operation of specialized stands with technical equipment. But, the best complex training can be implemented using integrated electronic training devices that are based on modern modelling technologies. These training simulators can be formed as specialized computer workstations [4]. Its features are only limited by assigned tasks.

2. Basic elements of the analytical complex

The informational complex is considered as a computer-based workstation that simulates an interaction between an attacker and a security system of the object, where both sides are controlled by the users. The automated workplace simulates functions of security systems to provide training opportunities for a training staff.

Main objectives of the complex are:

- training of skills of the operators of the control center;
- formation of abilities to observe and analyze the information by the security system staff;
- learning the decision-making methods caused by an unauthorized activity.

Work of the complex is based on the simulation of interaction "attacker - security system" with regards of characteristics of both [5]. The main functions of the complex are performed by a special software. It stores and archives data, executes algorithms, makes a connection between units and a fundamental mathematical model, visualizes the processes.

The software configuration is designed in the form of target blocks - units, which are allocated according to the performing tasks [6]. The units operate data that are represented by embedded or customer databases. Core units are presented in Table 1.

Table 1. Basic units of the analytical complex

	Function		Name of unit
1	First level	"Static" model	Construction Unit Evaluation Unit
	Description of basic characteristics of the object and the attacker		
2	Second level	"Dynamic" model	Imitation Unit
	Performing of interaction algorithms		
3	Third Level	Visualization	Visual Unit Interface Unit
	Displaying of the object, events and interface		
4	Support	Service	Information-analytical unit
	Legislative and methodical maintenance		

Constructive, evaluation and imitation units are the basis blocks of the complex. The first two are responsible for the imitation of the first level model – a "static" model. At this level, a user describes basic characteristics and the infrastructure of the object, for example: features of an object, properties of engineering and technical means of a security system etc. In addition, a user defines the general characteristics of both sides: the number of response forces and attackers, their "tactics" etc.

The imitation unit is responsible for the second level of the simulation. This level is determined by dynamic parameters of the model. At this stage, interaction algorithms of elements of the first level model are defined. Examples are the implementation of different "tactics" and versions of scenarios for response forces and attackers; the influence of an intruder; simulation modeling within an emergency situation etc.

The visual unit presents the third level of the simulation. Its function is to display the prototype of an object and events in a form that is convenient and realistic for a user. Another purpose is to visualize a graphical user interface. Along with the interface unit, the visual unit maintains a connection between a user and resources of the analytical complex [7]. The interface unit administrates a software interaction between databases and units. It is also responsible for linking organization of hardware, which is combined into an automatized workplace.

Visualization of security systems pursues a number of objectives. In turn, the computer simulation gives the ability to display the object of research as a combination of elements and their properties. In forming the initial requirements to the model of an object with considering the possibility of subsequent visualization, it is logical to determine the levels of detailing and connections of the visualization and computational parts of the model.

Imaging techniques in varying degrees are based on source data of the model. When forming the functional part of the security system model, its visualization is possible through the following options. The option one is when two separate units: visual and evaluation, are connected only at the logical level and their purpose is demonstration. An alternative option is a more "dense" interaction of the model elements and connections, up to the use of additional methods and visualization techniques.

In case of simulation of security systems, the "interactive environment' is in the spotlight. It is focused on computational models of physical and technical parameters of the technical means, response forces and variants of their complex interaction. Related tools provide construction activity at the information and documentary level that extends the functionality of such an environment. And the presence of a well-prepared graphical database converts this model in the framework of the environment at a significantly different level.

The information-analytical unit is responsible for informational support of the analytical complex. It is reference-methodical maintenance and a legislative and regulatory framework related to the operation and the construction of the security systems [8]. This unit presented as a database consisting of documentation related to security systems. It is an interactive documentary search engine to quickly find the desired information using a variety of filters. Besides, a user can add, remove and change the stored information. Extra tools for work represent analytical functions with the evaluation unit. This unit also allows choosing the modes of operation of the analytical complex depending on the demands in training of several types of the staff of the security systems.

3. Data sources

The databases form a resource base for describing parameters of the object and interacting elements. In general, databases are considered as systematized and categorized data. Their purpose is information support of the organization of the simulation process. Databases consists of numerical characteristics, independent or combined parameters as well as graphical and other elements of visualization. One or few units can use multiple elements of the databases.

Additional plug-resources - custom databases also can be used [9]. They give the opportunity to upload and download data for special calculations. An identical opportunity is presented for working with graphics and algorithmic resources. This greatly expands capabilities of the system as a whole.

All the data used by the informational complex are shown in Table 2 [10].

Table 2. Primary data sources

Type	Name	Description	Parameters
Active members	Attacker	characteristics of outside adversary	- number; - qualification; - type; - equipment etc.
	Guards	characteristics of response forces	
Description of the object	Object	description and features of secure object for simulation purposes	- characteristics of standard objects; - targets (object of defense); - operation mode; - threats; - operation mode; - site area parameters etc.
	Barriers And Sensors	parameters of engineering and technical means of security system	- type; - time of delay; - triggering probability; - strength and durability - location etc.
Support	Library	legislative and regulatory framework, supporting assistance	- information support; - fragmentary and full text documents; - visual material etc.
	Media	video, sound data, graphical primitives	- primitives and elements of graphical maintenance; - video elements; - acoustic elements etc.

A user can perform different types of training. He can observe and analyze data from video surveillance and sensors, performing a role of an "operator of the security system", while the other is trying to disrupt the facility as an attacker. An expert mode allows a synchronous organization up to four roles. Active Members are an "operator of the security system", an "attacker" and "response forces". Passive participants are "observer experts". The number of participants of the "training" determines the maximum number of personal devices.

The minimal configuration of the complex can be formed as a personal device. It is enough for training of the operator of security systems. The automatized workstation is organized with the simultaneous operation of the server and client parts of the complex.

The local configuration of each of the personal devices includes a specialized software, the subunits of the security staff and the adversary, the data on engineering and technical means [11, 12]. Thus, each workstation is a separate system and can be used independently from other workstations. The interaction between the users and subsystems of the instructor and the operator is implemented through the appropriate interfaces. The considered configuration suggests the possibility of a separate and independent work in training.

4. Conclusion

As a result of the work, the conception of the informational complex to simulate the work of security systems is developed. The proposed set of the core units enables modelling of a security system, and then, on this basis, performing the functioning of the used engineering and technical means. Thus, this allows one to simulate the interaction "attacker – security system". This development can be used for training of a security staff, effectiveness evaluation of used engineering and technical means.

5. References

[1] Garcia M L 2005 *Vulnerability Assessment of Physical Protection Systems* 1st ed (Butterworth-Heinemann) p 400
[2] Lovecek T, Ristvej J, Simal L 2010 Critical Infrastructure Protection Systems Effectiveness Evaluation *J. of Homeland Security and Emergency Management* 7 p
[3] Fischer R J, Hailibozek E P, Walters D C *Introduction to Security* 9th ed (Elsevier) p 544
[4] De Jong T, Linn M C, Zacharia Z C 2013 Physical and virtual laboratories in science and engineering education *Science* **6130** 305-308
[5] Bukovetskiy A V, Stepanov B P, Tatarnikov D A 2016 Initial Data Forming for Process Simulation in System "Intruder – Physical Protection System" *Key Engineering Materials* **685** 148-152
[6] Godovykh A, Stepanov B 2015 Development and Creation of Software and Information Environment for Simulation of Nuclear Facility *Advanced Materials Research* **1084** 652-654
[7] Darrah C 1995 Workplace Training, Workplace Learning: A Case Study *Human Organization: Spring* **54** 31-41
[8] Garcia M L 2007 *The Design and Evaluation of Physical Protection Systems* 2nd ed, (Butterworth-Heinemann) p 370
[9] O'Neil H F, Perez R S 2013 *Web-based learning: Theory, research, and practice* (Routledge) p 444
[10] Fennelly L 2012 *Effective Physical Security* 4th ed (Butterworth-Heinemann) p 384
[11] Lee W W, Owens D L 2004 *Multimedia-based instructional design: computer-based training, web-based training, distance broadcast training, performance-based solutions* 2nd ed (John Wiley & Sons) p 488
[12] Dennis J 1987 Computer Based Training *Education + Training* **29** 11 – 12

Tuning fuzzy systems parameters with chaotic particle swarm optimization

I A Hodashinsky[1], **M B Bardamova**[1]

[1] Tomsk State University of Control Systems and Radioelectronics, 40, Lenina Ave., Tomsk, 634050, Russia

E-mail: hodashn@rambler.ru

Abstract. In this paper, we investigate the application of chaotic particle swarm optimization (PSO) to fuzzy system parameter estimation. Unlike traditional PSO, chaotic PSO improves search capabilities of particles using chaotic transformations of random coordinates. Different mapping functions were investigated to generate chaotic transformation sequences. The efficiency of the algorithm was tested on approximation problems originating from data sets of the KEEL repository. The experimental study compared several existing techniques for fuzzy systems identification. The results indicate that chaotic PSO is a competitive approach to fuzzy system identification able to create fuzzy systems of high precision at an acceptable set of rules.

1. Introduction

Fuzzy Systems (FSs) have been successfully applied to problems in classification, clustering, modeling, control, decision making and other applications [1].

The main challenge in the development of FS is the generation of a fuzzy rule-base. The design of FS can be considered as a non-linear optimization problem in a multi-dimensional space featuring a complex (fitness landscape) structure. Due to the complexity of the problem, many authors proposed the use metaheuristics, such as evolutionary algorithms (EAs) and swarm methods, to tackle the problem. Whilst EAs, particularly genetic algorithms, have been applied successfully to the optimization of FS [2, 3], the application of swarm algorithms, such as ant colony algorithms [4, 5] and particle swarm optimization (PSO) [6, 7], is not so common. The focus of this paper is on the application of a variant of a PSO algorithm to address the problem of FS parameter tuning. PSO algorithms share similar features to EAs, e.g. both algorithm types do not require gradient information of the function being optimized, they can deal robustly with search landscapes of different topology, and are fairly easy to implement in any standard programming language. However, PSO tends to converge faster to high quality solutions than EAs, making it more suitable to solve optimization problems in real-time [8]. Similar to EAs, PSO can be hybridized easily with other approaches and has found application in various domains; the interested reader can be referred to [9] for a list of 26 application domains of PSO.

However, due to the fast convergence, the risk of getting stuck at local optima is higher than with many other EAs. Different approaches have been proposed to deal with the convergence speed vs solution quality trade-off including modifying the parameters of a PSO algorithm during the optimization process and the application of diversity-enhancing schemes. Recently, the chaos concept has been introduced to PSO to help control the rapid convergence and improve search capabilities of particles [10].

Chaotic PSO has found application in different types of problems including those with discrete and continuous decision variables. For example, in [11], the algorithm is used to plan the trajectory of a mobile robot, whilst in [12], it is used to solve the traveling salesman problem. In [13, 14], a binary chaotic PSO algorithm is used in conjunction with a nearest neighbor algorithm for selecting attributes in a classification problem. In this paper, we extend the list of applications of chaotic PSO by the optimization of FS parameters, particularly membership functions parameters.

The paper is organized as follows. Section 2 outlines the problem of FS identification and singleton-type FSs, which are used here as an example to demonstrate the performance of chaotic PSO. The basic idea of traditional PSO is described in Section 3. In Section 4, several mapping functions for the generation of chaotic transformation sequences are presented, and the chaotic PSO algorithm is described. In Section 5, the performance of the algorithm is validated using several approximation problems. Finally, Section 6 concludes the paper.

2. Singleton-type fuzzy systems

The main feature of FS is that they are based on the concept of fuzzy representation of information and operate with fuzzy sets instead of numbers.

We consider singleton type FSs, where the i-th rule has the following form:

$$\text{IF } x_1 = A_{1i} \text{ AND } x_2 = A_{2i} \text{ AND } \ldots \text{ AND } x_n = A_{ni} \text{ THEN } y = r_i,$$

where A_{ji} is a linguistic term estimating variable x_j , and r_j is a real number estimating output y.
The FS performs mapping $f : \Re^n \to \Re$ defined by the following formula:

$$f(\mathbf{x}) = \frac{\sum_{i=1}^{R} r_i \prod_{j=1}^{n} \mu_{Aji}(x_j)}{\sum_{i=1}^{R} \prod_{j=1}^{n} \mu_{Aji}(x_j)},$$

where $\mathbf{x} = (x_1, x_2, \ldots, x_n)$ is an input vector, n the number of input variables, R the number of rules, and $\mu_{Aji}(x_j) \in [0,1]$ is a membership function defining the membership degree of the j-th input variable to term A_{ji}.

The FS can be represented as follows:

$$y = f(\mathbf{x}, \boldsymbol{\theta}),$$

where $\boldsymbol{\theta} = (\theta_1, \ldots, \theta_N)$ is a vector of parameters, and N is a total number of parameters identified by the system.

Given a set of training data (a table of observations) $\{(\mathbf{x}_p; t_p), p = 1, \ldots, m\}$, where $\mathbf{x}_p$ is an input vector, t_p is the output value, and m is the number of observations, the mean square error (MSE) of a FS is calculated according to

$$E(\boldsymbol{\theta}) = \frac{1}{m} \sqrt{\sum_{p=1}^{m} (t_p - f(\mathbf{x}_p, \boldsymbol{\theta}))^2}. \quad (1)$$

FS identification includes two main steps: structure identification and parameter estimation. Structure identification is concerned with determining the structure characteristics of a FS, such as the number of fuzzy rules and linguistic terms for each variable. Parameter estimation involves the estimation of the parameters of the membership functions and consequents of fuzzy rules. The paper focuses on parameter estimation, in particular, on estimating membership functions parameters.

The problem of parameter estimation can be solved as a problem of finding a minimum of given function $E(\boldsymbol{\theta})$ in the multidimensional space defined by a set of coordinates corresponding to the FS parameters.

In order to solve the given minimization problem, it is suggested to use a chaotic particle swarm algorithm.

3. Traditional Particle Swarm Optimization

PSO is a population based stochastic optimization method developed by Kennedy and Eberhart [15]. It is based on observations of the social behavior of animals, such as birds or fish, searching for food.

The particles (conceptual mathematical objects) are located in the space defined by the parameters of the problem being solved. Each position of the particle is characterized by its fitness. In traditional PSO, there are three factors that define the movement of a particle in the search space: inertia, memory and cooperation. Inertia means that a particle cannot change its direction of motion immediately. Each particle has a memory storing its best position in the search space. Cooperation among particles is enabled by informing each particle of the position of the best particle of the swarm. Knowing the personal and global best position, a particle dynamically adjusts its speed according to its own experience and the experience of the movement of other particles. Typical criteria causing a swarm to stop moving include 1) reaching the equilibrium phase, 2) discovering a solution with a certain desired objective function value, or 3) reaching a certain number of iterations.

Here, the search space is $\boldsymbol{\theta} \subset \Re^N$, where N is the total number of membership functions parameters. Assuming a swarm of M particles, vectors $\mathbf{x}_i$, $\mathbf{p}_i$ and $\mathbf{v}_i$ define the current position, the personal best position, and the current velocity of the i-th particle in the population are, respectively:

$$\mathbf{x}_i = (\theta_{i1}, \theta_{i2}, ..., \theta_{iN}) \in \boldsymbol{\theta},$$

$$\mathbf{p}_i = (p_{i1}, p_{i2}, ..., p_{iN}) \in \boldsymbol{\theta},$$

$$\mathbf{v}_i = (v_{i1}, v_{i2}, ..., v_{iN}) \in \boldsymbol{\theta},$$

The velocity and the position of the i-th particle in the swarm at step $k+1$ are determined as follows:

$$\mathbf{v}_i(k+1) = w \cdot \mathbf{v}_i(k) + c_1 \cdot rand \cdot (\mathbf{p}_i(k) - \mathbf{x}_i(k)) + c_2 \cdot Rand \cdot (\mathbf{p}_g(k) - \mathbf{x}_i(k))$$

$$\mathbf{x}_i(k+1) = \mathbf{x}_i(k) + \mathbf{v}_i(k+1),$$

where $i = 1, 2, \ldots, M$; $\mathbf{v}_i(k)$ is the velocity vector of the i-th particle at iteration k; $\mathbf{x}_i(k)$ is the vector of coordinates of the i-th particle at iteration k; c_1 and c_2 are positive speedup factors; $\mathbf{p}_i(k)$ is the best position of the i-th particle so far (i.e. personal best position); $\mathbf{p}_g(k)$ is the position of the best particle in the swarm (i.e. global best position); w is the inertia factor; *rand* and *Rand* are random numbers drawn from interval [0, 1].

Coefficient c_1 is a cognitive parameter, which indicates the trust of a particle to its own past experience; this coefficient is responsible for the discovery of new areas in the search space. Coefficient c_2 is a social parameter, showing how the particle trusts the swarm; this coefficient is responsible for the exploration of promising search space areas.

4. Traditional Particle Swarm Optimization

Chaotic PSO extends traditional PSO with the principles of chaotic dynamics, which come into play when computing a particles new position $x(k+1)$. In more detail, a chaotic search behavior is induced by computing $x(k+1)$ based on chaotic sequences. The idea is that chaotic dynamics improves the diversity of a swam and thus reduces the risk of getting stuck at local optima [10, 16]. Given $cx_i(k)$ is the i-th chaotic variable at iteration k, then the set of commonly used mapping functions to generate chaotic sequences includes:

a) Logistic mapping [10]

$$cx_i(k+1) = 4\cdot cx(k)\cdot(1- cx_i(k)), \text{ where } cx_i(0) = 0.2027;$$

b) Sinusoidal iterator [17]

$$cx_i(k+1) = 2.3\cdot cx_i^2(k)\cdot \sin(\pi\cdot cx_i(k)), \text{ where } cx_i(0) = 0.7;$$

c) Tent mapping [17]

$$cx_i(k+1) = \begin{cases} \dfrac{cx_i(k)}{0.7}, \text{ if } cx_i(k) < 0.7 \\ \dfrac{cx_i(k)\cdot(1-cx_i(k))}{0.3}, \text{if } cx_i(k) \geq 0.7 \end{cases};$$

d) Gauss mapping [17]

$$cx_i(k+1) = \begin{cases} 0, \text{ if } cx_i(k) = 0 \\ \dfrac{1}{cx_i(k)}, \text{if } cx_i(k) \in (0,1) \end{cases}.$$

A chaotic transformation of variable x is performed using the following procedure.

Step 1. Calculate the value of chaotic variable cx_i by normalizing the value of x_i

$$cx_i(k) = \frac{x_i(k) - x_{i,lower}(k)}{x_{i,upper}(k) - x_{i,lower}(k)},$$

where $x_{i,lower}$, $x_{i,upper}$ are minimum and maximum values of x_i.

Step 2. Apply the chaotic transformation to variable $cx_i(k)$, based on the selected chaotic sequence.

Step 3. Re-normalize the value of resulting variable $cx_i(k+1)$

$$x_i(k+1) = x_{i,lower}(k) + cx_i(k+1)\cdot(x_{i,upper}(k) - x_{i,lower}(k)).$$

Factor $Score_i$ is introduced for each i-th particle in the chaotic PSO

$$Score_i(k) = \frac{E(\mathbf{x}_{i,worst}(k)) - E(\mathbf{x}_i(k))}{E(\mathbf{x}_{i,worst}(k)) - E(\mathbf{p}_i(k))},$$

where $\mathbf{x}_{i,worst}(k)$ is the worst position of the i-th particle within the first k iterations.

Factor $Score_i$ affects the inertia coefficient, which is calculated for each particle individually as follows (let us note that the inertia is fixed in the traditional PSO):

$$w_i(k)= w_{i,\text{low}}(k)+(w_{i,\text{high}}(k) - w_{i,\text{low}}(k))\cdot(1-Score_i(k)),$$

where $w_{i,\text{low}}(k)$ and $w_{i,\text{high}}(k)$ are the lowest and highest inertia values of the i-th particle within the first k iterations.

The chaotic PSO algorithm is presented below.

Step 1. Set the parameters of the algorithm: a maximum number of iterations K, swarm size M, c_1, c_2, $\mathbf{w}(0)$, $\mathbf{v}_i(0)=0$, $\mathbf{x}_i(0)$, $\boldsymbol{p}_g(0)$, $k=1$.

Step 2. Evaluate the fitness value of each particle in the swarm by calculating the MSE values (see Equation (1)). If $k = K$, then go to Step 6.

Step 3. Perform a chaotic transformation (shift) of randomly chosen coordinates of randomly selected particles using a given chaotic sequence.

Step 4. Define $Score_i(k+1)$, inertia $w_i(k+1)$, velocity vector $\mathbf{v}_i(k+1)$ and new coordinates $\mathbf{x}_i(k+1)$ for each i-th particle.

Step 5. Calculate the best new positions of each particle $\mathbf{p}_i(k+1)$. Redefine the best position in swarm $\mathbf{p}_g(k+1)$. $k = k+1$. Go to Step 2.

Step 6. Return $\mathbf{p}_g(k+1)$.

5. Experiments

5.1. Experimental setup.

The performance of the chaotic PSO algorithm was validated on four data sets of the KEEL (Knowledge Extraction Evolutionary Learning) repository [18]: *diabetes*, *ele*-1, *ele*-2, and *dee*. The objective of the experiments was to minimize the MSE (see Equation (1)).

The experiments were conducted according to the following procedure proposed in the repository: First, each data set from the repository was partitioned using 5-fold cross validation. For each split of training (*tra*) and testing (*tst*) data set, and each chaotic sequence type presented above, 5 algorithmic runs of the chaotic PSO algorithm were performed (i.e. in total, 25 runs were performed for each data set and chaotic sequence type).

FSs were learnt (by the chaotic PSO algorithm) using training data sets, and then validated on the testing data sets.

Triangular membership functions are considered for each input variable x_j. The following default parameters were used for the chaotic PSO algorithm: $c_1 = c_2 = 1.4$, $w(0) = 0.4$, $K = 1000$, $M = 20$. Average MSE values and their standard deviation were calculated for both training and testing data sets. The data sets considered, and FS setups employed for each data set, are detailed further.

diabetes: This data set has two real-type input variables and consists of 43 observations. The data was obtained from a study investigating the factors affecting the model of insulin-dependent diabetes mellitus in children. The study analyzes the dependency of the serum *C*-peptide level on various other factors to understand patterns of residual insulin secretion. The output is the concentration of *C*-peptide, and the inputs are alkalipenia and a patient's age [18]. The following FS parameters were used for this problem: 4 fuzzy rules, and 2 terms for each input variable.

ele-1: This data set has two input variables and consists of 495 observations. The data investigate the problem of forecasting the total length of low voltage lines (output) as a function of the number of inhabitants and the distance from the city center to the most distant customers (inputs) [18]. The following FS parameters were used for this problem: 16 fuzzy rules, and 4 terms for each input variable.

ele-2: This data set has 4 input variables and consists of 1056 observations. The data aim at estimating electrical maintenance costs of a city (output) as a function of the total length of all streets combined, the area of the city, built up area of the city, and the energy capacity required [18]. The following FS parameters were used for this problem: 81 fuzzy rules, and 3 terms for each input variable.

dee: This data set has 6 input variables and consists of 365 observations. The data aim at predicting the average daily consumption of electricity in Spain (output) as a function of the daily energy consumption of hydroelectric, nuclear electric, carbon, fuel, natural gas, and other sources of energy in Spain in 2003 [18]. The following FS parameters were used for this problem: 64 fuzzy rules, and 2 terms for each input variable.

5.2. Experimental results

Table 1 shows the average MSE and its standard deviation (for training and testing data sets) obtained by the two best-performing chaotic sequences on each data set.

The performance of the chaotic PSO algorithm has been compared against the methods described by [19]. Tables 2, 3, 4 and 5 compare the results of these algorithms.

For each data set, the chaotic PSO algorithm used the best-performing chaotic sequence as shown in Table 1.

From the tables, it can be concluded that chaotic PSO is able to create FSs of high accuracy using an acceptable number of rules. This indicates a good degree of interpretability of the obtained systems.

The good MSE results on testing data sets *MSE*(*tst*) indicate that chaotic PSO has a high predictive capacity. In fact, the algorithm performs best on all testing data sets, except for data set *diabetes*. Here, the algorithm featured the second-best average MSE, but this performance was achieved using fewer rules than the other algorithms. However, it is worthwhile mentioning that on the *diabetes* data set, chaotic PSO performed also poorly on the training data set. Conversely, on the data sets, *ele*-1 and *ele*-2, chaotic PSO achieves the best learning capacity, which translates also to the best predictive ability. Finally, chaotic PSO is also able to perform well on the *dee* data set: the outcome achieved was the third-best result on the training data sets, and the best on the testing data set, using the second fewest rules of all algorithms compared.

Table 1. Average MSE and its standard deviation (for training and testing data sets) obtained by the two best-performing chaotic sequences on each data set.

Chaotic sequence	*MSE*(*tra*) average	*MSE*(*tra*) σ	*MSE*(*tst*) average	*MSE*(*tst*) σ
		diabetes (4 rules)		
Tent Map	0.18525	0.0161	0.37926	0.0876
Sinusoidal Iterator	0.19000	0.0154	0.36000	0.0811
		ele-1 (16 rules)		
Logistic Mapping	317058	15905	389927	28075
Tent Mapping	309449	12306	388291	23602
		ele-2 (81 rules)		
Logistic Map	16616	1355	17832	1307
Tent Map	16806	1633	18409	2109
		dee (64 rules)		
Logistic Map	0.14023	0.0051	0.17003	0.0153
Tent Map	0.13901	0.0060	0.16855	0.0133

Table 2. Comparison of the number of rules and average MSE, and their standard deviation (for training and testing data sets) obtained by chaotic PSO and other existing algorithms on data set *diabetes*.

Algorithm	*number of rules* average	*number of rules* σ	*MSE*(*tra*) average	*MSE*(*tra*) σ	*MSE*(*tst*) average	*MSE*(*tst*) σ
Chaotic	4	0	0.18525	0.0161	0.37925	0.0876
Wang–Mendel	18.6	1.4	0.22836	0.0425	1.40241	0.6890
COR-BWAS	18.6	1.4	0.17496	0.0250	1.45869	0.7091
Thrift	46.2	0.7	0.07448	0.0098	0.87825	0.3575
Pittsburgh	15.0	2.9	0.10398	0.0182	0.95088	0.7881
Fuzzy-GAP	10.0	0.0	0.14292	0.0376	0.50141	0.3014
Pitts-DNF med	5.4	0.5	0.12958	0.0136	0.32134	0.1922
Pitts-DNF max	9.6	1.2	0.10656	0.0150	0.63396	0.5276

Table 3. Comparison of the number of rules and average MSE, and their standard deviation (for training and testing data sets) obtained by chaotic PSO and other existing algorithms on data set *ele*-1.

Algorithm	*number of rules*		*MSE(tra)*		*MSE(tst)*	
	average	σ	average	σ	average	σ
Chaotic	16	0	309449	12306	388291	23602
Wang–Mendel	22.0	1.4	423466	8069	455262	19943
COR-BWAS	22.0	1.4	354304	7065	417142	9823
Thrift	46.4	1.0	335086	5285	435373	57252
Pittsburgh	17.2	4.3	342464	19209	738691	543165
Fuzzy-GAP	11	0.0	481603	58989	548122	70968
Pitts-DNF med	8.2	0.7	344636	8999	415266	71200
Pitts-DNF max	14.0	1.1	330496	4815	440692	40370

Table 4. Comparison of the number of rules and average MSE, and their standard deviation (for training and testing data sets) obtained by chaotic PSO and other existing algorithms on data set *ele*-2

Algorithm	*number of rules*		*MSE(tra)*		*MSE(tst)*	
	average	σ	average	σ	average	σ
Chaotic	81	0	16616	1355	17832	1307
Wang–Mendel	65.0	0.0	112270	1498	112718	4685
COR-BWAS	65.0	0.0	102664	1080	102740	4321
Thrift	524.6	6.4	146305	12991	168472	20135
Pittsburgh	240.0	21.1	210717	32027	265130	30161
Fuzzy-GAP	33.0	0.0	279166	90017	290062	89155
Pitts-DNF med	18.6	1.4	86930	3955	99310	12996
Pitts-DNF max	32.4	6.6	70207	1658	88017	8968

Table 5. Comparison of the number of rules and average MSE, and their standard deviation (for training and testing data sets) obtained by chaotic PSO and other existing algorithms on data set *dee*.

Algorithm	*number of rules*		*MSE(tra)*		*MSE(tst)*	
	average	σ	average	σ	average	σ
Chaotic	64	0	0.13900	0.0059	0.16855	0.0132
Wang–Mendel	178	2	0.14117	0.0074	0.22064	0.0264
COR-BWAS	178	2	0.12463	0.0052	0.20513	0.0231
Thrift	13020	33	0.38778	0.0357	0.45830	0.0804
Pittsburgh	982	56	0.42111	0.0784	0.72109	0.3263
Fuzzy-GAP	89	0	0.17751	0.0130	0.20633	0.0172
Pitts-DNF med	57	3	0.13821	0.0060	0.27465	0.1366
Pitts-DNF max	98	5	0.11267	0.0035	0.21692	0.0359

6. Conclusion

In this paper, we validated an efficient algorithm applied to the optimization of FS parameters. Experiments conducted on several benchmark data sets, as well as a comparative analysis with various existing techniques showed that the proposed algorithm is able to create FSs that are highly accurate whilst using few rules.

The algorithm has a good learning ability, meaning MSE values on test data sets are not far off from the MSE values on training data sets. This fact indicates the absence of overtraining.

Acknowledgments

The project was carried out with financial support from the Russian Foundation for Basic Research (project №16-07-00034).

References

[1] A T Azar (ed) 2010 *Fuzzy Systems* (InTech)
[2] Fazzolari M, Alcala R, Nojima Y, Ishibuchi H and Herrera F 2013 *IEEE Trans. Fuzzy Syst.* **21** 45-65
[3] Herrera F 2005 *International Journal of Computational Intelligence Research* **1** 59-67
[4] Juang C-F and Chang P-H 2010 *IEEE T Fuzzy Syst* **18** 138-149
[5] Khodashinskii I A and Dudin P A 2012 *Optoelectronics, Instrumentation and Data Processing* **48** 54-61
[6] Maldonado Y and Castillo O 2013 *Appl. Soft Comput.* **13** 496-508
[7] Juang C-F 2010 Combination of Particle Swarm and Ant Colony Optimization Algorithms for Fuzzy Systems Design *Fuzzy Systems*, ed A T Azar (InTech) p 195
[8] Angeline P J 1998 *Proc. 7th Int. Conf. on Evolutionary Programming VII* (London: Springer) p 601
[9] Poli R, Kennedy J and Blackwell T 2007 *Swarm Intell.* **1** 33-57
[10] Liu B, Wang L, Jin Y H, Tang F and Huang D X 2005 *Chaos, Solitons Fractals* **25** 1261–1271
[11] Zhao Q and Yan S 2005 *Advances in Natural Computation*, ed L Wang, K Chen and Y S Ong (Berlin: Springer) p 632
[12] Yue C, Yan-duo Z, Jing L. and Hui T 2009 *Progress in Robotics*, ed Kim J-H et al. (Berlin: Springer) p. 372
[13] Chuang L-Y, Yang C-H and Li J-C 2011 *Appl. Soft Comput.* **11** 239-248
[14] Xu H, Lu S and Zhou S 2013 *Proc. 2012 Int. Conf. on Information Technology and Software Engineering* (Berlin: Springer) p 619
[15] Kennedy J and Ebenhart R 1995 *Proc. 1995 IEEE Inter. Conf. on Neural Networks* **4** 1942
[16] Cui Z, Cai X and Zeng J 2009 *International Journal of Innovative Computing, Information and Control* **5** 951-960
[17] Alatas B, Akin E and Bedri Ozer A 2009 *Chaos, Solitons Fractals* **40** 1715–1734
[18] Alcalá J, Fernández A, Luengo J, Derrac J, García S, Sánchez L and Herrera F 2011 *Journal of Multiple-Valued Logic and Soft Computing* **17** 255-287
[19] Casillas J, Martinez P and Benitez A D 2009 *Soft Computing* **13** 451-465

Revisiting evaluation of the integral module from the product of two real functions

V P Hranilov [1], L P Kogan [2], P V Gorev [3], P N Kuzmichev [4], P A Egorova [5]

[1] Nizhny Novgorod State Technical University, Nizhny Novgorod, 24, Minin Str., 603950, Russia
[2] Nizhny Novgorod State University of Architecture and Civil Engineering, Nizhny Novgorod, 65, Ilyinskaya Str., 603950, Russia,
[3] Giprogascenter JSC, 26, Nizhny Novgorod, Alekseevskaya Str., 603950, Russia,
[4] Giprogascenter JSC, 26, Nizhny Novgorod, Alekseevskaya Str., 603950, Russia,
[5] Nizhny Novgorod State Technical University, Nizhny Novgorod, 24, Minin Str., 603950, Russia

E-mail: L_kog@list.ru

Abstract. The article envisages the question of estimating the integral module from the product of two real functions. To solve this problem, it is required to transit to the integral by the range values of one of the multipliers that allows transiting to the probabilistic notation of the assumed integral. As a result, it is possible to obtain the required estimation in the form of the composite function product from one of multipliers by the integral of another factor.

1. Introduction

This work raises the question on evaluating the integral module from the product of two multipliers of type

$$\xi = \int_a^b s(x)g(x)dx\,, \tag{1}$$

where $s(x)$ and $g(x)$ are functions of real ones at $x \in [a, b]$ that can alternate within this interval. The similar integrals often occur in the problems concerning the mathematical modeling of the wave propagation in the heterogeneous media and directing systems. It is considered that dependence $s(x)$ is continuous throughout boundaries $x \in [a, b]$, as well as with the limited number of possible points of discontinuity of the first-order derivative, whereas $g(x)$ is regarded as the continuously differentiable function. Additionally, the nonoccurrence of the finite extent segments within the $[a, b]$ interval with the fulfilled condition $g(x) = \text{const}$ is required. It is considered that values $g(x)$ at $x \in [a,b]$ are changed within $g_{min} \le g(x) \le g_{max}$.

The authors set a goal to obtain the relevant estimations in the form that is close to that one occurring in the course of extended mean-value theorem application [1] – [6]. This is due to the advantage of the given theorem (in comparison with other integral inequalities (see [7] – [12]), which relates to the fact that the resulting equation is in the form of the product coefficient associated with one of the subintegral functions by the assumed integral of the second multiplier. Such property is significant, especially, when determining the degree of convergence of Neumann series in the defined problems. Moreover, in this paper some other relations and estimations, which are also important for the investigation of the properties of certain integrals, have been obtained. As a consequence of calculations presented below, the integral of equation (1) leads to the kind, which is associated with the integration by the range of values of one of the multipliers (see also [7]), that is equal to the integration in a probability sense. This allows, in particular, using the probability theory for $|\xi|$ when giving the estimation.

2. Materials and methods

It is necessary to write equation (1) as the sum of the integrals by the sections of function monotonicity.

$$\xi = \sum_{i=1}^{N} \xi_i ; \quad \xi_i = \int\limits_{X_{i-1}}^{X_i} s(x)g(x)dx . \tag{2}$$

Here, $X_0 = a$, $X_N = b$ and $x = X_i$, $1 \le i \le N-1$ are values of inner points of $g(x)$ function extrema within the interval of integration into equation (1); function $g(x)$ possesses values $Z_i = g(X_i)$ at $x = X_i$, $0 \le i \le N$. If changing variable $z = g(x)$, it is required to transform ξ_i from equation (2) to the integral of the ξ_i type of equation (3):

$$\xi_i = \int_{Z_{i-1}}^{Z_i} s[x_i(z)]A_i(z)dz , \tag{3}$$

where $x_i(z)$ is an argument of equation (3), which is the solution of equation $g(x) = z$ belonging to interval $[X_{i-1}, X_i]$. We consider that the number of such solutions is $M(z)$. It is obvious that the $x_i = x_i(z)$ dependence involves the $z \in [\mathrm{Min}\{Z_{i-1}, Z_i\}, \mathrm{Max}\{Z_{i-1}, Z_i\}]$ definition domain being a branch of the function that is opposite to $g(x)$. By condition, function $g(x)$ is differentiable and thus continuous at $x \in [a, b]$, as well as strictly monotone at any interval $[X_{i-1}, X_i]$. As a consequence, function $x_i(z)$ is available and continuous on the specified sections of the z axis according to the continuity theorem of the inverse function [1].

There is a derivative of the relevant inverse function branch in $A_i(z) = \frac{dx_i(z)}{dz}$ (3). Consideration of the differentiability of $g(x)$, the availability and boundedness of function $A_i(z)$ at the specified interval of axis z (except for infinitesimal neighborhood of values $z = \tilde{z}$ corresponding to the stationary points of function $g(x)$) is resulted from the theorem on the derivative of the inverse function.

Values $A_i(z)$ at $\tilde{z}$ points go to the infinity, that is why integral ξ_i of equation (3) (by values of z) is an improper integral of the second kind, which convergence is resulted from the convergence of equation (2). It is simply, if $Z_{i-1} < Z_i$, then $A_i(z) \ge 0$, and when $Z_{i-1} > Z_i$, then $A_i(z) \le 0$. Thus, it is convenient to rewrite it as $\xi_i = \int_{\mathrm{Min}\{Z_{i-1}, Z_i\}}^{\mathrm{Max}\{Z_{i-1}, Z_i\}} s[x_i(z)]\, |A_i(z)|\, dz$.

As a result we obtain:

$$\xi = \int\limits_{g_{min}}^{g_{max}} \left(\sum_{i=1}^{M(z)} s[x_i(z)]A_i(z) \right) zdz . \tag{4}$$

It is necessary to admit that index i of equation (3) and in any sum of the $\sum_{i=1}^{M(z)}$ kind means the branch number of inverse function $x_i(z)$ responding to reaching value z by function $g(x)$ while this index is equal to the number of the monotonicity interval of function $g(x)$ in equation (1).

Let us multiply and divide the right-hand side of equation (3) by function $f(z) = \frac{\sum_{i=1}^{M(z)} |A_i(z)|}{b-a}$. Then, the following equation is obtained:

$$\xi = (b-a) \int_{g_{min}}^{g_{max}} T(z)zf(z)\, dz, \tag{5}$$

where $T(z) = \frac{\sum_{i=1}^{M(z)} s[x_i(z)]|A_i(z)|}{\sum_{i=1}^{M(z)} |A_i(z)|}$ is a multiplier available in equation (4). From the definition of $f(z)$, it follows that $f(z)dz = \frac{\sum_{i=1}^{M(z)} |dx_i(z)|}{b-a}$. Here, the right member is equal to the ratio of sums of $M(z)$ paths of infinitely small intervals of the X-axis where $z \leq g(x) \leq z + dz$ is a relation valid for every interval ($b - a$ content function of the whole integration domain of equation (1)). Such relation is equal to elementary probability $P(z \leq g(x) \leq z + dz) = P(z, z + dz)$ calculated by the enormous number of tests for the event consisting in entering function values $g(x)$ into interval $[z, z + dz]$ of its range of values under condition of the proportional and random distribution of argument x within interval $[a, b]$. Thus, dependence $f(z)$ is identically equal to the probability density of random variable $g(x)$ relating to the defined distribution of argument x.

It is to be admitted that if $g(x)$ is a sinusoid function (i.e. when the module of $| g'(x)|$ is the first-order derivative, which is the same in all points, where $g(x)$ possesses the same values, which are peculiar for function $\sin(x)$), so equation $T(z)$ is too simplified. Whereas all values $|A_i(z)|$ with this provision in $T(z)$ become the same at all i indexes (the same for every z, in general), the following equation is obtained:

$$T(z) = \frac{S(z)}{M(z)} = \bar{s}_{g(x)=z} , \tag{6}$$

where designation $S(z) = \sum_{i=1}^{M(z)} s[x_i(z)]$ is introduced and symbol $\bar{s}_{g(x)=z}$ means the averaging of function $s(x)$ by the set of points of the X-axis that belong to interval $x \in [a, b]$ with $g(x) = z$. Therefore, equation (5) is transformed into the equation of the (7) kind:

$$\xi = (b - a) \int\limits_{g_{min}}^{g_{max}} \bar{s}_{g(x)=z}\, z f(z)\, dz . \tag{7}$$

The relation of equation (7) allows obtaining the number of estimations. Firstly, equation (7) is obtained when applying the first law of mean [1] to the integral on the right-hand side of the equation and changing the 'mean value' that is nearly always unknown into the largest value of the subintegral function:

$$|\xi| \leq (b - a)\, Max \left|\bar{s}_{g(x)=z}\right| Max\{|g_{min}|, |g_{max}|\} . \tag{8}$$

It is considered here that $\int_{g_{min}}^{g_{max}} f(z)\, dz = 1$.

Let us assume that M_s is a number of points that belong to interval $x \in [a, b]$ and in which $|s(x)|$ reaches the largest value within this interval. Thus, if $Min\,\{M(z)\}_{z \in [g_{min}, g_{max}]} > M_s$, then $Max\left|\bar{s}_{g(x)=z}\right|_{z\in[g_{min}, g_{max}]} < Max|s(x)|_{x\in[a,b]}$. As a result, the estimation of equation (8) for $|\xi|$ becomes more precise than that oriented immediately to the integral of equation (1) due to the first law of mean (as well as changing the 'mean' value of the subintegral function into the largest one according to the modulo).

It is also admitted that if g_{min} and g_{max} are variables with the same sign, then the equation will be obtained in accordance with the extended mean value theorem:

$$|\xi| \leq (b - a) Max \left|\bar{s}_{g(x)=z}\right|_{z\in[g_{min}, g_{max}]} \left| \int\limits_{g_{min}}^{g_{max}} z f(z)\, dz \right| = Max \left|\bar{s}_{g(x)=z}\right|_{z\in[g_{min}, g_{max}]} . \tag{9}$$

It is considered here that $\int_{g_{min}}^{g_{max}} z f(z)\, dz = \frac{\int_a^b g(x)dx}{b-a}$. It must be emphasized that it will be necessary to use multiplier $Max\, |s(x)|_{x\in[a,b]}$ in most cases of evaluating $|\xi|$ if directly applying the extended mean value theorem to integral of equation (1) instead of coefficient $Max\left|\bar{s}_{g(x)=z}\right|_{z\in[g_{min}, g_{max}]}$ proceeding the integral sign in equation (9). This will practically always result in the increasing of the upper bound of values $|\xi|$ compared to the results obtained in this paper. Additionally, it is to be admitted that relation $(z) = \frac{1}{2\pi} \int_{-\infty}^{+\infty} L(p) \exp(-ipz)\, dp$, where $L(p)$ is a characteristic function corresponding to random variable $g = g(x)$ at the random distribution of argument x within interval $[a, b]$ specified above, is valid for probability density $f(z)$. Thus, considering equation (5), integral ξ may be written as

$$\xi = \frac{1}{2\pi}(b-a)\int_{g_{min}}^{g_{max}}\int_{-\infty}^{+\infty} L(p)\exp(-ipz)\, dp\, T(z)zdz = (b-a)\int_{-\infty}^{+\infty} L(p)\Phi(p)dp =$$

$$= i(b-a)\int_{-\infty}^{+\infty} L(p)d\Phi_1(p), \qquad (10)$$

where $\Phi(p) = \frac{1}{2\pi}\int_{g_{min}}^{g_{max}} T(z)z\exp(-ipz)\, dz = i\Phi'_1(p)$; here $\Phi_1(p) = \frac{1}{2\pi}\int_{g_{min}}^{g_{max}} T(z)\exp(-ipz)\, dz$.

3. Conclusion

The relations of equations (5) and (9) are the major results of this paper. Using equation (5), it is possible to write the assumed integral of equation (1) as that allows using the machinery of probability theory for its estimating. Particularly, it allows $|\xi|$ to reach inequation (9). Upon conditions specified above and frequently realized for sinusoid functions, this estimation is more precise than that which would be valid when directly applying the extended mean value theorem. Besides, inequation (8), which affords to define more precise results of the first law of mean at set conditions, as well as equations (10) are actual for evaluating the integrals of equation of the (1) kind.

References

[1] Fikhtengolts G M 1970 *Kurs differentsialnogo i integralnogo ischislenia* (Moskva, Fizmatlit) pp 151-153
[2] Iliin V A, Pozniak E G 1982 *Osnovi matematicheskogo analiza* (Moskva, Nauka) **1** 175-176
[3] Khinchin A I 1948 *Vosem lektsiy po matematicheskomu analizu* (Moskva, OGIZ,) 347-348
[4] Hobson E W 1909 Proc. London Math Soc. **7 (1)** 14–23
[5] Gorev P V, Anisimova L N, Koldanov A P 1985 Vsesoyuznaya nauchno-tekhnicheskaya konferentsia «Razvitie i vnedrenie novoi tekhniki radiopriemnykh ustroistv» 18-19
[6] Bubukin I T, Kogan L P 1999 *Izv. Vuzov. Radiofizika* **42 (5)**, 438-451
[7] Il'in V A, Sadovnichij V A, Sendov Bl H 1985 *Matematicheskij analiz. Nachal'nyj kurs* (Moskva, MGU) 367 – 369
[8] Mironenko L P., Petrenko I V, Rubcova O A 2012 *«Iskusstvennyj intellekt»* **2**, 46–53
[9] Smirnov V I 1974 *Kurs vysshej matematiki* (Moskva, Nauka) **5** 188 – 193
[10] Chernov AV 2013 Vestnik Nizhegorodskogo universiteta **6(1)** 151–161
[11] Shagidullin R R 2013 *Integralnie uravneniya* (Kazan: KFU) 29-42
[12] Piskunov N S 1985 *Differentsialnoe i integralnoe ischislenia dlia vtuzov* (Moskva Glav. redaktsia fiz.-mat. lit.) **2** 260-262

The information system for enterprise visitors' verification, identification and authentication utilizing modern identification features

A Y Iskhakov[1], **R V Meshcheryakov**[1]

[1] Tomsk State University of Control Systems and Radioelectronics, 40, Lenina Ave., Tomsk, 634050, Russia

E-mail: iay@security.tomsk.ru

Abstract. The problem of distant enterprise visitors' verification, identification and authentication in case of the pass regime is investigated. In order to solve the problem, a structure of the information system utilizing modern identification features is suggested. The article presents the access control systems identifiers classification and an overview of modern identification features and data transmission technologies.

1. Introduction

Lately, the access control system (ACS) has been one of the most effective methods to solve the problem of complex security of the different property objects [1]. ACS allows preventing nonauthorized (illigal) access to the object territory, into the building or some floors. At the same time, the systems of such a kind do not prevent the staff and visitors from entering the allowed areas. ACSs are installed at the objects of different kinds: ranging from small offices to big plants and technoparks with several buildings. Wide application of such systems leads to the fast development of technologies used.

One of the problems solved with ACS is identification of every visitor (the subject of access) and the subsequent opportunity to enter the allowed areas [2]. This procedure utilises comparison of the identification features attached to some particular subject with the data base information. When the members of the staff is it concerned there are used contact or contactless cards [3] of different forms and standards. Table 1 shows the classification of the modern access identifiers.

There are many enterprises which functional peculiarities need a path regime with the opportunity for outside people to visit them. The goal of such visits could be excursions, exhibitions, scientific and practical seminars [4]. Individual electronic passes for the outside people are expensive, therefore the most popular way of the visitors idetification is inspection of the document certifying the person followed by registration in the registry. This way is not effective, because identification is done only at the entrances controlled by the security staff. In addition, the policy of total inspection is not effective, because it slows the business process.

The alternative way to solve the problem is the on-line service when the visitor can order the pass to the object himself. After registration, the visitor would receive the one-time pass with a bar code, print it himself and use at the electronic equipped entrance. But in this case, it is impossible to identify the visitor personality, impossible to check the validity of the data written by the visitor. Besides, such method does not allow one to check legitimacy of the identifier, in other words, does not allow identifying the visitor. Thus, the information system with the following functions becomes actual:

- An opportunity of the distant registration of the visitor without preliminary visit.
- Verification of the visitor personality at the stage of registration.
- A possibility to use modern identifications.
- Realization of the authentication procedures.
- An opportunity to install the system into modern ACS.

In order to realize these functions, it is necessary to define the procedures for the following stages: registration, identification and authentification of the visitor.

Table 1. Classification of the modern ACS identifiers

Identification features used	Basic element for use	Examples of identifier	Examples of reading devices
Mechanic	Elements of the identifier construction	Mechanical keys with perforating holes	Abloy SL905, Mottura 54.793
Magnetic	Magnetized parts of the surface or magnetic identifier elements	Cards with magnetic strips, Wiegand cards	Cipher MSR210U-33, Posiflex MR-2106U
Optic	Marks put on the surface or inside identifier having different optical characteristics in case of the reflected or transmitted illumination	Cards with a bar code, topographical labels	AP-IRC, СБН–4_KMKS
Electronic contacts	Electronic code written into the identifier microchip	Electronic keys	ACR38U-I1, OMNIKEY® 3121
Electronic radiofrequency	Radio channel used to transmit the data	Contactless cards of excess	Bolid PROXY–3A, Matrix II IronLogic
Acoustic	Coded acoustic signal	Device to generate acoustic signals	Stelberry M-70, Invensense NMP621
Biometric	Individual person physical features	Fingerprints, palm geometry, voice, signature dynamics	Smartec ST-FR030EMW, BioSmart PV-TS
Combination	Several identification features	Contactless access card and fingerprints	Smartec ST-FR031EM, ACR83

2. Registration and verification of the visitors

To realize the distant registration of a person, it is suggested to work out the system utilizing web-oriented architecture not needed addition soft at the visitor side. The traditional scheme of a visitor registration in ACS is shown in Fig. 1.

The distant verification of a registered person includes delegation of the mechanism M2 authority to mechanism M1. To do this, the author suggests using the framework based on the mechanism of trusted persons in the ACS registration sub-system. In this case, other people who previously controlled the legacy of the operation and also have needed privilege should affirm the personality of the visitor. The John Branard group was the first to suggest this mechanism.They realised and tested the system utilizing the social authentication when the visitor personality was proved by other people.

The successful example of this system application is working up to the present system of renovation of access to the account «Trusted Contacts», used in the social network, Facebook, since 2013. A visitor can name several friends to prove his personality using a special key in case the account is blocked. It is considered that these people can telephone the visitor or meet him personally to prove the legality of the access restoration. The administration of the social network thinks that this way is safe, secure safe and secure for the access restoration. In case of Facebook, the mechanism of trusted people is applied to restore the access to the existing accounts. The author is going to apply the method for personality verification before the account is written.

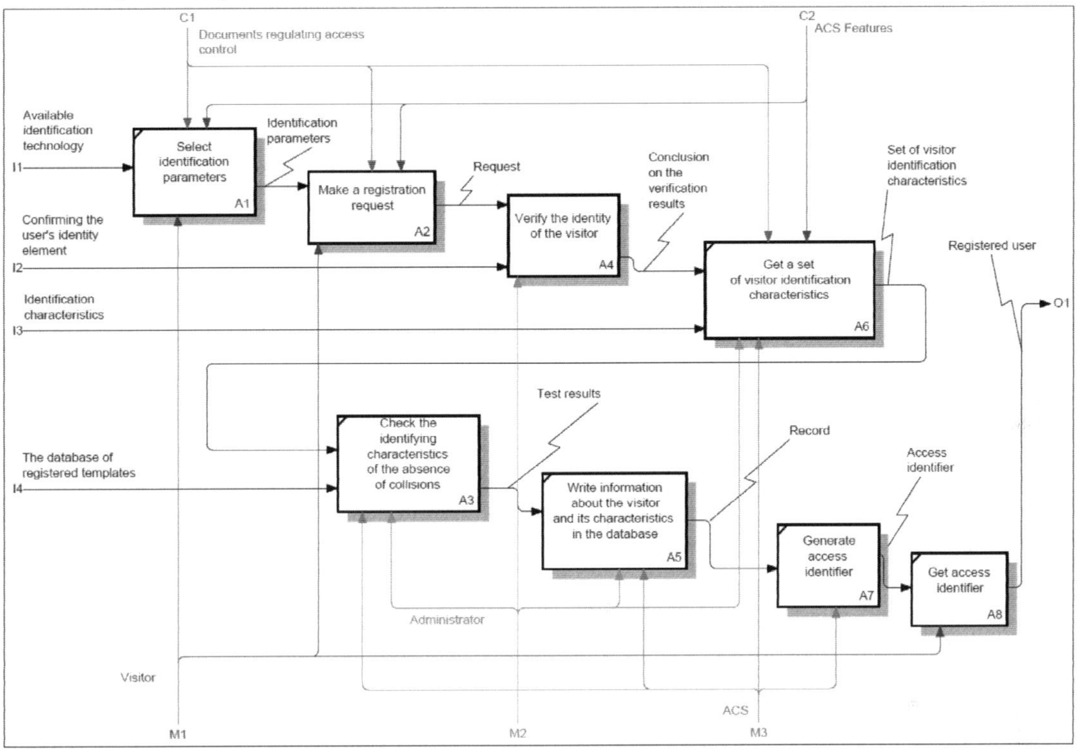

Figure 1. The traditional scheme of an ACS user registration

3. Identification and authentication

3.1 The modern identification features

According to statistics, presented by the International Telecommunication Union by 2016, there were used 7 377 million of mobile communication devices (the population of the Earth is about 7 448 million). A cellular telephone came into the everyday life and one cannot imagine life without it. So, the author suggests integrating the fast developing mobile technology into the industry of security to use the mobile device (smartphone) as the access identifier.

The modern smartphones, having powerful calculating opportunities, large memory and fast access to the Internet can not only provide mobile communication, but at the same time also keep several virtual passes allowing access to different objects with the help of a telephone [5]. Further, the main advantages of the method compared to electronic passes are given.

Advantages of application

- There is no need to use several identifiers (electronic passes) - each for every object.
- An opportunity of operational distant identification control including issue, change and recall.
- An opportunity of hidden reading devices (RD) installation (not entering the room), vandalism protection.
- An opportunity to use in case of longer distances (as an example, a barrier at the entrance of the parking place or the garage).

Security

- Mobile operation systems controlling the modern smartphones provide reliable security. In addition to the OS security service, it is also possible to use devices of electronic signature and coding. Besides, mobile applications work within the isolated program area (Sandbox) without the possibility of access and data variation by other programs.
- Basic functions of a screen automatic blocking with the demand of several passwords provide an opportunity to put in the additional authentication factor.
- Modern telephones allow using a functional that would inform the owner about the device position in case it is lost.
- A probability to find the lost telephone is much higher than the access with the card.
- The probability to transfer the smartphone to other people is lower than in case of the access card.

Convergence opportunities

A mobile communication device used for identification can be also used to restrict the access to the physical objects and to the organization electronic resources. The following examples can be mentioned: authentication of the means of wireless communication, VPN, corporative portals and others.

3.2 Data transmission technology

Among different technologies of data transmission used in mobile devices, Near field communication (NFC), Bluetooth Smart (BLE) and processing the coded information on the smartphone screen are most convenient for the idetification of data transmission in ACS [6],.

NFC is a technology of wireless high frequency communication for not large distances having three modes:

- passive (emulation of a smart-card). A passive device works as a contactless card of one of the standards;
- data transmission by the devices equal in rights;
- an active mode (reading or writing).

In order to emulate access with the mobile device, it is necessary to have the Secure Element (SE), a SIM-card for example. For the centralized SE, an ecosystem using the Trusted Service Manager (TSM) was worked out. This leads to the complicated technical processes of integration and to business models making application of contactless NFC more difficult [7]. In 2013, the Google company introduced the technology of Host-Based Card Emulation into operational system Android (HCE), allowing the emulatation of the contactless card into the mobile application without SE. The drawback limiting the application of NFC technology is absence of its support in some mobile OS dissimilar Android.

The technology of Bluetooth Smart (BLE), presented in 2010, allows transmitting data between the devices at the distance of several tens of meters, having low energy consumption and needing no compulsary procedures of the devices conjugation (compared to mating Bluetooth). BLE is the only contactless technology which is supported by many mobile OSs. The obvious advantage of the technology of a larger radius adds some drawbacks and needs additional protection of the transmitted identity data.

Displays of the modern smartphones have rather good resolution, allowing one to use them to show the data on the screen. So, the reading device should be a video camera which signals are processed by the access controller. For transmitting the identity data, it is better to use the barcode coding algorithms. One of the examples is the OR-codes technology having a big volume of the transmitted information and a high speed of recognition [8,9]. The drawback of this data transmission method is low immunity against such factors as low illumination and presence of the alien bodies between the transmitter and the reader.

3.3 The system structure

Fig. 2 shows the structure of the suggested identification system. To get universality and to widen the system potential users, it was decided to utilize three technology of the data transfer.

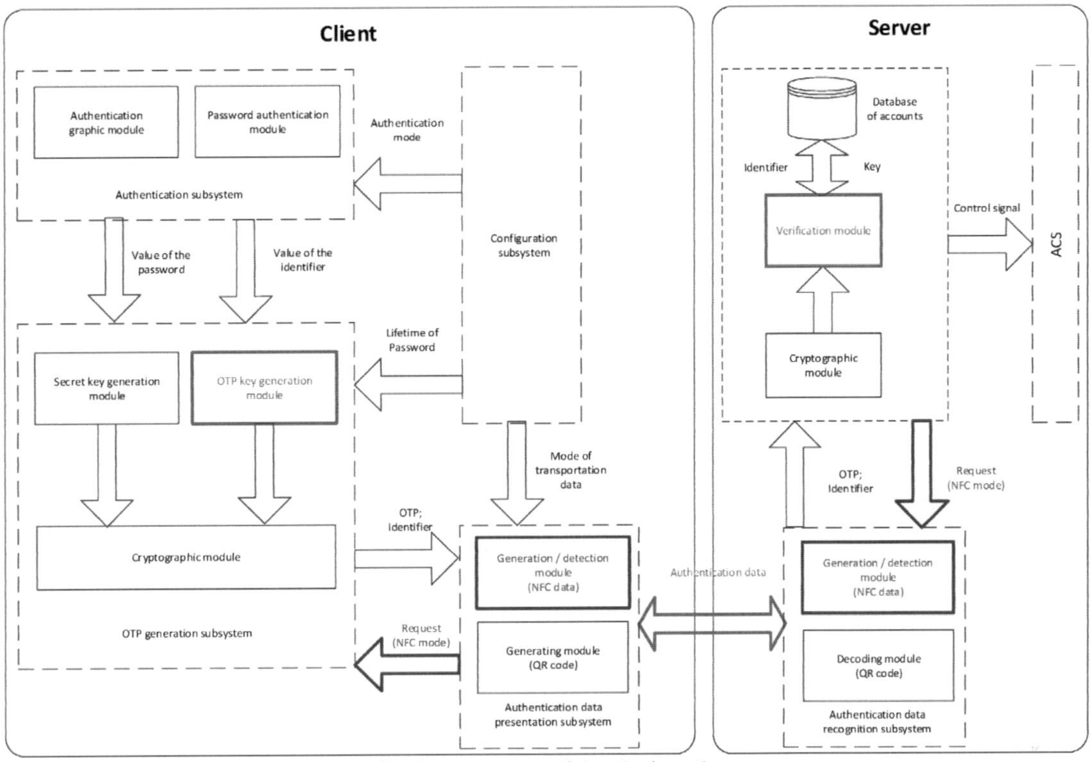

Figure 2. The structure of the designed system

To protect the transmitted data from compromentation, the autentification mechanism based on the one-time authorization code technology in the " synchronization in time" mode is suggested. Function *F (1)* describes the process of the authentication code receiving [10]:

$$F(u,p,t) = Q[u * ER[H(p),l](u \oplus t)], \tag{1}$$

where u – the user identifier;

p – a group from the secret key components;

t – the current temporal mark;

$Q(x)$ – the message x coding operation to pass to the reading device;

$*, \oplus$ – the operation of concatenation with some dividers;

$E_k(x)$ – the operation of encoding message x at key k;

$R(x,y)$ – the operation of the secret key formation using input line x according to demands y to the secret key in the process of the utilized mode of encoding;

$H(x)$ – the operation of the x line hashing;

l – demands to the secret key in the process of the utilized mode of encoding.

The second factor of the authentication is protection from the unauthorized access to the application using the modern method of graphic pass keys. The graphic pass key has two components complementing each other. It includes an image from the user multi-media library and a group of signs (gestures), put above the image. Using the private images makes security more reliable and helps to remember the pass-key.

4. Conclusion
The main advantage of the discussed system is the author's idea of realizing the procedure of identification without providing the subject of access with additional physical objects still keeping the opportunity of the authentication parameters variation to meet the demands of the object security policy. The author suggests using mobile platforms and utilizing their technologies. Certainly, the author does not suggest the full replacement of the traditional electronic passes by the mobile access at the moment. Still such systems could be the convenient addition to the ACS, providing new opportunities and advantages.

5. Acknowledgments
This work was supported by the Ministry of Education and Science of the Russian Federation within the 1.3 federal program research and development in priority areas of the scientific-technological complex of Russia for 2014-2020 (grant agreement 14.577.21.0172 of October 27, 2015; identifier RFMEFI57715X0172).

References
[1] Bezzateev S V, Zharinov R F, Petrov V I and Zybin V A 2015 *Information Security Problems. Computer Systems* **4** 17–22
[2] Desnitsky V A and Kotenko I V 2016 *XIX IEEE Int. Conf. on Soft Computing and Measurements (St. Petersburg)* pp 481-83
[3] Munch-Ellingsen A, Karlsen A, Andersen A and Akselsen S 2015 *First Conf. Mobile and Secure Services (Gainesville)* pp 1-6
[4] Kuliesius F and Dangovas V 2016 *Eighth Int. Conf. on Ubiquitous and Future Networks (Vienna)* pp 894-99
[5] Birichevskij A R 2016 *SPIIRAS Proc.* **46** 128–38
[6] Fan W, Huang W, Zhang Z, Wang Y and Sun D 2015 *Trustcom/BigDataSE/ISPA* pp 1324-28
[7] Fortat F, Laurent M and Simatic M 2015 *Int. Workshop Network and Systems Support for Games (Zagreb)* pp 1-3
[8] Yin L R, Zhou J and Hsu M K 2015 *Computer Software and Applications Conf. (Taichung)* pp 678–79
[9] Markman A, Javidi B and Tehranipoor M 2014 *IEEE Photonics Journal* **6** 1–9
[10] Iskhakov A Y 2014 *Bezopasnost Informatsionnykh Tekhnology* **3** 97–102

Simulation modelling as a tool to diagnose the complex networks of security systems

S Y Iskhakov[1], **A A Shelupanov**[1], **R V Meshcheryakov**[1]

[1] Tomsk State University of Control Systems and Radioelectronics, 40, Lenina ave., Tomsk, 634050, Russia

E-mail: frosty86@mail.ru

Abstract. In the article, the questions of modelling of complex security system networks are considered. The simulation model of operation of similar complexes and approbation of the offered approach to identification of the incidents are presented. The approach is based on detection of uncharacteristic alterations of the network operation mode. The results of the experiment allow one to draw a conclusion on possibility of the offered model application to analyse the current status of heterogeneous security systems. Also, it is confirmed that the application of short-term forecasting methods for the analysis of monitoring system data allows one to automate the process of formation the criteria to reveal the incidents.

1. Introduction

The prompt development of security system market over the last years is caused by the increase of topicality of international terrorism threat, anthropogenic and natural disasters and, also, the growth of cyber-crime and industrial espionage. In practice, the problem of development of a uniform security system complex often arises. A widespread approach to its solution is the integration of heterogeneous equipment on the basis of local area networks (LAN). The main units of the involved systems connect in the complex security system network (CNSS).

The CNSS is a heterogeneous complex of software and hardware which differs in principles of operation and types of the provided information, but performs a common task - to provide security of people's lives and health and to decrease the probability of material damage to a company.

The development of a security system sector is impossible without implementation of science intensive technological solutions. Scientists all over the world work on the development of new ways to detect threats, to simulate the behaviour of operations of malefactors etc. The majority of modern research is directed to the solution of theoretical and practical tasks from allied sciences. The interconnection of scientific research and technical solutions are thus clearly seen. Despite the obvious topicality and common points of contact, there is no research on the analysis of the CNSS operation and on forecasting incidents [1] in similar systems.

At the same time, there are no models of CNSS functioning and methodological support of the incident identification process in security networks. The objective necessity of methodological generalisation and development of methods and technologies to detect the incidents in the security system operation follows from this unsolved problem.

Thus, modelling of the CNSS operation and process of incident identification in operation of similar complexes is an actual scientific problem.

2. Problem statement

The application of LAN technologies in the CNSS leads to appearance of some essential vulnerability besides obvious advantages in the form of development of the consolidated control tool for all components of security. As the exploitation of this vulnerability can threaten security of people's lives and health, the task of providing the protection of the CNSS obtains a high priority status.

In most cases, the risks connected to external intrusions are taken into account at the design stage of security networks. At the same time, the probability of internal attacks to accessibility, integrity or confidentiality of CNSS elements is usually assessed as low. The internal attacks include: video camera failure, video recording archive failure, blackout of alarm systems, failure of access card readers, malfunction of communication channels etc.

Susceptibility of information processes of CNSS to security threats defines necessity of constant monitoring and status control of such complexes for the purpose of identification of operation incidents. For the solution of the given problem, it is necessary to introduce a monitoring system of the object status in the CNSS. In such system, data on the equipment status should be collected automatically [2]. The analysis of such information will allow one to provide opportunity of constant development and perfection of a network reliability control system.

As mentioned earlier, the nature of incidents can be various [3]. However, during their occurrence there are always events that are uncharacteristic for the standard operation scenario. This fact allows us to combine various incidents and study them by the nature for consideration in a unified manner. At the same time, regular changes of operating modes can occur in the CNSS. Consideration of all possible scenarios and identification the general laws in practice is not possible not only owing to considerable labour-intensity of the process, but also because of peculiarities of the CNSS functioning.

Nevertheless, recording or identification the moment of uncharacteristic change of the operation scenario during a specific period of time allows one to make an assumption that an incident approaches [4]. The use of mathematical models allowing one to describe the CNSS operation seems logical. Thus, for the solution of the problem of identification of the CNSS operation incidents, it is necessary to obtain the model of the CNSS functioning.

3. Construction of a security network model

Generally, the CNSS structure can be represented in the form of a set of devices and communication channels. Let us represent the CNSS in the form of graph *K*, apexes of which are the devices belonging to a security network, and arches are the communication channels connecting the devices.
The present status of the security network can be defined as a set of statuses of all devices composing its structure. Taking into consideration possible scales of the CNSS, modelling a similar complex in the form of a certain set of objects is logical within the limits of a practical task. Such approach allows us to simplify the modelling process and to abstract away from the redundant data that do not influence the result.

Then, the security network can be presented in a form of graph $K`$, apexes of which are the devices chosen to control the network status, and arches are the communication channels connecting chosen devices.

$$K` = (U`,M`) \tag{1}$$

where $U`$ is a set of the devices chosen to control the network status; $M`$ is a set of the communication channels connecting the controlled devices.

Device u_i possesses a certain set of parameters. Every parameter is characterised by a value that can change in time. Then, the device status in the present moment can be defined as a set of values of all its parameters. Considering the significant number of characteristics of each device, modelling a specific device in the form of a limited set of parameters is logical within the limits of a practical task.

At the stage of the choice of devices and their parameters it is necessary to try to obtain a list of parameters necessary and sufficient for the solution of a certain practical task. If this condition is met, the constructed model and the further decision on classification of the found denial of the system operation are adequate.

Let us define the set of all parameters of device u_i as set P_i.

$$P_i = \{p_{i1}, p_{i2}, \dots, p_{ik}\} \tag{2}$$

where k is the number of parameters of device u_i.

Let us represent device u_i in the form of the form of set $P_i`$ – the set of parameters chosen for the control of its status within the limits of the practical task. At that, ratio $P_i` \subseteq P_i$. Thus, $|P_i`|$ defines the number of controllable parametes.

At each moment of time t, parameter $p_{ij} \in P_i'$ is characterised by some value $z \in Z_{ij}$, where Z_{ij} is a set of all values which can be taken in by the j-th parameter of an i-th device.

To define the regular operating mode for each p_{ij}, let us introduce a set of acceptable values $Z'_{ij} \subseteq Z_{ij}$. The problem of formation of set Z'_{ij} will be considered further.

To define the status of device u_i in present time t, it is necessary to check whether all obtained values enter set Z'_{ij}.

$$\text{Criterion } (a, B) = \begin{cases} TRUE, \; if \; a \in B \\ FALSE, \; if \; a \notin B \end{cases} \tag{3}$$

Let us substitute a with value z in function (3). Value z was obtained when measuring parameter p_{ij} at time t. B is also substituted with a set of acceptable values Z'_{ij}.

Then device status u_i at time t is defined by logic function (4).

$$S_i = \text{Criterion } (z_{i1}, Z_{i1}`) \wedge \dots \wedge \text{Criterion } (z_{ij}, Z_{ij}`) \tag{4}$$

where S_i is device status u_i; z_{ij} is the obtained value of parameter p_{ij}.

Let us assume that device u_i functions normally if S_i is *TRUE*. If S_i is *FALSE*, it means that there is a possible incident in the operation of this device. This moment is marked as *DENIAL* and the process to define whether this *DENIAL* is caused by an incident is sterted.

In turn, the condition of the normal operation of a security network at time t will be defined by logic function (5).

$$S = S_1 \wedge S_2 \wedge \dots S_i \tag{5}$$

where S is the status of the CNSS.

Thus, we believe that at time t the CNSS works in a regular mode if S is *TRUE*. In cases when S is *FALSE*, there is an assumption of a probable incident in the security network operation.

4. Application of the model in practice

For application of the above described model, it is necessary to solve the problem of the set formation of admissible values Z'_{ij} for each device u_i. The authors tried to find an alternative way of criterion formation for the solution of the given tasks, which was considered in detail in [5]. The offered technique of information obtaining allows the operator to choose from the set of devices and parameters that are the most important within the limits of the practical problem being solved to make the choice easier.

Understanding the structure and principles of the CNSS operation allows us to draw the conclusion that the similar complexes represent environment where, during each moment of time, a large quantity of heterogeneous data is generated and processed [6]. Change in time of any of the CNSS parameters forms a time series. Each of such time series represents a certain component of the general status of a network. Denials and failures in the operation directly influence the values of time series [7]. It was defined that at the moment of an incident approach the parameters of observed processes change. In particular, probability of observed events before and after the approach of incident varies [8]. Hence, to reveal an incident, it is necessary to define the moment of uncharacteristic change of the scenario operation with certain probabilistic characteristics.

The offered approach to the solution of the problem of incident identification consists in construction of time series on the basis of data on the values of CNSS parameters and in the application of the forecasting method to the obtained time series. It is supposed that to reveal an incident it is necessary to form a forecast on the basis of the estimation of the previous values of the investigated parameter and to compare it to the current value.

The developed software "SOWA" [9] was introduced in open joint stock company "Special Technology and Innovation Economic Zone" of Tomsk and was used to control the CNSS consisting of more than 1000 active devices. The video surveillance subsystem is one of the most dynamical components of the investigated CNSS, which is caused by a number of special features of its operation. Besides, the information from the video surveillance cameras is fed to the workstations of the duty staff in a real time mode and is constantly processed by operators. It allows strengthening the control of the investigated object of research and, consequently, increasing the accuracy of the experiment. 3 video surveillance cameras digitalizing the analogue signal from connected video cameras were selected and data were transferred to the server.

The described above features of the system operation connected to the use of various record modes allowed us to make an assumption that the traffic generated by the video surveillance cameras has specific character. There are both constant and variable components of the traffic profile. The constant component is provided by means of data transmission to the server from the video cameras recording constantly. The variable component is specified by data transmission from the video cameras with the "Record on Movement Detection" mode. The intensity increases unevenly due to aperiodicity of the recording.

5. Results of the experiment

The main objective of the experiment is to evaluate the possibility of use of the offered methodological groundwork and of the developed software for identification of the incidents in the CNSS operation on an example of a video surveillance subsystem. The experiment was being carried out during 30 days and consisted in the following. In each watchpost there is a register where operators of video surveillance record all incidents connected to the system operation (irregularities of information display, interference etc.). The register is kept according to strict instructions.

As an alternative to operator's work, there was an attempt to reveal the incidents on the basis of the forecast of traffic in the network interface of the video surveillance camera to control the system status. As a result of application of the authors' technique, the traffic generated by the video surveillance camera was controlled. The monitoring system worked irrespective of video surveillance system. To reduce the human factor influence, security guards were not notified about the experiment and continued working according to duty regulations. Inquiries about the amount of traffic in the interface of the video surveillance camera were made every 5 minutes. The data was collected in the storage and was processed by the Holt and Winters method [10].

In case of probable occurrence of an incident the monitoring system sent notifications to the administrator by means of e-mail.

The special feature of the investigated system of video surveillance is the logging subsystem. Besides, the system provides the means to operate the register of events. Within the limits of the experiment, at the end of each day the system administrator possessing the expertise in network equipment and also aware of the features of the investigated object compared the security log entries with the data logging subsystem.

The data received by video surveillance operators was checked by the administrator every day and was verified with the logging system. The obtained result was recorded. To make the analysis complete, the administrator used archive of video surveillance. Similar actions were also carried out for the introduced system of monitoring. The administrator checked the messages of monitoring system about the found incidents every day as well. The incidents confirmed by the administrator were summarised every day and recorded.

The estimation of the received results consisted of two stages:

- the estimation of the work of video surveillance operators as regards the logging subsystem;
- the estimation of monitoring system operation concerning the logging subsystem.

A dispersive analysis was applied to estimate the effect of the means of gathering information on incidents in a CNSS on the result of monitoring with the use of the Fisher criterion. The principal cause of application of the given aspect of the analysis is the possibility to estimate the effect of nonmetering qualitative factor on number indicators.

5.1 Estimation of operators' of activities

Evaluating factor *A* - a way of data acquisition about incident occurrence in a CNSS was defined. Factor *A* accepts the following values at an estimation of activity of operators of video surveillance: 1) the work of operators of video surveillance; 2) logging system incidents in the CNSS. Also two statistical hypotheses were generated:

- Zero hypothesis H_0: work of operators does not influence the number of the revealed incidents in comparison with logging system (distinctions in the number of the revealed incidents between are not more definite than casual distinctions in each group);

- Alternative hypothesis H_1: the effect of work of video surveillance operators has an essential effect on the number of the found incidents in the operation of the CNSS (the number of the incidents revealed by the operator essentially differs from a "reference" indicator - system logging).

Comparison of calculated and tabular values of the criterion showed that in all cases the calculated value of the criterion is more than critical. It means that the difference between dispersions is significant and zero hypothesis about absence of influence of operators work on the number of revealed security incidents should be rejected. In other words, the number of the incidents revealed by the operators essentially differs from data of the logging system processed by the system administrator.

5.2 Estimation of monitoring system functioning

Evaluated factor A - the way of data acquisition on occurrence of incidents in a CNSS was defined. Factor A accepts the following values at the estimation of "SOWA" monitoring system functioning: 1) the application of the monitoring system; 2) the system of logging incidents in the CNSS. Also two statistical hypotheses were produced:

- Zero hypothesis H_0: application of the "SOWA" monitoring system does not have an essential effect the number of the revealed incidents in comparison with the logging system;

- Alternative hypothesis H_1: the effect of application the "SOWA" monitoring system has an essential effect on the number of the revealed incidents in the CNSS operation in comparison with the logging system.

Comparison of the calculated and tabular values of the criterion showed that in all cases the calculated value is less than critical. It means that the difference between dispersions is insignificant and zero hypothesis about absence of monitoring system effect on the number of the revealed incidents in comparison to the regular logging system cannot be rejected. That is, the difference in the number of the revealed incidents between two specified ways are not more expressed than random difference in each group.

6. Discussion of results

The obtained results show that the number of the incidents revealed by the monitoring system slightly differs from the records in the logging system. At the same time, the number of the incidents recorded by operators considerably differs from a regular logging system.

Considering that the logging system was accepted as a standard and all results of experiment proved to be true by the system administrator of the video surveillance who has an expertise in its operation, it is possible to draw the conclusion about the possibility of use of the developed monitoring system for identification of the incidents.

It is determined that the developed toolkit allows using the offered technique of gathering the information to the full. At the same time tests showed that the monitoring system can be used as an alternative of the detailed studying of the given system logging. In practice, the involvement of the system administrator to deeply study the logging system data will undoubtedly lead to increase in labour content and consequently to the higher cost of the system operation. The automated system of monitoring that has similar indicators to the ones of an expert should become an alternative. An additional factor characterising the monitoring system positively is the considerable reduction of time needed to detect an incident. In other words, the developed software allows us to improve the status control of the CNSS of an enterprise qualitatively.

7. Conclusion

Construction of complex security system networks has recently become one of the basic directions of people's life and health protection at enterprises. Despite the topicality of such complexes, the problems of the status control and monitoring remains unstudied. Absence of both models of the CNSS functioning and methodological support of incident identification of operation process is observed.

The purpose of the given research is to fill the gap in the scientific and technical literature lacking the research in the field of modelling of security systems of heterogeneous objects, including the models of functioning and methodological support of the process of incidents identification in operation of such complexes.

The authors defined the features of the CNSS functioning which cause the limited applicability of the status control methods and incident identification methods to them, which are used in a typical LAN. A model of the regular operation of a CNSS is offered and approbation of the approach offered earlier based on detection of uncharacteristic changes in the operating mode of a network was carried out. This approach allows one to generate a logical and time structure of the incident identification process.

The "SOWA" program toolkit is developed. The monitoring system of complex security systems networks allows one to use the offered methodological support of the process of incident identification for strengthening the subsystem status control and to increase the speed of reaction to arising incidents.

8. Acknowledgments

The work was partially supported by the Ministry of Education and Science of the Russian Federation in terms of the basic part of the state order for Tomsk State University of Control Systems and Radioelectronics in 2015–2016 (project № 3657); RFBR (project № 16-47-700350).

References

[1] Lavrova D S and Pechenkin A I 2015 *Information Security Problems. Computer Systems* **2** 80–85
[2] Basov G B, Ostrovskiy E O and Uraskov Y L 2015 *Telecommunications* **12** 16-20
[3] Baranov V A 2011 *Information Security Problems. Computer Systems* **2** 24–31
[4] Fedorchenko A V, Levshun D S, Chechulin A A and Kotenko I V 2016 *SPIIRAS Proc.* **47** 5-27
[5] Iskhakov S Yu, Shelupanov A A and Mescheryakov R V 2014 *Proc. Int. Sci. and Tech. Conf. (Omsk)* 1–4
[6] Sun S *2015 Int. Conf. Computation on Intelligence and Communication Networks (Jabalpur)* pp 1183-87
[7] Kalashnikov A O, Yermilov Y V, Choporov O N, Razinkin K A, Barannikov N I 2013 *World App. Sci. J.* **25 (3)** 399–403
[8] Homenda W, Jastrzebska A and Pedrycz W 2014 *Fuzzy System (FUZZ-IEEE): IEEE Int. Conf.* S **1** 2055–62
[9] Iskhakov S Yu and Shelupanov A A 2013 *Telecommunications* **6** 16-21
[10] Ekberg J, Ylinen J and Loula P 2011 *Int. Conf. for Internet Technology and Secured Transactions (Abu Dhabi)* 627-31

Key management schemes using routing information frames in secure wireless sensor networks

V A Kamaev[1], A G Finogeev[2], A A Finogeev[2], D S Parygin[1]

[1] Volgograd State Technical University, 28, Lenina Ave., Volgograd, 400131, Russia
[2] Penza State University, 40, Krasnaya Str., Penza, 440026, Russia

E-mail: kamaev@unix.cad.vstu.ru, alexeyfinogeev@gmail.com

Abstract. The article considers the problems and objectives of key management for data encryption in wireless sensor networks (WSN) of SCADA systems. The structure of the key information in the ZigBee network and methods of keys obtaining are discussed. The use of a hybrid key management schemes is most suitable for WSN. The session symmetric key is used to encrypt the sensor data, asymmetric keys are used to encrypt the session key transmitted from the routing information. Three algorithms of hybrid key management using routing information frames determined by routing methods and the WSN topology are presented.

1. Introduction

The protection of corporate information systems from security threats is the basis for the implementation of any IT project, including Supervisory Control And Data Acquisition (SCADA) systems [1]. SCADA systems use wired or wireless sensor networks (WSN) as a transport medium for collecting telemetry data and sending commands to actuators [2]. Most such systems are not directly connected to the Internet, but they are connected to industrial business and information systems, to the communication maintenance manufacturers and consultants from the external networks. The uses physical isolation of the critical SCADA system will be at risk of need to receive information from the outside world. The introduction of such measures generates new ways of information security violations, which are more difficult to manage [3]. For example, the Stuxnet virus passed through APCS firewalls using indirect ways, such as USB keys and CDs, or via protocols that firewalls were configured to miss. Therefore, the purpose of the security of SCADA systems is to implement architecture that increases resistance the sensor network, communications, devices and dataframes [4].

The current trend of building a transport network for SCADA systems data acquisition determines the use of wireless self-organizing networks with features of the equality of nodes, dynamically changing topology, self-repair after failures, etc. The ZigBee technology provides a good basis for the construction of reliable WSN for data collection [5]. WSN are gradually replacing the wired network and are used in industry for control of technological equipment [6], in the housing sector to control the heat supply [7], lighting, air-conditioning and ventilation [8], to commercial account of energy and water consumption, in the fire security systems, home automation systems, medical monitoring, etc.

For the protection from other types of impacts, experts use hardware and software methods for a multi-layer protective model for the components of the SCADA system and security information interaction with public wireless data transmission. The task of ensuring the sensor networks security is shifted to prevent data leaks from the system [9], intrusion detection and attack [10], etc.

2. Key management for data encryption in the WSN SCADA systems

Modern cryptographic data protection is based on the encryption using a symmetric key or an asymmetric private/public key. Special codes are used for authentication of the elements of the SCADA system and nodes of the sensor network, the hash functions are used to control the integrity of transmitted data. In any complex information system, there is a lot of secret information that requires protection against compromising that leads to the implementation of key management systems.

If cryptographic algorithms to protect information are developed well enough, the procedures for secure creation, keys use and management are problematic tasks. There are two problems:

- How to generate keys with the necessary cryptographic properties?
- How to send the keys safely to the participants of the information interaction in WSN?

The complexity of key management in wireless sensor networks is determined by the absence of any fixed routes data due to self-organization, spontaneous connections during routing and random nature of information interactions. The main objective of key management is to provide participants of information interaction with key data in wireless sensor networks for implementing the confidential exchange of information via a secure communication channel. The key-management procedures that should be implemented in the control system are as follows:

1. Registration of network nodes as the interaction participants.
2. Synthesis of cryptographically strong keys.
3. Transmission and distribution of keys between nodes of a WSN of the SCADA system.
4. Managing connections between the exchange participants and the keys.
5. Keys replacement.
6. Key recovery in case of accidental destruction.
7. The planned or compromised destruction of the keys.

3. The extended keys structure of the WSN ZigBee technology

The confidentiality mechanism in the WSN ZigBee specification is the encryption and protection of key data when establishing a trust between interacting partners, both at the stage of installation of keys and data transfer process [11]. The security framework is governed by the IEEE 802.15.4 standard, where security is provided by means of special profiles. Specification ZigBee Pro Feature Set supports data encryption, determines changes in the keys distribution and encryption. An additional encryption protocol can be used at the application level when exchanging data cannot be decrypted by any other node in the network, despite the fact that they all have a common network key.

The security system is based in accordance with the ZigBee specification on the AES symmetric encryption algorithm with 128-bit keys, which may be associated with the network (network key – NK) or channel (link key – LK). The key synthesis is based on the use of the master key (MK), which controls their compliance. The initial master key must be obtained through a secure environment.

The control centre keys to which other nodes trust the distribution of keys is assigned to the ZigBee network. Each node in the network must be pre-loaded with the address of the control center keys to get the NK and session keys for the LK connection. During configuration or reconfiguration of the network, the center control key enables/disables the connection for new devices, i.e. working with access control lists (ACLs). Typically, the control center also serves as the coordinator of the WSN, but it may be associated with the server. In the WSN ZigBee standard uses three types of keys:

1. The master key, which is used as an original shared secret code between two nodes in the procedure of generating the session link key.

2. The network key (NK) provides security at the network layer and each node of the network has its own one. These keys are used when disconnecting and re-connecting nodes to the network. The center may periodically update the network key, and broadcast to all nodes in the new key.

3. Session LKs provide a secure unicast transmission of frames between nodes at the app level.

As the ZigBee security is based on symmetric keys, the sender and the recipient of the data frame must have the same shared key. There are three methods of transfer switches for the participants of the information exchange: (i) pre-installation, (ii) transmission from the centre of keys management, (iii) the synthesis of keys by the participants of the interaction. In case of pre-setting, the keys are placed in the nodes or PLC in advance in the process of firmware of the device. In the second case, the centre of key management sends the keys to the devices. In the third case, one of the participants generates its own keys before information exchange and sends it to the partner. Using the symmetric exchange, the participants are sent the same key (shared key) to encrypt/decrypt that causes the following problems:

- the need for secure transmission of keys to each subscriber via a secret or secure channel;
- the complexity of key management, which means the quadratic growth of the number of keys that are to generate, transmit, store and destroy for each pair of nodes in the sensor network.

To solve these problems, an asymmetric encryption scheme with public key is used in network systems. The use of asymmetric algorithms eliminates the problem of key distribution in the system, but raises the problem of validating the received keys and their source authentication, especially in wireless networks. The technology of an electronic digital signature is used when the message previously is subjected to the hash, using the private key, and the other party, using the public key, can verify the authenticity of the recipient signature. Such a scheme of joint application of asymmetric encryption is used in the RSA cryptosystem, where the sender is first added to the message's digital signature.

Although this method solves the problem of symmetric schemes associated with the initial transfer of the key to the other party, such systems are demanding in the length of the keys, computing resources and the performance of the whole network, which does not allow it to be applied in sensor networks. Therefore, the hybrid encryption system for the use in WSN is of greater interest, as it combines the advantages of an asymmetric system with the performance of symmetric cryptosystems. The key structure used in the ZigBee standard, should be supplemented by a special type of key (asymmetric key – AK), which will be used to encrypt a symmetric session key connection (figure 1).

The MK, NKs and special AK keys are long term and session keys have typically a short lifespan.

A key information control scheme and a key exchange procedure in WSN can be implemented in the routing process using frames of routing information and the receipt confirmation to save energy.

4. Key management protocols in WSN

In SCADA systems with a small number of monitoring objects WSN with centralized control is the most common network, which has a topological structure of the "star" or "cluster tree." This network uses a coordinator associated with the server, where it is logical to install the key management system.

The task of key management is more complicated in networks with a large number of controlled objects. Such networks include full or partially decentralized structures where multiple coordinators are responsible for separate areas of monitoring and interact with each other through routers. To provide secure information interoperability of sensor nodes, industrial controllers and a zone coordinator, there are tasks of ensuring secure communications between coordinators and routers.

Key management protocols can be divided into three groups: protocols of pre-placement keys; arbitration protocols with a third trusted party; autonomous protocols. Protocols of pre-placement keys can reduce the service traffic in a sensor network, since the keys are placed in the firmware in advance when configuring sensor nodes. In arbitration protocols, the third-trusted party is used for generating, distributing, installing and maintaining the keys, established by the coordinator or the associated server. In the process of information exchange, the third-trusted party plays some roles (see figure 2). The key controlling system generates, stores and distributes keys, produces accounting, network addressing and configuration of sensor nodes, and is responsible for their authorization.

The autonomous protocols work on the scheme of a self-distribution pair of identical keys between communicating parties. In the first case, the disadvantages are the keys transmission to other party via an wireless communication channel. In the second case, the drawback is the computational complexity of the algorithms of generating the keys pair, the encryption/decryption complexity and the need for a key generation node authentication that leads to additional time and energy of sensor nodes.

Therefore, the system of a hybrid encryption is the most effective one, where pairs of asymmetric keys are used to encrypt the symmetric key before passing it to the interaction participants with the transmission initiator authentication by signature. But it does not exclude the growth of routing traffic.

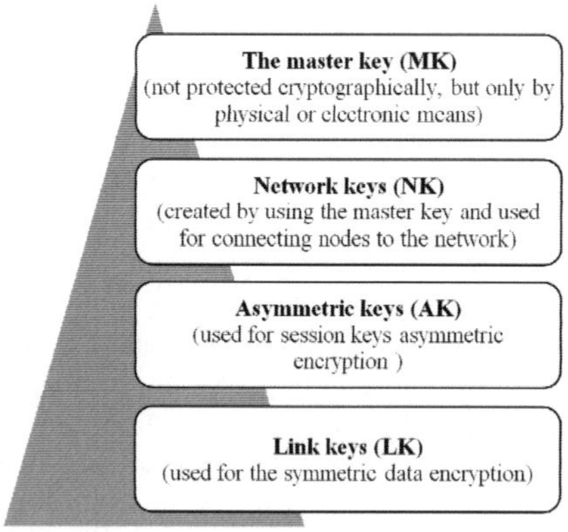

Figure 1. The key structure to protect the sensor data in WSN.

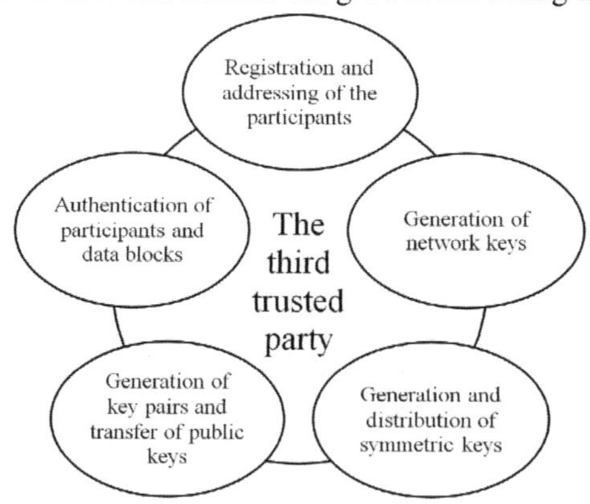

Figure 2. The third trusted party roles.

5. Hybrid key management in the WSN ZigBee technology

In traditional wireless networks, the problem of data protection is ensured by the services at the program level. Sensor networks do not differ from other types of wireless networks in terms of security. They are vulnerable to passive listening attacks and active falsification attacks, as the wireless network is available to the public. Limited energy, computational power and memory nodes are not capable of providing powerful data protection. These restrictions narrow the use of cryptographic mechanisms and protocols at the data link and physical layers of the network model that requires the implementation of an architectural component security at the network and application levels.

6. Autonomous hybrid key management with dynamic routing

Reactive dynamic routing protocol Ad hoc On Demand Distance Vector (AODV) is used in mesh topology sensor networks and sets a route from the source to the destination by broadcasting queries [12]. When one of the touch nodes is going to send data, it sends a broadcast request to create a route (Route Requests – RREQ). The WSN routers broadcast frame relay and make an entry for the node in their routing table. "Logical distance" from the requester to the current position is also written in the frame. The recipient will receive some RREQ frames with different "logical distances" and sends a reply (Route Reply – RREP) to the device, from which the package with a minimum "logical distance" came. RREP is transmitted on the shortest chain until the source. Thus, the response is returned to the optimal path, and generates a vector of the direct route for the frame transmission. If the connection is unreliable, then the node may send a receipt confirmation of a route to the destination (RREP-ACK).

The key-management procedures should be integrated into the routing protocol to reduce the service traffic. It needs to add the appropriate fields in the route frames RREQ and RREP to write the keys and hash functions in them. The encryption and authentication methodology will be as follows:

1. The sender generates a random session key, which encrypts and prepares the data frame to send.
2. The sender sends a broadcast request to create the RREQ route and obtain the public key from the recipient to encrypt the session key.
3. The receiver generates a random pair "public key-private key" for the RSA algorithm and sends the public key to the sender together with the RREP route reply. To authenticate the receiver, the hash of the frame with the public key is computed, encrypted by a key known to both parties.
4. The sender encrypts the session key with the public key and sends it to the recipient along with the encrypted session data frame key. To authenticate the sender and verify the data in the frame, the

hash function of the encrypted frame is computed, which is passed along or together with the frame data, or together with a confirmation receipt of the RREP-ACK route with a bad link quality.

5. The recipient decrypts the session key and the hash function, checks the authenticity of the sender and the integrity of the encrypted frame. Further, he decodes the frame data and deletes the key.

7. The arbitration scheme of hybrid key management in the hierarchical routing process

Another routing method of cluster topology in ZigBee networks is hierarchical routing. When building a cluster tree of the ZigBee network [13], the coordinator, and then the attached routers assign address ranges to child devices in a hierarchical manner. Each node can determine whether the recipient address of the data frame belongs to its "child" branches or is in the other part of the network.

In such sensor network topology and method for hierarchical routing, it is advisable to use the arbitration key management protocol, where on a network coordinator or a related server of the SCADA system, the role of the trusted certification center in a hybrid encryption is implemented. The arbitration method of a hybrid key management for the transmitted data encryption will be as follows:

1. The joined nodes to sensor networks receive addresses for the branches of the cluster tree from the coordinator or the router in accordance with the ranges.

2. Each newly joining node generates a random pair "public key-private key" by the RSA algorithm and sends frames with the public key, the address and the calculated hash as a digital signature the center key management that records the public keys and digital signatures sensor nodes.

3. The source sends a request to the key management center for the generating of the session key for the symmetric data encryption and the frame receiver address for the same key transfer.

4. The management centre authenticates the source, generates a session connection key, finds the public keys of the source and destination in the keys database, encrypts the session key with.

5. The encrypted session key is sent to the source and destination, where the authenticity of the key management center is also verified and the session key is decrypted using the stored private key.

6. The source encrypts the frame using the session key, destroys the key and sends the frame to the recipient who decrypts it with the same key and then destroys it.

8. The arbitration scheme of the hybrid key management in the WSN Many-to-One routing

The third type of routing in the ZigBee network takes into account the specificity of information flows, which are transmitted from a plurality of end nodes to one or more coordinators. This type of routing is called Many-to-One Routing. When using this mechanism, the central coordinator periodically sends a broadcast request (SINK_ADVERTISE) to all nodes. When the node receives the SINK_ADVERTISE request, it sends the Route Record frame back and waits for a receipt confirming the route. Thus, the coordinator receives the full information about the route to the source node and uses it to send a receipt confirmation of the route and the subsequent receipt of a data frame. With the receipt, the coordinator can send any additional information to the node.

The technique for the encryption of data frames and the sender authentication is as follows:

1. The central coordinator sends the SINK_ADVERTISE broadcast request and waits for the Route Record staff in response.

2. The sensor node receives the SINK_ADVERTISE request, generates a random pair "public key-private key" by the RSA algorithm, forms the Route Record frame adding a public key, then calculates and adds a hash function to authenticate the frame and sends the frame to the coordinator.

3. The coordinator receives the frame with the routing information, the source address, its hash function and the public key. Then it generates a session connection key and encrypts it by the received public key. Then, the coordinator adds the session key to the route supporting receipts, calculates the hash function to his authentication using the public key and sends the frame back to the source.

4. The sensor node decrypts the session key using the private key, authenticates the sender by calculating and comparing the hash function.

5. Then the frame is encrypted, using the session key, and sent to the coordinator.

6. The coordinator decrypts the received frame by the same key and destroys it.

9. Conclusion

The advantage of these approaches is the use of existing routing procedures for the simultaneous exchange of key information that allows reducing energy consumption during the information transmission. However, it is required the consumption of network nodes as well as the additional energy for the generation, storage and destruction of keys, the calculation of hash functions, etc. The size of transmitted frames with routing information also increases.

The main disadvantage of key management methods in the hybrid and asymmetric encryption is the possibility of a successful attack to spoof the public key or nodes, where the pair keys for the asymmetric encryption are generated, which leads to a compromise of the entire sensor network. The process of the asymmetric public key obtaining is vulnerable to attack, in which the attacker interferes with the interaction between the sender and the receiver, and can modify the traffic between them. Therefore, the open asymmetric key must have a digital signature to authenticate its sender. Today, there is no such system, in which it would be possible to guarantee the authenticity of the public key.

In the next step, the session key is encrypted using the asymmetric encryption algorithm and the asymmetrical public key of the recipient. The encrypted session key is attached to the frame routing which also includes the added electronic signature. The entire data packet is transmitted to the recipient via the unprotected WSN, and, of course, it is also subjected to sniffer attacks.

To solve these problems, it is possible to abandon the cryptographic encryption of session keys by high computational complexity algorithms, but instead carry out covert transfer of open or encrypted key information by the steganographic methods. Despite the fact that the cryptographic security mechanisms are a prerequisite for the security of sensor networks today, other methods also require the intensive study. The examples are steganography to hide the fact of classified information transfer, the use of the timestamps and synchronization technologies in the generation and disclosure of key information, the intrusion detection and prevention in the sensor network, etc.

Acknowledgments

The reported study was funded by RFBR according to research projects No. 16-07-00031_a, No. 15-07-01720_a, No. 16-37-50017_mol_nr, No. 15-57-54033_Vet_a and the Russian Ministry of Education in scope of the base part (project 2586 of task N 2014/16).

References

[1] Zhao G 2011 *Network protocols and algorithms* (Penza: PSU) **3(1)** 46–63
[2] Kim HJ 2012 *Int. J. of Distributed Sensor Networks* (Hindawi Publishing Corp.) 8(11) pp 1–10
[3] Kravets A G, Bui N D and Al-Ashval M 2014 *Communications in Computer and Information Science* (Springer IPS) **466** 371–382
[4] Bolshakova O N and Chusavitina G N 2016 *Proc. 3d Int. Conf. on Information technologies in science, management, social services and medicine* (Tomsk: TPU) pp 122–126
[5] Finogeev A G, Maslov V A, Finogeev A A and Bogatyrev V E 2011 *Proc. VSTU (Actual problems of management, computer science and informatics in technical systems)* (Volgograd: VSTU) **3(10)** 73–81
[6] Kizim A, Matokhina A and Nesterov B 2015 *Communications in Computer and Information Science* (Springer IPS) **535** 354–366
[7] Finogeev A G, Dilman V B, Maslov V A and Finogeev A A 2011 *Applied Informatics* (Moscow: Market DS) **3(33)** 83–93
[8] Sokolov A, Tyukov A, Sadovnikova N, Zhuk S, Khrzhanovskaya O and Brebels A 2015 *Communications in Computer and Information Science* (Springer IPS) **535** 462–473
[9] Banokin P I 2016 *Proc. 3d Int. Conf. on Information technologies in science, management, social services and medicine* (Tomsk: TPU) pp 106–109
[10] Frolov S G and Demin A U 2016 *Proc. 3d Int. Conf. on Information technologies in science, management, social services and medicine* (Tomsk: TPU) pp 74–76
[11] *ZigBee Specification Overview* http://www.zigbee.org/Specifications/ZigBee/Overview.aspx

[12] Rivest R L, Shamir A and Adleman L 1978 *Communications of the ACM* (New York, NY: ACM) **21(2)** 120–126
[13] *ZigBee cluster library specification* http://www.zigbee.org/zigbee/en/spec_download/spec_download.asp?AccessCode=1351395201

Attacks and intrusion detection in wireless sensor networks of industrial SCADA systems

V A Kamaev[1], **A G Finogeev**[2], **A A Finogeev**[2], **D S Parygin**[1]

[1] Volgograd State Technical University, 28, Lenina Ave., Volgograd, 400131, Russia
[2] Penza State University, 40, Krasnaya Str., Penza, 440026, Russia

E-mail: kamaev@unix.cad.vstu.ru, alexeyfinogeev@gmail.com

Abstract. The effectiveness of automated process control systems (APCS) and supervisory control and data acquisition systems (SCADA) information security depends on the applied protection technologies of transport environment data transmission components. This article investigates the problems of detecting attacks in wireless sensor networks (WSN) of SCADA systems. As a result of analytical studies, the authors developed the detailed classification of external attacks and intrusion detection in sensor networks and brought a detailed description of attacking impacts on components of SCADA systems in accordance with the selected directions of attacks.

1. Introduction

The effectiveness of solving problems of APCS and SCADA systems information security depends mainly on data transmission protection technologies applied to transport environment components.

SCADA systems use a wired or wireless sensor networks as a transport environment for gathering telemetry data and sending commands to executing devices [1].

Because of the transition from wired to wireless network technologies for the construction of sensor networks for gathering telemetry data, the quality of such protection [2] is determined not only by hardware and software solutions for industrial controllers and sensor nodes, but also by the chosen principles of their information interaction in the process of synthesis of network topology, routing determination and data transfer.

Traditional information security measures (the use of sophisticated encryption algorithms, multi-factor authentication, antivirus programs, firewalls, etc.) are not always applicable due to the limited computational and energy resources of sensor nodes and WSN as a whole. In addition, manufacturers of industrial automation and execution devices are developing proprietary protocols, which do not allow implementing security technologies using IPSec, SSL, VPN, etc.

If the SCADA system is set in a large area, for example, for monitoring and management of distributed engineering services (heat, water, electricity and gas supplies) [3], then the network of mobile operators (GPRS/3G modem connections) is used as a transport environment with the possibility of public access [4]. This effectively provides a channel for attacks.

Therefore, to build the effective ways of protecting information in wireless sensor networks, it is necessary to analyze the possible types of attacks, methods of their detection, and reasons of system vulnerabilities. Solving these problems is the purpose of this article.

2. Attacks classification

The modern trend towards transport environment of SCADA systems defines the use of self-organizing wireless networks with peer equality, dynamically changing topology, the possibility of network reconfiguration, self-recovery, dynamic routing, etc.

Currently used principles of data transmission in wireless networks provide the possibility of making the four types of impacts: interception, alteration, destruction and code injection. In accordance with the definition of security, all attacks in WSN of SCADA systems can be divided into the following categories [5]:

- Access attacks, which include attempts to gain an unauthorized access to system resources.
- Attacks on privacy, which represent attempts to intercept the data transfer in the transport environment.
- Attacks on integrity, which include the generation and transfer of frames to capture the control over the SCADA system, to cause faults and failures in its work or to prepare other attacks.

We consider the classification of attacks in detail by the directions of impacts and give a detailed description of the main types (see figure 1).

1. Attacks on the sensor network of the SCADA system.

1.1. Creating active interference in the work area of the SCADA system. To create permanent noises, "white noise" generators are used. They operate on the same frequency as the SCADA system. A source of that noise can be determined using spectrum analyzers and it is possible to stop the attack by locating and eliminating its source [6]. The most dangerous are natural (lightning) or artificial impulse noises, that can lead not only to a system failure, but also damage the sensor nodes and industrial controllers.

1.2. Attacks on the human-machine interface (HMI) of the SCADA system. An unauthorized access to the web-interface from a mobile device can be carried out in the case of open wireless networks or networks with weak authentication.

1.3. Attacks on WSN addresses spoofing aimed at Denial of Service (DoS) initiating. We can distinguish two types of such attacks:

1.3.1. The interception of sensor nodes frames with the purpose of spoofing MAC source and destination addresses, which leads to the failure or malfunction of the SCADA system.

1.3.2. The replacement of the central coordinator to change the address space of the sensor network configuration.

2. Attacks on sensor network nodes and related devices.

2.1. Changing the firmware, drivers and software of industrial controllers (PLC – Programmable Logic Controller) and terminal sensor nodes (RFD – Reduced Function Device). The attack is conducted by PLCs and RFDs scanning to identify the opportunities of the preset operating system, firmware, drivers and controllers changing.

2.2. Injection attacks by spoofing or replacement of the WSN nodes, responsible for collecting and relaying data in the network (FFD – Fully Function Device) to intercept and redirect network traffic. The main purpose of such attack is to redirect the network traffic to the injected or replaced node. We will consider the variety of such attacks:

2.2.1. Compromising the node by replacing routes confirmation tickets to redirect traffic from the end source-nodes to the injected receiver-node. As a result of such replacement, the real coordinator stops collecting data from the PLCs and sensors, and the dispatch service loses control of technological processes.

2.2.2. The router (the FFD node) replacement in a sensor network is aimed to violate the correct operation of routing algorithms. The attack can be carried out by:

- the creation of a "false" tunnel (on the injected router there runs a program that copies retransmitted frames to transfer them to another sensor network, or, conversely, a program of frame transmitting with control commands from another network);
- setting filters (on the injected router there runs a program that filters and destructs retransmitted frames on the specified criteria or content);

- changing routes (on the injected router there runs a program that changes the contents of "Route Record" packets by a given algorithm or at random.

2.2.3. The replacement of the WSN central coordinator to the organization of run- up broadcast storm and to achieve service denial or to power supplies fast discharge.

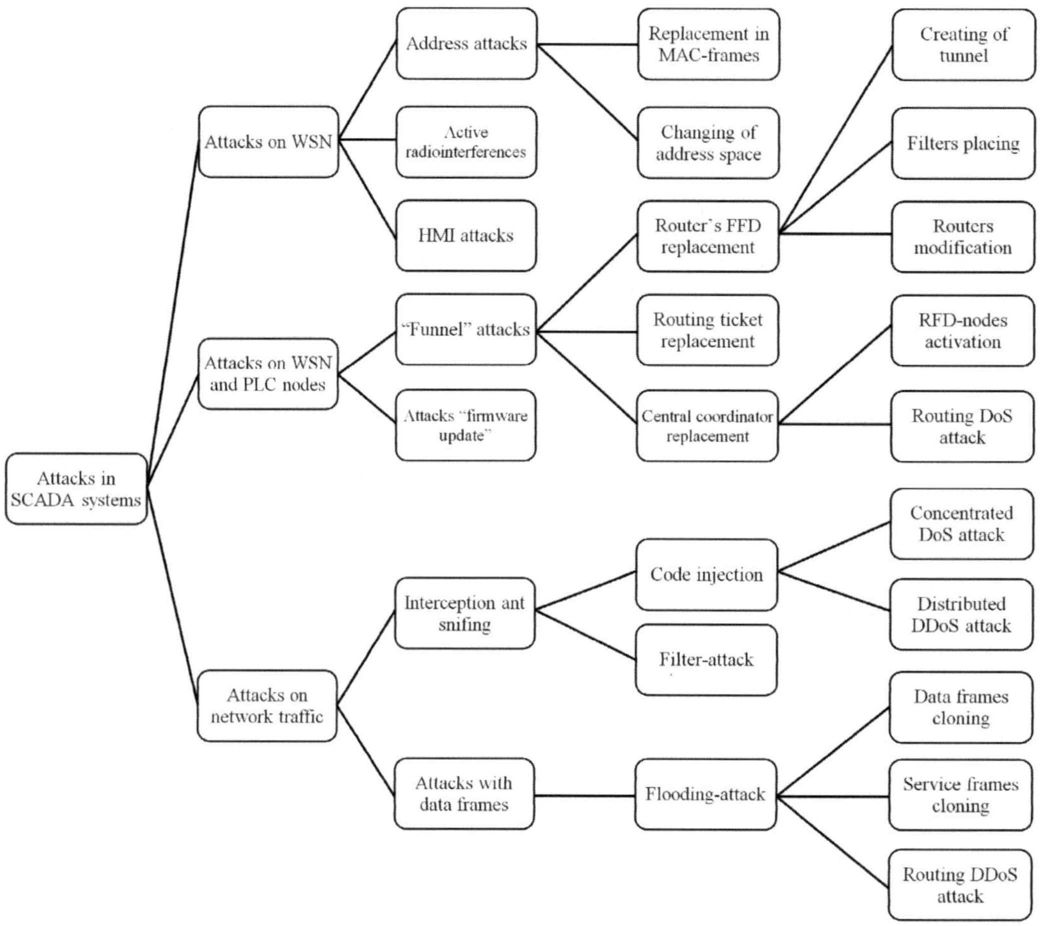

Figure 1. Classification of attacks by impacts directions.

3. Attacks on the sensor network traffic.

3.1. Listening to data transmission channels. It is produced by network traffic intercepting and decoding with special utilities (sniffers) for the subsequent frames analysis for extracting the required information.

3.2. Attacks with data frames. It is performed by flooding or by generating "false" service or data frames or replacing the contents of captured frames and the subsequent injection into the network [7]. Let us consider the basic options of such attacks.

3.2.1. The injection of a malicious code. It focuses on bringing malfunction to the executing devices, the entire network or on changing the parameters of technological processes. The injection of a self-replicating worm into the network routers leads to infection and transformation of all nodes to the botnet, which nodes generate data frames to increase the network reaction time, producing faults and failures (distributed DoS attack [8, 9]).

3.2.2. Frames filtering and selective broadcast. It is produced by injecting into the network a special type of software or hardware filters that intercept data frames, filter them, and may perform a selective broadcast. The effectiveness of the attack increases with its integration with the "funnel" attack.

3.2.3. Flooding attacks by generating "false" frames (service or data) and broadcasts:

- cloning and broadcast of data frames are performed by intercepting and reproducing repeatedly the same data frames followed by broadcasting in the network to achieve input buffers overflow and network failures;
- generation and broadcast of polling units frames and HELLO-frames to achieve failures of network resources; creating and sending a set of HELLO-frames with non-existent addresses of nodes in the network, it is possible to make an image of a "non-existent" area of the sensor network;
- synthesis of "virtual" source-nodes to broadcast route packets from them (routing DdoS attack); here the weakness of Source-Routing technology is exploited if it is used in centralized SCADA systems with one coordinator and a gateway, namely, the excessive network load with the broadcast routing traffic.

3. The results of intrusion detection in WSN

There are few general traditional techniques aimed to detect attacks in transport network media [10]. All of them include the following procedures:

- identification and validation of a non-standard network traffic;
- periodic inspection of privileges and authorizations for the personnel access to specific information resources of the SCADA system;
- disabling of the unused protocols and services;
- disabling of the remote access and control of the network nodes and applications;
- a periodic scan of network interfaces and drivers;
- timely updating of nodes software from the trusted sources.

There are three ways to detect attacks in networks:

1. Detection by the signatures. The signature defines the characteristics (profiles) of previously committed attacks. During the scanning, a coincidence of signatures is revealed and notification is made. However, this method does not reveal the attack with new (unknown) signatures.

2. Detection of the anomalous behavior. The attack detection occurs when identifying an abnormal behavior of the network node or deviations from its normal operation. The disadvantage of this approach is that an incorrect behaving node may be affected by other factors that are not related to the attacks, such as software, hardware or sensor failure.

3. Combined detection by the specifications. This method combines the two previous ones to reduce their shortcomings. WIDS (Wireless Intrusion Detection System) is a software and hardware solution, which consists of software agents that perform the function of collecting, processing and analyzing network traffic packets. Agents interact with the server, transmitting captured packets to it. The server processes the received data for detecting attack signatures and the anomalous behavior of network nodes, as well as responding to events. Thus, WIDS combines signature and behavioral ways, and relates to the third method. In operation, WIDS performs monitoring and analysis of traffic in the sensor network. Its functionality includes the following standard procedures:

1. The analysis of the WSN topology.
2. The determination of WSN vulnerabilities.
3. The compilation and maintenance of network nodes lists. These lists are generated on the basis of network traffic analysis and retrieval of MAC-addresses of the network nodes from the captured frames. In the future, the resulting lists will actually allow detecting the appearance of new "foreign" potentially dangerous nodes in the network.
4. Detecting and countering attacks in WSN. At the moment, the number of detected attacks in WSN is far less than the number of detected attacks in wired networks, as it is only limited by the OSI

model data link layer traffic analysis. The result of the attack detection is the administrator's notification on potential problems in different ways (via email, SMS messages, etc.) and event logging for auditing.

5. Locating the source of the attack and its suppression. WIDS can use such mechanisms of repression as the implementation of DoS attacks on the attacker's node, blocking the attacking agent by active network equipment. Locating the source of attacks means the detection of the coordinates of the device that violates security policy by the trilateration, multilateration or triangulation technologies.

6. Control of security policy. It is based on the analysis of the network nodes list in order to detect changes in the policies set by the administrator. An audit can detect the appearance of unauthorized nodes and applications, violations of traffic protection policy.

7. Performing controlled invasion tests through the existing vulnerabilities of the SCADA system and its components by specific exploits.

8. Monitoring of the wireless network capacity and the network response time. In the process of monitoring, WIDS can monitor the physical and data link layers of the network, and identify problems such as:

- overload of a channel, node or network;
- a sharp increase in the number of data frames received by the coordinator, routers and end nodes;
- reduction of radio signals power;
- a sharp increase in the broadcast service or routing frames;
- overlapping with the neighbouring networks;
- reduction of the network bandwidth for no apparent reason;
- a dramatic increase of the route search time;
- a sharp increase of the server applications' reaction time to client requests;
- an increase of collisions in data channels;
- an appearance of new network nodes;
- reduction of the data transmission rate;
- overload of the network nodes and the network as a whole;
- overflow of nodes' buffer memory, denial of service, etc.

On the basis of such monitoring results analysis by responsible persons for the information security, the necessary decisions may be made and appropriate operational and long-term measures may be implemented.

4. Conclusion

Despite of rather a large number of possible attacks in wireless sensor networks and SCADA systems, the internal anthropogenic threats are the most dangerous for information security, which include [11]:

- unintentional personnel actions that create the auspicious conditions for external attacks by hackers;
- intentional ignoring the requirements of information security by the staff serving the SCADA system;
- the lack of personnel qualification in the field of information technologies and implementation of information security methods.

Unlike the external intruder, the staff of the enterprise has great opportunities for attacks to infect and spread a malicious code on the sensor network. Information security problems are often caused not so much by external attacks, but by the staff non-compliance of regulations and rules of the enterprise information security policy.

Managers and other staff of the enterprise may ignore their duties and in their "free" time are busy with "surfing" the Internet, social networking, and playing computer games.

It may result in an unauthorized PC infection by computer viruses, Trojan horses and worms, which then may penetrate into the sensor networks. This explains the fact that viruses and worms like Stuxnet are often present in industrial systems, and the fact that their presence is normally hidden by staff and managers, as the disclosure of this information will lead all the staff and management to the detailed inspection and then to the subsequent negative consequences for them. In addition, the finding of the infection in the SCADA system may cause a need of hard reset to clean the virus and will stop the most of the enterprise's processes, but it is not always feasible from the economic standpoint.

The lack of qualifications of the personnel, who work with PLCs and SCADA systems, also requires the involvement of outside experts to identify and correct software changes in controllers, because after cleaning the system, it is necessary to be ensured that the programs and settings in the controllers correspond to the values required for the proper functioning of the industrial automation complex.

It is well known that the human factor is the main reason of deviations from the normal operation status in various technical systems. This requires special attention to the establishment and maintenance of appropriate technical regulations.

Acknowledgments
The reported study was funded by RFBR according to research projects No. 16-07-00031_a, No. 15-07-01720_a, No. 16-37-50017_mol_nr, No. 15-57-54033_Vet_a and the Russian Ministry of Education in scope of the base part (project 2586 of task N 2014/16).

References

[1] Tyulkina E, Vassiliev P, Janovsky T and Shcherbakov M 2014 *Communications in Computer and Information Science* (Springer IPS) **466** 542–555
[2] Kravets A G, Bui N D and Al-Ashval M 2014 *Communications in Computer and Information Science* (Springer IPS) **466** 371–382
[3] Kamaev V, Finogeev A, Finogeev A and Shevchenko S 2014 *Proc.11th Joint Conf. Knowledge-Based Software Engineering* (Springer International Publishing Switzerland) pp 1–15
[4] Finogeev A G, Dilman V B, Finogeev A A, Nefedova I S and Finogeev E A 2014 *Mater. Int. Sci.-Practic. Conf. on Innovative information technologies* ed S U Uvaysov (Moscow: HSE) part 3 pp 109–116
[5] Frolov S G and Demin A U 2016 *Proc. 3d Int. Conf. on Information technologies in science, management, social services and medicine* (Tomsk: TPU) pp 74–76
[6] Sokolov A, Tyukov A, Sadovnikova N, Zhuk S, Khrzhanovskaya O and Brebels A 2015 *Communications in Computer and Information Science* (Springer IPS) **535** 462–473
[7] Banokin P I 2016 *Proc. 3d Int. Conf. on Information technologies in science, management, social services and medicine* (Tomsk: TPU) pp 106–109
[8] Beitollahi H and Deconinck G A 2011 *The 10th IEEE Int. Conf. on Trust, Security and Privacy in Computing and Communications (IEEE TrustCom-11)* pp 11–20
[9] Beitollahi H and Deconinck G 2012 Analyzing Well-Known Countermeasures against Distributed Denial of Service Attacks *Elsevier J. of Computer Communications*
[10] Beitollahi H and Deconinck G 2012 *Elsevier J. of Computer Communications* **35(7)** 829–841
[11] Bolshakova O N and Chusavitina G N 2016 *Proc. 3d Int. Conf. on Information technologies in science, management, social services and medicine* (Tomsk: TPU) pp 122–126

Monitoring of sucker-rod pump units as a result of the analysis wattmeter cards

M I Khakimyanov[1], I N Shafikov[1], F F Khusainov[1], R A Semisynov[1] and E A Bezryadnova[1]

[1] Ufa State Petroleum Technological University, 1, Kosmonavtov ave., Ufa, 450062, Russia

E-mail: hakimyanovmi@gmail.com

Abstract. The problems of automation of oil wells operated by sucker-rod pumps are considered. The authors have developed a method of mathematical processing of wattmeter cards for diagnosis of downhole pumping equipment. Analysis of wattmeter cards allows controlling the condition of the ground equipment of the pumping unit. Using wattmeter cards, the balance and efficiency of the pumping unit can be calculated. Wattmeter cards spectrum analysis allows determining the vibration and impact loads, which makes it possible to diagnose defects of the reducer and bearings.

1. Introduction

When automating sucker-rod pump facilities (SRPF), two methods are mainly used: dynamometry and wattmetering. The first method allows controlling the operation of the downhole equipment, and the second – that of the ground equipment. Ground equipment of SRPF includes an electric motor, a V-belt drive, a reducer and a pumping unit with balancing weights.

The advantages of wattmetering include the ease of measurements and the ability to keep a record of consumption of the electric power drive [1].

Wattmeter cards analysis allows determining the degree of balance and efficiency of SRPF, diagnosing breakage of belts and rods, detecting blows in the kinematic scheme of the pumping unit.

Historically, the wattmetering method began to be used before dynamometry in the analysis of wells operating modes, since for measuring the force on the polished rod complex, sensors were required. In his works [2, 3], Krichke V.O. developed electronic devices that implement wattmeter cards algorithms. However, the use of wattmetry until now was restrained by computing abilities of controllers [4, 5].

2. Diagnosis of SRPF according to the wattmeter cards

The method of mathematical processing of SRPF wattmeter cards specifically targeted for use in the controller of control station was developed by the authors [6].

The initial data for the analysis are the instantaneous values of the active power in one period of SRPF oscillation. It should be noted that from the arrays of instantaneous current and voltage values, the following parameters can be obtained: active, reactive and full power, power factor, effective values of voltage and current in each phase.

Wattmeter card processing begins with the construction of the smoothed graph (Figure 1). AT the next stage, the average value, variance and standard deviation are calculated for the evaluation of shock and vibration components.

3. Mathematical processing of wattmeter cards

The mean power value of the oscillation cycle is the average arithmetic meaning of the wattmeter card array:

$$W_{mean} = \frac{\sum W_i}{n}$$

Here, n is the number of array points.

The variance can be found from formula

$$S^2 = \frac{\sum (W_i - W_{SM\_i})^2}{n},$$

Where W_{SM_i} – the array point of the wattmeter card after filtration.

The mean square deviation is found from the variance.

For the vibration component, the mean square deviation is accepted. For shock component, the instantaneous value of the maximum deviation of the original and smoothed wattmeter cards is taken.

Next, the construction of the spectrum wattmeter card is made. To construct a spectrum constant, the component is removed from the array by passing through a highpass filter. The constructed wattmeter card spectrum is shown in Figure 2. In the analysis, a maximum frequency of the spectrum, frequency of the spectrum's maximum, the presence and number of peaks are of importance. Defects of gear reducers, bearings, pumping unit elements, the low oil level can be diagnosed by the spectrum.

4. The construction of the wattmeter card spectrum

In Figure 3, the original, smoothed graphs and their differences are built. The power line of the noise component of the signal is drawn. By the nature of the noise component of the signal, the dynamics of impacts in SRPF drive elements can be detected.

To calculate the balance of SRPF, let us compare the energy emitted during the course of the up and down movement of the rod. The energy equals the area under the power changing curve. The balance factor is the ratio of these energies with the ideal balance ratio equal to one.

For the calculation of the facility's efficiency the following energies are found: consumption of the network is going to overcome the friction forces in the ground equipment and transfer them to polished rod. The energy consumption of the network is the area of the figure under the wattmeter card curve for the oscillation cycle. The energy expended to overcome friction forces is equal to a constant power for the oscillation cycle [7-10]. The energy expended to lift the rod string and fluid, as well as to lift the counterweight is defined as the area of the "bulges" of the wattmeter card or as the energy difference between the consumed energy from the network and the expended energy to overcome friction.

Efficiency of the facility is the ratio of the energy transferred on the polished rod to the energy consumed from the network.

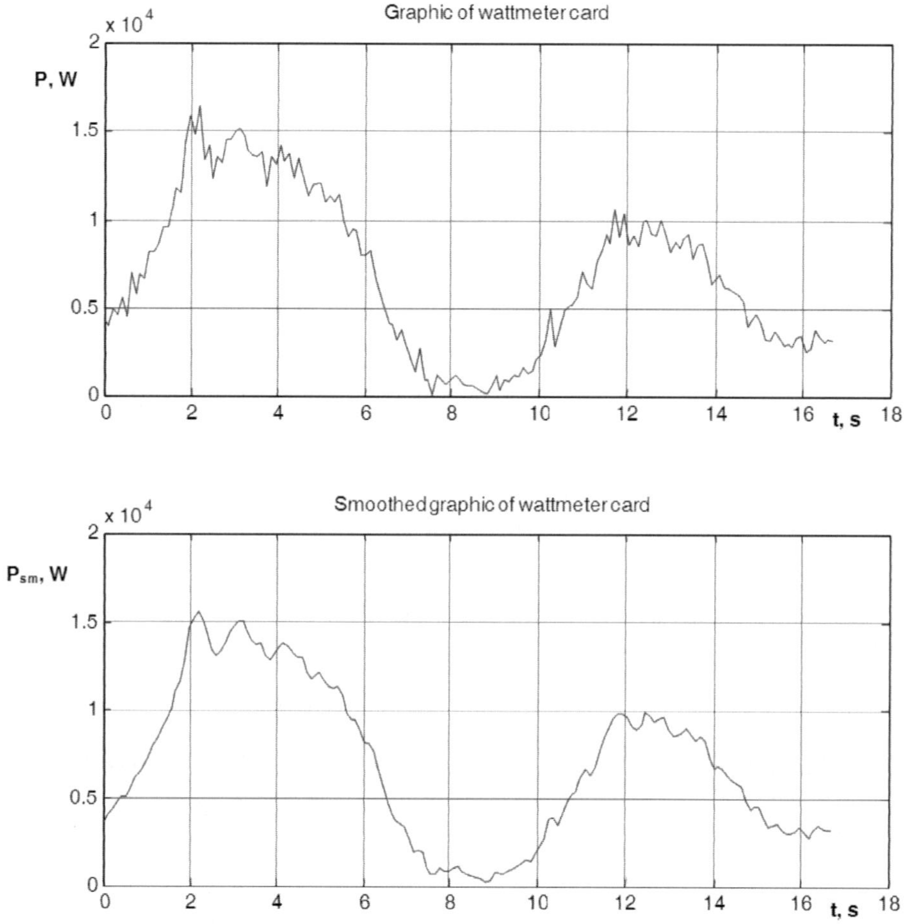

Figure 1. The source and smoothed graphics of the wattmeter card.

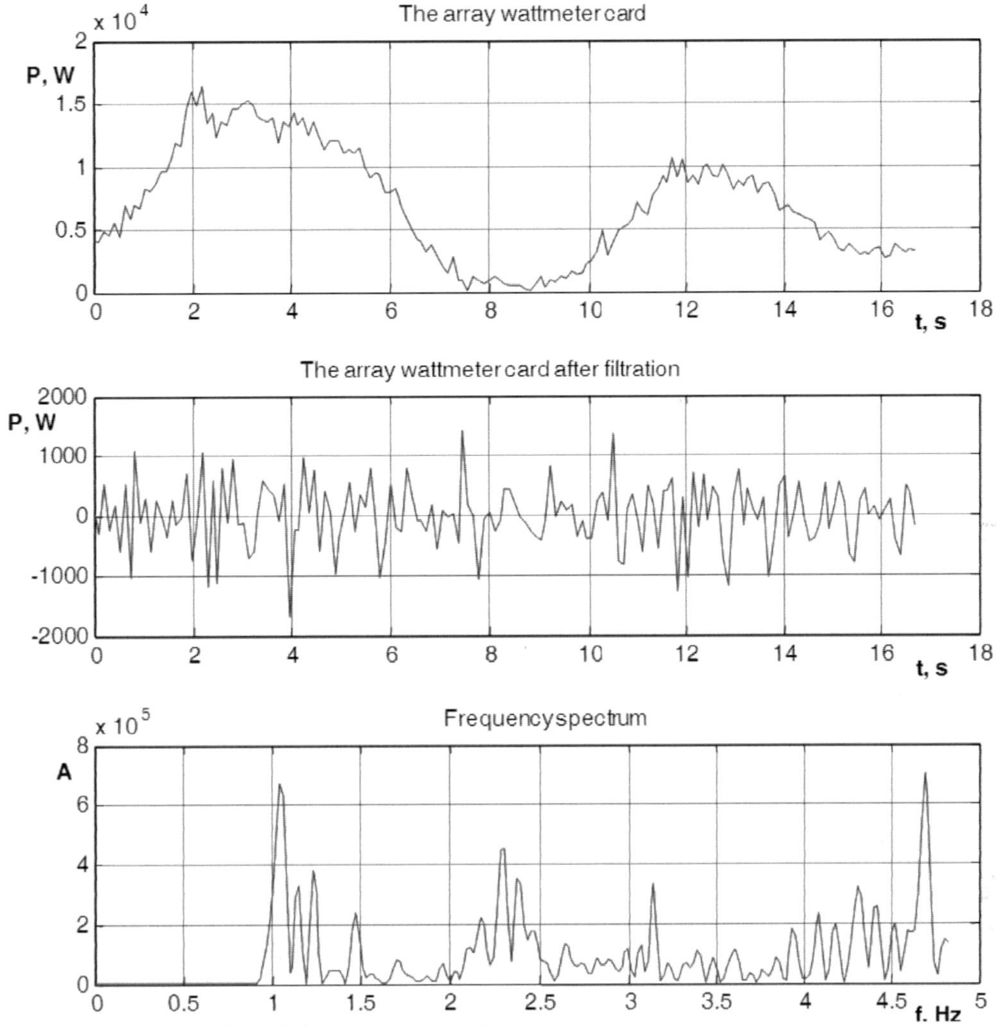

Figure 2. The construction of the wattmeter card spectrum.

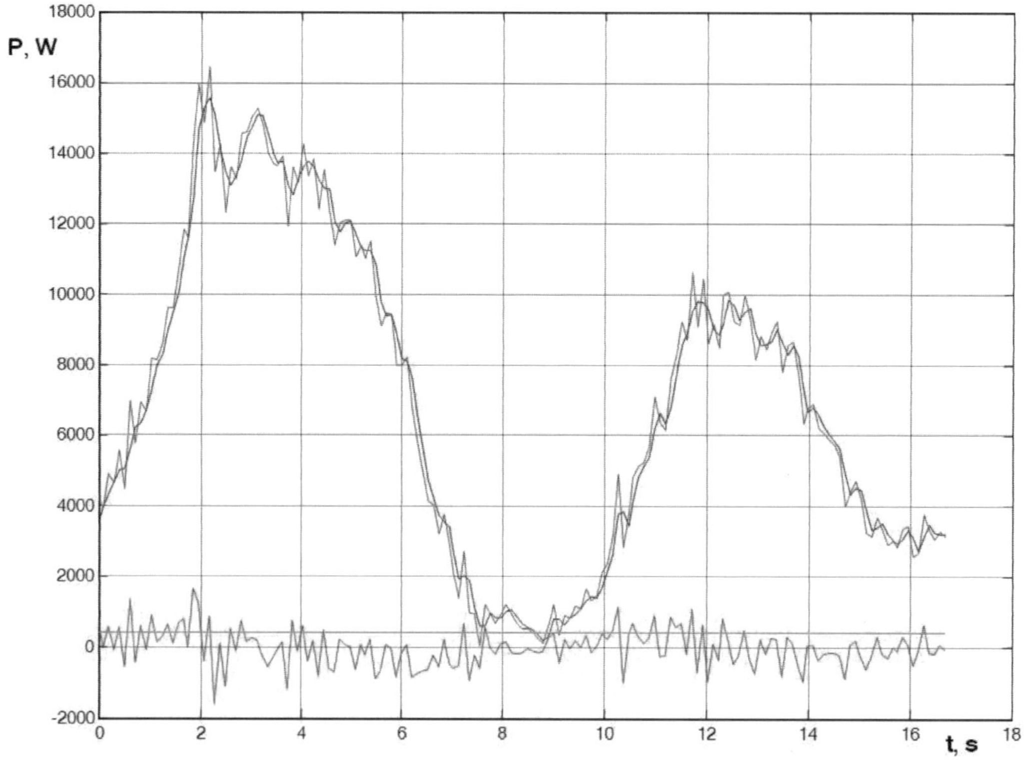

Figure 3. The difference between the source and the smoothed graphs of the wattmeter card.

5. The difference between the original and the smoothed graphics of the wattmeter card

A three-dimensional graph of the spectrum can be built to enhance the visibility (Figure 4).

A three-dimensional image of the wattmeter card spectrum allows evaluating the vibration and impact loads fluctuations. At present, in the wattmeter card, there is a blow at the end of the period of oscillation. On the spectrum, it is clearly seen that at the end of the period there are oscillations with a frequency of 2-5 Hz.

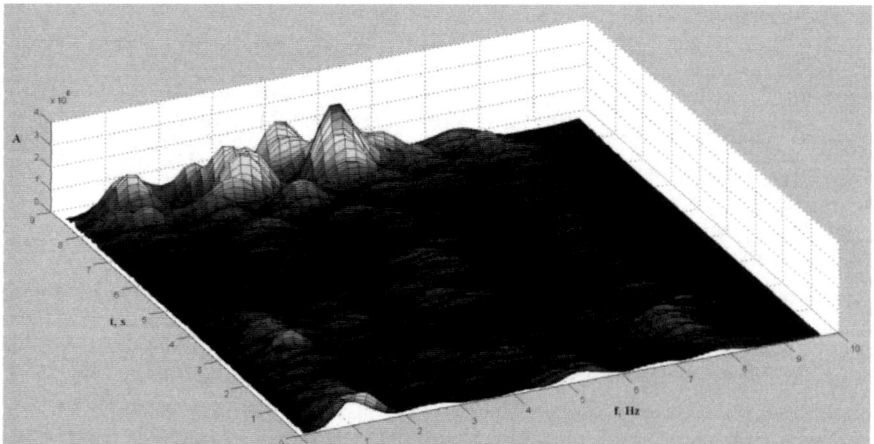

Figure 4. The graph of the three-dimensional spectrum of the wattmeter card.

6. Conclusion

Thus, the following conclusions can be made based on the results of the conducted research:

- control station should provide measurements of dynamometer and wattmeter cards to implement integrated monitoring of the condition and the mode of operation of SRPF;
- wattmetering is an effective way to control the condition of the ground equipment of SRPF;
- wattmetering allows a high degree of accuracy to evaluate balance and calculate the efficiency of SRPF;
- wattmeter cards spectrum analysis allows determining the vibration and shock loads, which makes it possible to diagnose bearing and gear defects.

References

[1] Ordonez B, Codas A and Moreno U F 2009 *Control Applications, (CCA) & Intelligent Control (ISIC)* 1259-1264
[2] Lindh T, Montonen J H, Grachev M and Niemela M 2015 *IEEE 5th International Conference* pp 416-420
[3] Ordonez B, Codas A, Moreno U F and Teixeira A 2008 *IEEE International Conference* pp 282-289
[4] Guluyev G, Pashayev A, Pashayev F, Rzayev A and Sabziev E 2012 *Proc. of IV International Conference "Problems of Cybernetics and Informatics"* 19-21
[5] Aliev T, Nusratov O, Guluyev G, Rzayev A, Pashayev A and Pashayev F 2010 *Proc. of The Third International Conference "Problems of Cybernetics and Informatics"* pp 219-222
[6] Rzayev As H, Rezvan M H and Khakimyanov M I 2013 *Electronic Science magazine "Oil and Gas business"* **5** 19-29
[7] Khakimyanov M I, Khusainov F F and Shafikov I N 2015 *Electronic Science magazine "Oil and Gas business"* **1** 533-563
[8] Xu P, Xu S, Yin H 2007 *Journal of Petroleum Science and Engineering* **1** 43-48
[9] Guoqing H et al 2003 *Oil Drilling & Production Technology* **5** 21-22
[10] Bezerra M A D et al 2009 *ICEIS International Conference* 351-355

Environment as a basis for the design of advertising structures by forming

Y P Khmelevsky[1], **V A Seryakov**[1], **G Y Mamontov**[1], **D T Tsarenko**[1]

[1] Tomsk Polytechnic University, 30, Lenin avenue., Tomsk, 634050, Russia

E-mail: hmelevskiy@tpu.ru

Abstract. A few different neighbouring styles of architectural forms are quite frequent in the cities of great historical past. As a result, a designer or architect has to solve the complex problem while designing the objects within such environment, i.e. one has to fit them naturally into the existing site development. Often, form making is found to be hard, due to the fact that the existing architectural forms of totally different stylistic execution coexist in the visual proximity. Presently, placement of the advertising bills in urban environment is both an urgent and debatable issue. On the one hand, advertising providers are keen to present their product bigger and brighter, on the other hand, the overall and eye-catching exhibition stands can be disharmonious with the surrounding architectural ensemble of the city. This situation is relevant for every cultural city.

1. Introduction

The emerging contradictions require a conceptual analysis based on the study of the historical city center and the necessity for placement of the major advertising and notice boards therein. Review of the existing structures which place the advertising bills in Tomsk city showed that they are executed in most cases against the style of the surrounding architectural ensemble, which is not infrequent. The typical examples are promotional structures of the stadium "Trud" (Figure 1a) and the Italian speciality restaurant "POMIDOR" (Figure 1b).

a b

Figure 1. a - promotional structures of the stadium «Trud», Tomsk; b - promotional structures of the Italian specialty restaurant «POMIDOR», Tomsk.

Examples of promotional displays that consider the historical architectural space while forming the favorable cultural background site development are placed in Tomsk city near the main buildings of Tomsk State University (Figure 2a) and Tomsk Polytechnic University (Figure 2b).

a

b

Figure 2. a - promotional displays near the main building of TSU Tomsk; b - promotional displays at the main building of TPU, Tomsk

2. Form making of advertising structures

The advertising banners that hinder perception of the building architecture are also placed on the front of the grand concert hall in Tomsk city, Figure 3. The building of the grand concert hall has an architectural concept of the distinct geometric patterns with the elements of the modernist style in large frontal lanterns, Figure 4; the benches are in the immediate vicinity, Figure 5, and the elements of classicism are in street lights.

Figure 3. The building of the grand concert hall of Tomsk city

Figure 4. Street lights a la modernist style

Figure 5. A bench a la modernist style

The principal task of the project was the development of the multiple solutions of the promotional displays to use them for placement of the different sized poster canvases, which generally made it possible to combine the ads, depending on the repertoire abundance. Thus, the primary advertising field can be presented in several ways, Figure 6.

First of all, it was necessary to present the strict geometric image of the constructivist building in the advertising structures. The solution to the task was found in using two square tubes being parallel to each other in the pillar construction, while forming a geometric frame integrating the general composition of the object. The picturesque elements in the form of the stylized curls are placed in the pillars' corners to present the characteristics of the historic environment a la classicism and modernist style [1,2].

The overall dimensions of the structure are shown in Figure 7. The basic elements of promotional displays are metal profiled square tubes of 40x40x3 mm, metal profiled square tubes of 100x100x3 mm are used as bearing pedestal of the structure.

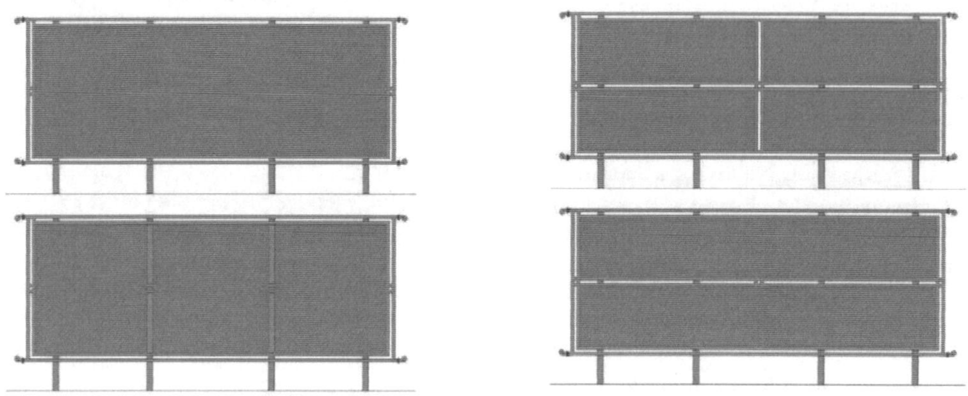

Figure 6. Multiple solutions of the poster canvases placement.

Two stylistic solutions of the structures for advertising bills were used in the project. Both structures are made a la modernist style with the specific picturesque elements: a fancy version with the forging elements, Figure 8, and a version which stylistically most fully conforms with hardscaping (benches and street lights) made a la modernist style, Figure 9.

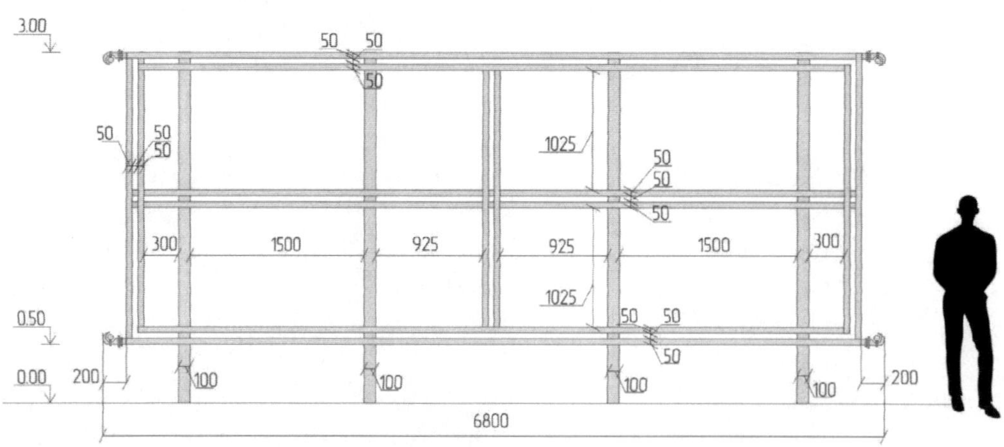

Figure 7. The overall dimensions of the structure

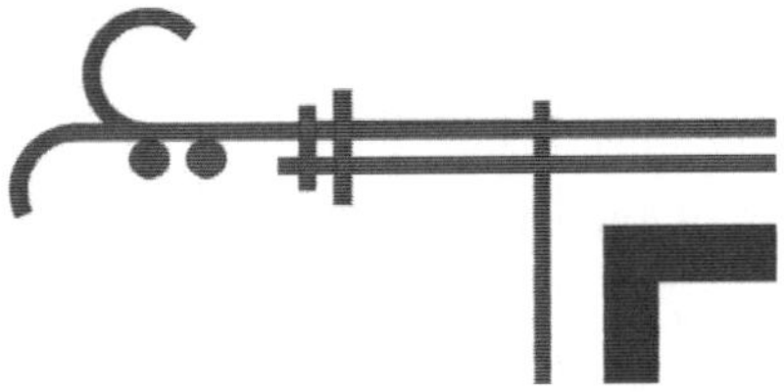

Figure 8. A fancy version with the forging elements

Figure 9. A version made a la modernist style

While designing the object, it was also important to determine its location in the current architectural environment. A characteristic feature of this project is distribution of advertising structures along the territory in front of the grand concert hall. The requirement was identified to locate the sufficient number of advertising structures provided that the significance of the building architecture is maintained, Figure 10.

Figure10. The building of the grand concert hall of Tomsk city with the proposed promotional displays.

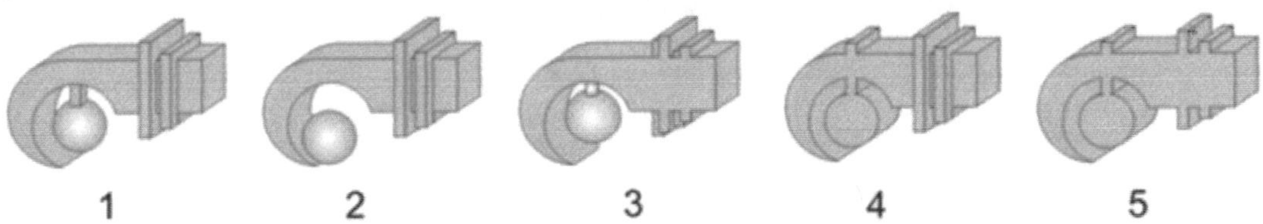

Figure 11. Variants of picturesque elements

During the stylistic analysis of five picturesque elements arranged in the corners of the promotional displays, variant №4 was selected as the one most fully conforming with a la modernist style.

Three options were considered when analyzing the manufacturing methods for picturesque elements. The first option is cutting of three-millimeter steel sheets with laser, bending of some elements of the product and joining by manual arc welding and grinding. Laser cutting is done by means of flow-through burning of sheet metal with laser beam. Advantages of this manufacturing method are that the laser cutting allows working with the profile of any complexity, a perfect line of cut-out is produced without a requirement for any additional cleaning, while saving considerable time, which contributes to high performance, and high-priced die molds or casting molds are not required [3,4].

The second option of the picturesque elements manufacturing is waterjet cutting of the basic element out of fifty millimeter steel sheets and joining them to a decorative bend by manual arc welding (Figure 11-4). They are cut-out with laser out of ten millimeter steel sheets. With waterjet cutting, the particles of a very hard material – an abrasive material are added to increase the destructive force of the water jet. The waterjet cutting has the following distinctive advantages as compared to the technologies of heat treatment (oxygen, plasma, laser, etc.):

- better quality of the cut because of the minimal thermal effect on the workpiece (without melting, fusing and burning the edges);

- cutting opportunities for heat-sensitive materials (a number of fire and explosive, laminated, composite materials etc.);
- ecological cleanliness of the process, a total absence of harmful gas emissions;
- explosive and fire safety of the process .

The third possible option of the manufacturing concept of picturesque elements in the advertising structures is the traditional preparation of molds and casting. The picturesque elements are manufactured by metal casting and welded to the basic structure, followed by cleaning and grinding. The stands are treated with anti-corrosive primer and stained with paint for exterior use on metal surfaces [5,6].

The best possible and cost-effective method of all the above-mentioned methods is manufacture of the picturesque elements with water-jet cutting. Colouring of advertising structures is proposed to be black, because the hardscaping (benches and street lights) located in this territory are of this colour already.

3. Conclusion

Design of advertising objects is relevant and quite a challenge, which necessitates the solution of a wide range of major issues such as a design concept, stylistic solution to maintain the cultural characteristics of the environmental perception, placement within the urban site development and economically feasible design concepts manufactured of the structural components. All these issues are to be solved by a designer, on condition of preservation of individual characteristics of every object.

4. References

[1] Kukhta M S, Sokolov A P, Pelevin E A 2015 *Semantics and technologies in modern design of interior stairs* IOP Conference Series: Materials Science and Engineering **93 (1)** 1-5
[2] Prokopyeva I A 2012 *The selection problem of forming methods in design* p 38
[3] Runge V F, Yu P Manusevich 2005 *Ergonomics in Environmental Design*. 328
[4] Seryakov V A, Khmelevsky Yu P, Mamontov G Ya *The design features of the body of the portable electrocardiograph "ECG-EXPRESS"* **93** 1-7
[5] Solovyev R, Seryakov V A. 2014 *Advertisements and information objects' positioning technologies based on the territorial zoning of the city of Tomsk* **66** 1-6
[6] Sokolov A P, Kukhta M S, Pelevin E A 2014 *Modern technologies of decorative surface treatment* Mechanical Engineering, Automation and Control Systems: Proceedings of the International Conference, Tomsk, October 16-18, National Research Tomsk Polytechnic University (TPU) pp 1-4

Diagnosis and restoration of Li-Ion batteries

I M Kirpichnikova[1], D V Korobatov[1], A S Martyanov[1], E A Sirotkin[1], E V Solomin[1]

[1] South Ural State University (National Research University), 76, Lenina Ave., Chelyabinsk, 454080, Russia

E-mail: nii-uralmet@mail.ru

Abstract. The paper presents the results of testing the LT-LFP.300.11.01 Li-Ion cells under different conditions of charge by stable and pulse width modulation current. The shown study proves the possibility of approaching the diagnosis of battery cell before multi stage charging cycles. This approach helps to determine the real cell state of charge, which is important for the charger in turn for the determination of the initial current value at the very first stage of charge. Using the trickle charge algorithm, each battery cell may be charged individually in accordance with its initial state of charge, chemical condition and temperature. This approach should use a special charger acceptable for multi-cell charge or diagnosis of individual cells with a balancing apparatus. The same approach could be used for battery restoration.

1. Introduction

Compared to other rechargeable batteries, Li-Ion power sources have much higher energy density for weight and volume, providing longer running time between charges. In general, it is rather simple to charge such type of the battery: just charge it, applying 1C current (A) in a few hours [1], where C is capacity of the battery, measured in Ampere-hours. However, if the temperature of the charging battery will be higher than acceptable, the battery/cell will be soon more dead than charged [2,3]. Or even it may explode [4]. If one of the cells has very low voltage and others are well, anyway the battery will not be charged because it is "unbalanced" [5]. And finally if charge will be approached with 1C current, then the time of charge will be acceptable from the commercial point of view. To avoid the problem of run-free-charge, the charger has to be equipped with the apparatus and a software system of diagnosis. Otherwise, it will conduce the degradation of the battery charge by charge. Especially, when not all, but one of the cells is almost or fully dead. Therefore, before the charging procedure it is hard to chose between a trickle charger and maintenance battery charger [6]. However, using the below described approach, it is relatively simple to make a new or the third version of the charger using a special algorithm of the initial state of charge (SOC) determination. In addition to that, the Li-Ion cells have to be charged in full each time, as they also have a "memory effect" [7].

This paper presents experiments of the state of the charge increase, which in turn shows the possibility of diagnostics, its importance and investigation of the influence of trickle charge parameters on the battery state of the charge level. The presented material is the continuation of the research published in [8].

2. Diagnostics of battery SOC

Any battery starts losing its SOC sooner or later, basically from the date of production [9]. This process is called a "degradation", which is impossible to prevent. However, it is real to extend the lifetime of the battery significantly. Getting older, the battery SOC decreases step by step (or actually charge/discharge by charge/discharge). If the periods of time between discharge/charge cycles are longer, the battery degradation they cause is greater [10]. Especially when the battery is not charged in full each time [7]. When the battery loses its SOC up to 70% or even less, it is considered that the battery is almost unworkable [11]. The main reasons, causing battery degradation, are well known: temperature (extremely high or low) [12], fast charge-discharge cycles (big currents) [13], unbalance of cells (different degradation speed of cells). The approach described below may overcome all three problems.

To slow down the degradation, several approaches could be used. We have shown these algorithms of multi-stage charge in our earlier publications [2] and [8]. First of all, each battery should be a subject for diagnosis every charging cycle, as it is recommended in [14]. It means that each cell of the battery should be tested and determined if it still has normal SOC. Second, each cell in battery should be charged individually with the control of temperature. Different scientific and commercial groups propose many different methods and approaches for diagnosis of battery SOC and charge cycling for the cells [15-17]. And many experiments have shown that the individual cell charge increases the lifetime of battery in times, especially when in the beginning of charge procedure, the process starts with the trickle charge or charge-discharge cycles. The testing of the pulse charge on the trickle charge stage shows very good results, because it can make alive apparently dead Li-Ion cell [18]. The difference in SOC, when charging either stable or pulse width modulation (PWM) current, is not big, but essential to make an assumption of PWM preferable usage [19].

We used Li-Ion cells LT-LFP.300.11.01 in experiments [20-21]. Nominal voltage is 3.2V, nominal capacity is 300 A-hours. In accordance with manual, it is not recommended that the current would exceed 2C. On the base of this information, we assumed that the charging current pulse width modulation would have amplitude 1C (or 300 (A)), but the porosity should be variable.

In the beginning, we tested the charge with stable current I_3 = 300 (A).

The first approach was to vary the pulse duty cycle and switching frequency (period), as it is shown in Table 1, in accordance with Fig.1. The similar method is proposed in [22]. However their approach is too expensive and long, compared with the presented below.

All variants of charge start from 80% of SOC. The time of each charge experiment is 30 minutes. The first charge has the duty cycle of PWM = 10 msec, switching period or time *f* between pulses is 5, 10 and 15 msec. The second, accordingly, is the duty cycle of PWM=100, *f* is 3, 10, 20 msec. The third one is the duty cycle of PWM=500, *f* is 3, 10, 50 msec. And the fourth is the duty cycle of PWM=1000, *f* is 3, 10, 50 msec.

Table 1. Dependence of SOC on PWM (duty cycle) and switching frequency

PWM (duty cycle)	stable	10	10	10	100	100	100	500	500	500	1000	1000	1000
f ($_m$sec)	0	5	10	15	3	10	30	3	10	50	3	10	50
SOC final (%)	84	83	82	81	83	82	82	84	84	84	85	85	84

The final SOC under the stable charging current was 84-85%, which is worse than when charging by the duty cycle of PWM = 1000 msec. It is obvious that the SOC is getting better when charging by PWM. However, the PWM is different for different battery types and even, as the experiment showed, for different cells. From the economic and technical point of view, it is worth doing the extra research in determination of optimal PWM for a specific battery. Then the charger can remember the cell and the battery and an approach to individual charge. This approach may save a lot of time and power losses compared to [23].

IOP Conf. Series: Journal of Physics: Conf. Series **803** (2017) 012070 doi:10.1088/1742-6596/803/1/012070

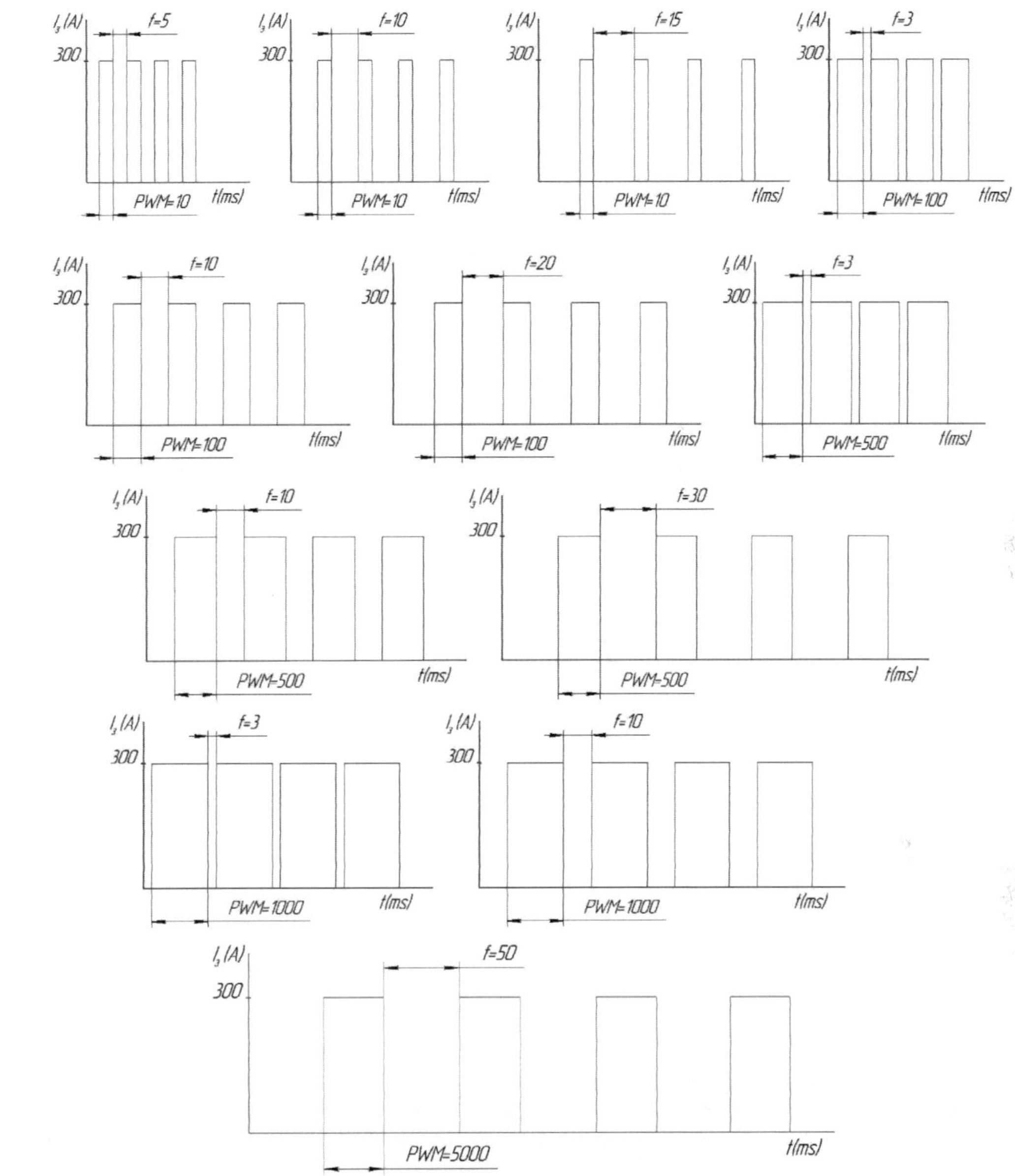

Figure 1. Trickle charge options of $LiFePO_4$ cells (charge only).

The next experiment involves the same PWM charging approach, but with discharge cycles with the current of $0.1I_3$. The results are shown in Fig. 2 and Table 2.

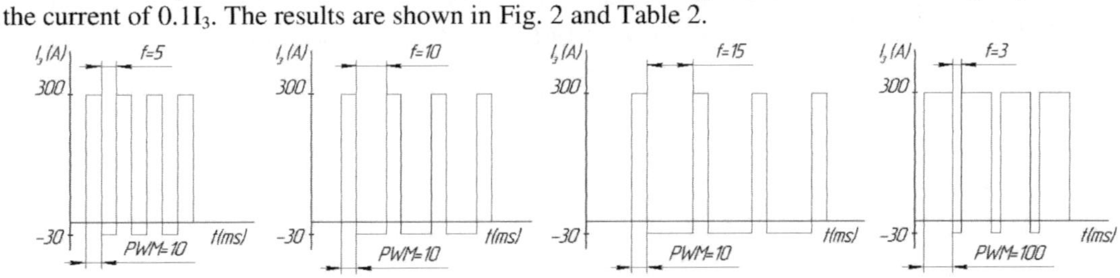

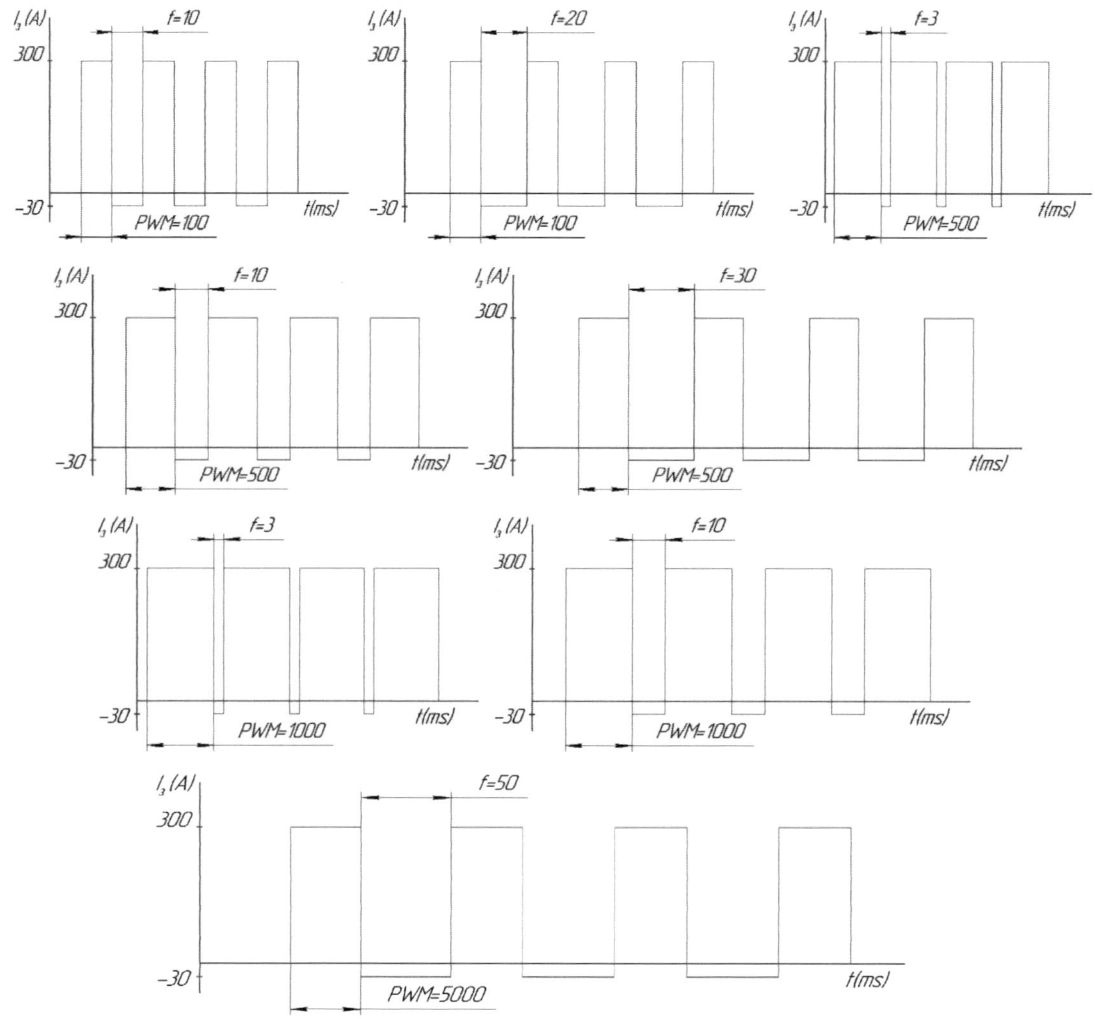

Figure 2. Trickle charge options of $LiFePO_4$ cells (charge and discharge alternation).

All variants of charge start again from 80% SOC. The time of each charge experiment is 30 minutes. The first charge has the pulse duration of PWM = 10 msec, time *f* between pulses is 5, 10 and 15 msec. The second, accordingly, is the pulse duration of PWM=100, *f* is 3, 10, 20 msec. The third one is pulse duration of PWM=500, *f* is 3, 10, 50 msec. And the fourth is pulse duration of PWM=1000, *f* is 3, 10, 50 msec. The discharge PWM in each case was equal to charge PWM in time, but the amplitude was 1/10 of charge PWM. From the technical point of view, this approach is not more complex than the previous one.

Table 2. Dependence of SOC on PWM (duty cycle) and switching frequency

PWM (duty cycle time)	stable	10	10	10	100	100	100	500	500	500	1000	1000	1000
f ($_m$sec)	0	5	10	15	3	10	30	3	10	50	3	10	50
SOC final (%)	84	83	82	81	84	83	82	86	85	85	88	87	87

Table 2 shows that the final SOC in this case of charge-discharge is better even when we discharge the battery. It shows that the battery is "training" and may absorb more energy than in the first experiment. The final SOC in the end of trickle charge stage is absolutely real and can be used as the basic value for the charger.

Another hybrid approach could be used with the determination of voltage at this stage and disconnecting each cell by the individual schedule and/or voltage and/or the current value. It means that for good cells the main charge can be started individually and immediately after SOC determined. For other cells, the main charge should follow the trickle charge. This approach could be used in the grids with alternative energy sources when the peak power is sensitive. Anyway, the hybrid approach could be automated and actually remove the so-called balancing, eleminating the extra apparatus required.

3. Restoration of battery

It is obvious that the similar approach could be used for restoration of the battery. If the charger "learns" to dignose and charge with control each cell of the battery, then the problem of balancing will be solved as it is shown in [24]. And the problem of restoration is to be solved as well. If we apply smaller current 0.1C – 0.2C for trickle charge and alternate it with the discharge for each cell, we will get the desired "training" of the battery when each cell will be charged to initial SOC acceptable for the normal charge. The similar experiments were presented in [24]. In this case we can apply even up to 6C current controlling the temperature every 100 ms. This time is enough to trace the temperature changes. This approach opens very wide perspectives for the restoration of Li-Ion batteries.

To develop this topic we are planning to make a series of experiments with different Li-Ion batteries studying the possibility of restoration in the field conditions. Since the currents required are very small, the charge could be made using almost any small power source like the autonomous renewable energy plant or small charger. Using this approach and small inexpensive device, anybody can "treat" the dead Li-Ion battery and then use it in full.

To analyse the idea related to the charger software for individual determination and charge of the cell, it is required to develop the algorithm of trickle charge, preliminary charge and normal charge. The charger may determine the voltage and current levels as well as PWM frequency for each cell individually.

4. Conclusion

The shown experiments prove the possibility of approaching the diagnosis of battery cell before multi stage charging cycles. Using the trickle charge and/or training charge algorithm, each battery cell may be charged individually in accordance with its initial state of charge, chemical condition and temperature. This approach should use a special charger with the software acceptable for multi-cell charge or diagnosis of individual cells with balancing apparatus. The same approach could be used for battery restoration. The important condition for trickle charge stage is testing a temperature in accordance with the manual requirements of the seller. The first steps were already made by [23] relatively the chargers of future using special algorithms.

5. Acknowledgments

The study was awarded by the RF Ministry of Education and Science under Subsidy Agreement # 14.577.21.0154 of 28.11.2014 (Unique Identifier of Agreement RFMEFI57714X0154).

References

[1] Hoffart Fran, *Linear Technology Corporation* 2015 p 242
[2] Long Cai, Ralph E. White 2011 *Journal of Power Sources* **196** 5985–5989
[3] Liu S, Wang J, Zhang F and Wang L 2016 Applied Surface Science **390** 86–91
[4] Pramanik A, Maiti S and Mahanty S 2016 *Chem. Eng. Jour.* **307** 239–248
[5] Leite M, Ruzmetov D, Bendersky L, Bartelt C 2014 *J. of Mat. Chem.* A **2(48)** 20552–20559

[6] Wu H, Du N, Zhang H and Yang D 2014 *J. of Mat. Chem.* A **2(48)** 20510–20514
[7] E V Solomin, D V Topolsky, I G Topolskaya 2015 *Proc. Eng. Jour.* **129** 213–218
[8] David Linden, Thomas B. Reddy 2002 *Handbook of Batteries,* Third Edition (New York: McGrawHill, Inc.) p 123
[9] Yang F C, Chen C C, Hwang Y S and W T 2006 *IEEE* pp 2723–2726
[10] Guan D, Li J, Gao X and Yuan C 2014 *J.l of All. and Comp.* **617** 464-471
[11] Chen J, Yang F, Lai C, Hwang Y and Lee R 2009 *IEEE Trans* **56 (7)** 1538–1542
[12] Hwang Y, Wang S, Yang F, Chen J and Lee W 2008 *BRC* pp 2259 – 2262
[13] Ke Y and Chuang Y 2007 *IEEE Trans* **22 (4)** 848–854
[14] Chen L R, Han J Y, Jaw J L, Chou C P and Liu C S 2006 *Proc. IEEE* pp 1087–1092
[15] Hwang Y S, Wang C C, Yang F C, and Chen J J 2007 *IEEE Trans* **54 (4)** 705–712
[16] Mohamad N, Kadir A, Mekhilef S 2010 *Jour. of Pow. Elec.* **10 (2)** 164
[17] USTUN T, Mekhilef S 2010 *Int. Rev. of Elec. Eng. (IREEE)* **5(4)** 1347–1353
[18] Hussein AA 2013 *IEEE ECCE* pp 677-681
[19] Liu J, Tang K, Song K, Aken V, Yu Y, and Maier J *2013 Physical Chemistry Chemical Physics* **15(48)** 20813-20818
[20] Lee Y and Park S 2013 *IEEE ECCE* pp 4725-4731
[21] Kumar P R and Mitra S 2013 *RSC Advances* **3(47)** 25058-25064
[22] Zhou Z, Shi Z, Ai G, Lu Y 2013 *Proc. of 2013 Int. Conf. on Qual., Reliab., Risk, Maint., and Saf. Eng.* pp 1938-1941
[23] Chen M and Rincón-Mora G 2006 *IEEE Trans., Exp. Briefs* **53 (11)** 1180–1184

Stability Study of Anthropomorphic Robot Antares under External Load Action

A S Kodyakov[1], N A Pavlyuk[1], V Yu Budkov[1], R A Prakapovich[2]

[1] St. Petersburg Institute for Informatics and Automation of the Russian Academy of Sciences, St. Petersburg, Russia
[2] United Institute of Informatics Problems, Minsk, Belarus

E-mail: ronzhin@iias.spb.su

Abstract. The paper presents the study of the behavior of the major structural elements of the lower limbs of anthropomorphic robot Antares under the influence of different types of loads (torsion, fracture). We have determined the required values for actuators torques for motion of the robot in space. The maximum values of torques are 5 Nm and 5.2 Nm respectively, and are able to withstand the upper and lower leg structures.

1. Introduction

Modern developments in the field of humanoid robotics are aimed at making motion characteristics of the robot closer to the kinematics of the human body [1,2]. The problem of high energy consumption in the realization of the kinematics of the anthropomorphic robot is partially solved by making the construction lighter. Auxiliary batteries (an installation site is provided) will be installed in the upper legs to reduce the electrical load on the main battery of Antares as well as to control actuator powering, which will help to avoid power supply problems. In this regard, the study of the stability of structures and materials and their ability to withstand a given load is of particular interest [3,4].

One of the simplest structures of the biped robot is described in [5]. It is made up of a two-millimeter aluminum sheet, includes six servos operated by an EyeBot controller, and weighs 1.11 kg. When walking, the robot reaches a speed of 120 m/h at a maximum angle of 60 degrees between the hips. Such robot architecture with six servos is used in [6] to study the operating angles of the joints of the knee, ankle and hip.

The anthropomorphic robot in the HanSaRam series, which regularly participates in the FIRA league since 2000, is discussed in [7]. The HanSaRam-VIII (HSR-VIII) robot has 28 servos, weighs 5.5 kg and can move at a speed of up to 12 cm/s.

In [8], anthropomorphic robot Lola, having 7 degrees of freedom per leg, weighs 55 kg with the 180 cm height. The problems of stability of the robot after stopping as well as a gradual contact of foot parts with the surface when walking are discussed. Elastic materials of the toe and heel of the robot foot ensure the reduction of impact force on the robot structure during touching the surface.

A home assistive robot with 14 servos and 16 degrees of freedom, presented in [9], has an anthropomorphic upper body architecture and a wheelbase. The paper mostly focuses on two-handed manipulation of objects by robots during transferring domestic objects.

Anthropomorphic robot SWUMANOID, 92 cm high, having 24 servomechanisms of the series Dynamixel in its construction, is designed in [10] to simulate motions and swimming in water.

Unstable rolling of the robotic body in water complicates the calculation of the kinematics of movements of a floating robot that consists of 21 composite components.

After reviewing the humanoid robot structures [11, 12] and focusing on the literature [13, 14], we have developed a model of robot Antares (Fig. 1). Antares is a robotic system designed for research activities. Its construction, as that of robot DARwin OP [12], consists of parts made from Duralumin D16T plates (2, 4, 8, 12 mm thick). The developed model of Antares has a number of advantages compared with robot Poppy [11]. These include an increased operating range of motors installed in the legs; the use of a more durable material; more optimal distribution of the internal space of the robot.

Details of anthropomorphic robot Poppy are created based on 3D-printing, which allowed much cheaper production of the robot parts and made it possible to use less powerful servos, but reflected on robot stability when walking. The robot has 25 servomechanisms Dynamixel MX-64 and MX-28 [12], which ensure unobstructed movement of the limbs with a given accuracy and strength margin as gears are made of metal.

A modular robot design helps the researcher to change the movement of any robot's limb by isolating the desired limb from the rest of the body, almost without affecting the performance. The structure is specially designed for the installation of additional sensors and connection cables. In addition, such a design facilitates periodic robot maintenance service.

Robot Darwin-OP is a robot platform intended for research and development within the framework of an educational process. DARwin OP has high performance and dynamic characteristics and a wide range of sensors.

A modular robot design helps the researcher to change the movement of any limb of DARwin OP. The structure also allows the installation of additional sensors.

In this paper, we study the stability of the basic structural elements of anthropomorphic robot Antares (Fig. 1) when subjected to external forces such as linear force and torsion. During the research, we have analyzed the movement of the lower limbs and modelled the impact of the external load on the constituent elements of robot's structure. Due to the use of lightweight materials, the engines consume less energy. Dynamixel actuators MX-64T (engines Nos. 1-12) and MX-28 (engines Nos. 13-22) are used in the construction of Antares; the gears are made of metal, which ensures movements with a given accuracy and a margin of safety [12].

The modular construction of the robot allows the researcher to change the movement of any robot's limb by isolating the desired limb from the rest of the body almost without affecting the performance. The structure allows for the installation of additional sensors and connection cables. Furthermore, such a design facilitates periodic maintenance service for the robot. The estimated height of the robot is 100 cm and the weight is 6 kg.

2. Motion analysis of lower limbs

Motion analysis is regarded as the movement of a three-dimensional model in the conditions close to real ones and the reading from sensors. The model of the robot, shown in Fig. 1, displays numbering of servomotors, for which the motion analysis is carried out using the method proposed in [15,16]. The purpose of the simulation is to determine the values of the torque and linear force acting on the robot structure. Data from the obtained results can be used to simulate weight loads on various positions of the robot in space. In the simulation process, we have determined the torques of the engine output shaft in the joints of robot's leg (ankle, knee and pelvis joints).

To analyze the motion, robot's foot was blocked while the leg was moving. Force of gravity G equals 9.81 m/s^2. The contact between the motor flanges and upper and lower leg plates was simulated. Then, equal parameters were set for four engines Nos. 2-5 (Fig.2). In the engine parameters, the function "segments" was selected, where engine's running time is 3 seconds; the motor shaft is moving at 85°; the type of segment is cubic. We used the force that replaced the weight of robot's torso of 20 N as the external load. To determine the torque on the engine shafts Nos. 2-5 (Fig. 2), sensors were installed. The engines worked synchronously and their running time was 3 seconds.

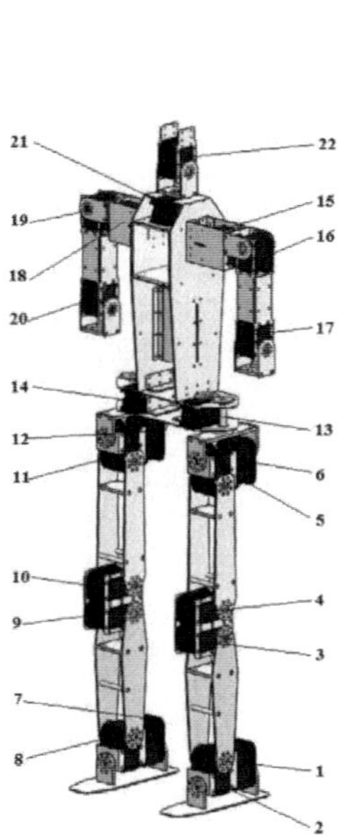

Figure 1. Engines of the generalized model of anthropomorphic robot Antares

As we can see clearly from the graphs, shown in Fig. 2, the engines need different torques to raise the torso. Engine No.2 only maintains balance, so it has a small torque in comparison with engine No. 3. The latter, rotating in the knee joint, moves the lower leg (which abuts against the foot), thus elevating the leg. The situation is similar with engines No. 4 and No. 5. Engine No. 4 lifts robot's leg, abutting against engine No. 3, while engine No. 5 maintains balance.

For engine No. 2, the torque is $\tau = 0.3$ N·m, and for engine No. 3 the torque is $\tau = 4.5$ N·m, so the total torque in the initial period of time approximately equals 0.3 + 4.5 = 4.8 N·m. The torque per one engine is 2.4 N·m. The moment of force applied to the assembly of the lower leg when robot's leg is in a "sitting position" is as follows:

$$F = \frac{\tau}{l} = \frac{2{,}4}{0.17} = 14{,}11 \approx 15 \text{ N},$$

where l is the length of the arm on which the force is acting.

For engine No. 4 the torque is $\tau = 4.5$ N·m, and for engine No. 5 the torque is $\tau = 0.5$ N·m, therefore, the total torque in the initial period of time is equal to 4.5 + 0.5 = 5 N·m. The torque per one engine is 2.5 N·m.

The moment of force applied to the upper leg when robot's leg is in a "sitting position" is equal to:

$$F = \frac{\tau}{l} = \frac{2{,}5}{0.17} = 14{,}7 \approx 15 \text{ N}.$$

Thus, based on the motion analysis of two leg assemblies, we determined moments that act on the parts of leg when lifting it from the "sitting position". These moments served to model the application of external loads to the assemblies of the lower and upper leg.

3. Simulation of weight loads on the lower limbs of Antares

Load modeling is testing if robotic parts and joints can withstand different loads as well as identifying weak spots in the structure. The essence of conducting the analysis of structural assemblies is to see what stresses occur in robot's details. On the basis of this analysis, we can evaluate if the developed model of the robot is suitable for carrying out different tasks.

We have studied the assembly of the upper and lower leg. The material selected was duralumin alloy D16T and brass. Duralumin was used for making plates for a lower and upper leg and a transverse strut. Brass was used for making a stiffening rib for an upper and lower leg. Duralumin is destroyed at a lower load than brass is [15], so the simulation process is terminated after reaching the threshold of destruction of duralumin. The "global contact (no movement)" was selected as the type of attachment. Surfaces of working holes served as a place of attachment. In addition to the "global contact", we have used a type of the connection "bolt-connector" for even load distribution. The assemblies are fixed with the help of "fixed geometry" which does not allow feet assemblies to move in servo flanges' attaching point. The linear force and torque were used as the external load at fixing points of servo flanges. Estimated weight of the robot assembly is 6 kg, with each leg weighing about 1 kg. Half the weight of robot's body (excluding legs) accounts for the assembly of one leg, which is approximately 20 N. This value was calculated for the vertical and lateral (along the normal to the front plane) load according to the following formula:

$$F = P/2 = 40/2 = 20 \text{ N},$$

where F is the force applied to the lower leg assembly; P is weight of legs; the denominator is 2, since all weight is distributed over 2 legs.

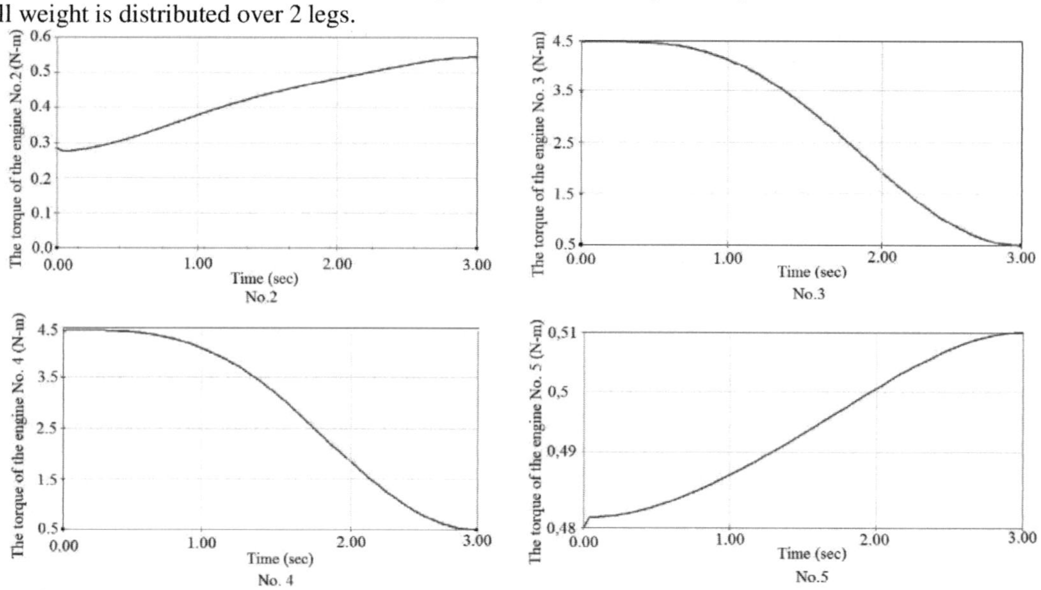

Figure 2. Dependence of torques of engines Nos. 2-5 on their running time for the leg construction

During the simulation process, assemblies were exposed to 2 types of loads - torque and linear force aimed at three directions: vertical (perpendicular to the transverse plane), lateral (along the normal to the front plane) and lateral (perpendicular to the sagittal plane).

Vertical load and lateral load (sitting position) on the assembly of the lower leg (or upper leg) accounts for 25 H, as 20 H is torso's weight and 5 H is the weight of the upper leg; only torso's weight (20 H) accounts for the upper leg.

For lateral load (position of the legs in cross splits), the force value was calculated based on the leg "motion analysis". It is equal to 20 H, since 15 H is torso's weight, and 5 H is the weight of the upper leg; only torso's weight (15 H) accounts for the upper leg.

Since motor's torque varies from 0 to 6 N·m, the simulation was carried out for the maximum torque value with the application of external vertical and lateral loads. The direction of the applied force is selected based on the fact that this force will have its maximum value when the foot is in the "sitting" position. Therefore, the direction will be along the normal to the sagittal plane of the leg.

To test the leg for torsion, we have modeled such a scenario when the robot is in a prone position and the inner part of its feet rests against the floor surface. Torsion occurs throughout the whole length of the leg (0.5 m), therefore, the torque is the same for the assemblies of the lower and upper leg. For the assembly of the leg (with the foot length equal to 0.0825 m and the weight of the leg - 1 kg), the torque has the following value:

$$\tau = F \cdot l = 10 \cdot 0{,}0825 = 0{,}825 \text{ N} \cdot \text{m}.$$

Figures 3 and 4 present the results of load modelling in the form of stress distribution diagrams, on the basis of which it is possible to draw conclusions about the behavior of the upper and lower legs assemblies under different kinds of applied load. The diagram of internal stresses reflects the ability of a part or assembly to withstand the applied loads.

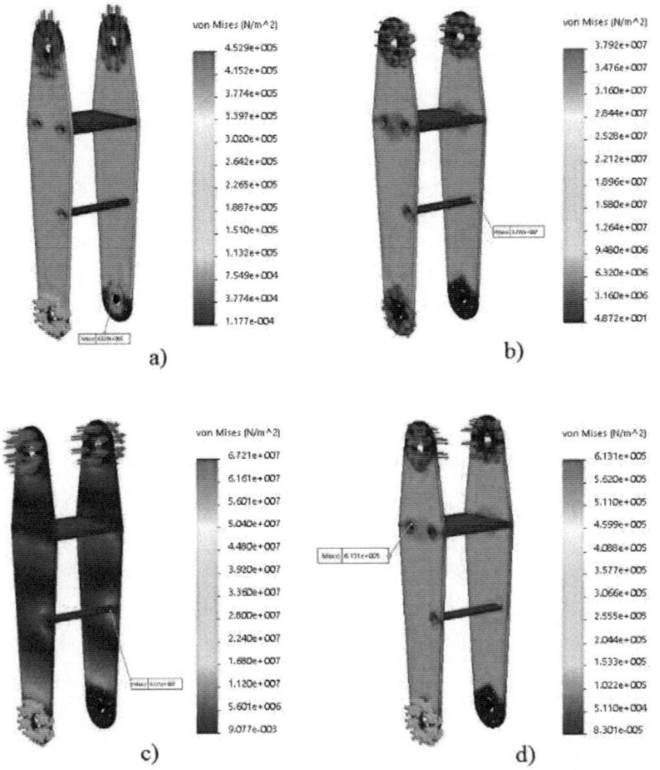

Figure 3. The simulation results of the lower leg model: a) vertical load; b) torque load and lateral load (along the normal to the transverse plane); c) lateral load (perpendicular to the sagittal plane); d) torsion

Diagrams of the maximum internal stresses are built according to von Mises and evaluated according to equivalent internal stresses in comparison with the yield strength of the used material. The material used for making the product begins to weaken when the stress according to von Mises is equal to the stress limit value. In most cases, yield strength corresponds to the stress limit value. A stress limit is a stress at which the deformation continues to increase without increasing the stress inside the product. Units of measurement of yield strength are N/m^2. Thus, the results are shown in the form of equivalent stress diagrams and a graphical representation of color palette directly on the assembly. The location of the maximum internal stress is painted in red and marked with a footnote to the current stress value.

3.1. A spatial model of robot's lower leg

During the simulation process, assemblies were exposed to 2 types of loads - torque and linear force aimed at three directions: vertical (perpendicular to the transverse plane), lateral (along the normal to the front plane) and lateral (perpendicular to the sagittal plane). To study the lower leg model, we have conducted four experiments described below:

1) Vertical load of 25 N is applied to the lower leg assembly and is distributed evenly throughout the whole volume of plates of the lower leg. The maximum internal stress of the assembly is concentrated in the plates of the lower leg (Fig. 3 a); the stress value is 4.53×10^5 N / m^2.

2) The torque of the engine flange (6 N·m) and lateral load of 25 N are applied to the lower leg assembly (Fig. 3 b). The load is almost equally distributed over the entire surface of the lower leg plates. The maximum internal stress of the assembly is found in the stiffening rib (in the attachment

point of a stiffening rib and a lower leg plate); the stress value is 3.79×10^7 N/m^2. The maximum internal stress of the lower leg plate is in the working hole of a stiffening rib and equals 2.29×10^7 N/m^2.

3) Load of 20 H is applied to the lower leg assembly (perpendicular to the sagittal plane). The maximum internal stress of the assembly is located in the stiffening rib (Fig. 3) and equals 6.72×10^7 N/m^2. The maximum internal stress of the lower leg plate is in the working hole of attachment of the stiffening rib and equals 2.45×10^7 N/m^2.

4) The torque equal to 0,825 N·m, acting along the axis that passes through the center of the transverse plate, is applied to the lower leg assembly. The maximum internal stress of the assembly is on the plate that connects plates of the lower leg (Fig. 3 g); the stress value is $6.13 \cdot 10^5 \cdot$ N/m^2.

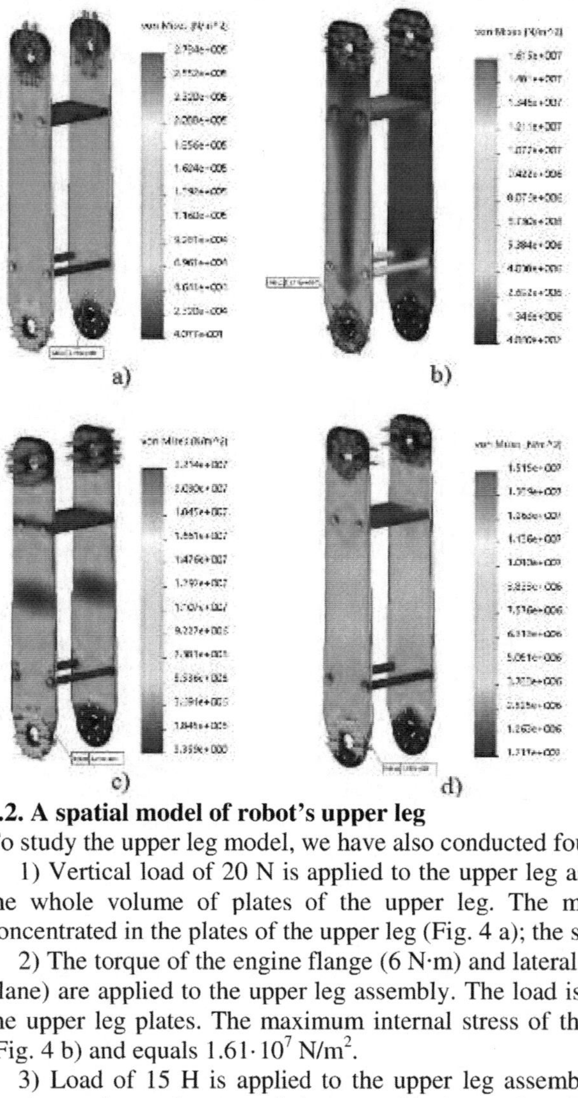

Figure 4. The simulation results of the upper leg model: a) vertical load; b) lateral load (along the normal to the front plane); c) lateral load (perpendicular to the sagittal plane); d) torsion

3.2. A spatial model of robot's upper leg

To study the upper leg model, we have also conducted four experiments described below:

1) Vertical load of 20 N is applied to the upper leg assembly and is evenly distributed throughout the whole volume of plates of the upper leg. The maximum internal stress of the assembly is concentrated in the plates of the upper leg (Fig. 4 a); the stress value is $2.78 \cdot 10^5$ N/m^2.

2) The torque of the engine flange (6 N·m) and lateral load of 20 N (perpendicular to the transverse plane) are applied to the upper leg assembly. The load is evenly distributed over the entire surface of the upper leg plates. The maximum internal stress of the assembly is on the plates of the upper leg (Fig. 4 b) and equals $1.61 \cdot 10^7$ N/m^2.

3) Load of 15 H is applied to the upper leg assembly (perpendicular to the sagittal plane). The maximum internal stress of the assembly is on the plates of the upper leg (Fig. 4 b) and equals $2.21 \cdot 10^7$ N/m^2.

4) The torque, equal to 0.825 N·m, acting along the axis that passes through the center of the transverse plate is applied to the upper leg assembly. The maximum internal stress of the assembly is on the plates of the upper leg (Fig. 4 d); the stress value is $6.13 \cdot 10^5$ N/m^2.

On the basis of the yield strength for the developed leg model, it can be concluded that the assembly of the upper and lower leg can withstand the given types of external loads, since the yield strength for the leg assembly lies within the permissible range [12, 15].

4. Conclusion

In this paper, we have presented motion analysis of lower limbs of robot Antares, and, on the basis of the obtained results, we have simulated application of external loads on the composite parts of robot's legs. According to the data presented in Figures 3 and 4, it can be concluded that by applying various external forces to the legs, the internal stress is close to the yield point for the given material.

Since the assemblies of robot's structure can withstand the load, calculated previously, without being destroyed, the stability of the leg's design is kept. During the simulation process, we have also identified some of the parameters of the future robot model. The lifting of the robot from the "sitting" position requires torque which is 2.4 times less than the maximum torque developed by the engine, and the construction of the leg can withstand the maximum torque. Thus, the robot is able to perform a vertical jump, which has been successfully confirmed experimentally.

The robot under development is intended for use for educational purposes, such as participation in robot football competitions, as well as for the development of assistive technologies of human-computer interaction based on multimodal interfaces [17-22].

5. Acknowledgments

This work is partially supported by state research 0073-2015-0005 and RSF (grant No. 16-19-00044).

References

[1] Gradeckij V G, Ermolov I L, Knjaz'kov M M, Semenov E A, Suhanov A N 2014 Kinematicheskaja model' jekzoskeleta ruki cheloveka i opredelenie oshibki pozicionirovanija *Mehatronika, Avtomatizacija, upravlenie. Teoreticheskij i prikladnoj nauchno-tehnicheskij zhurnal.* **5** 37-41

[2] Zhidenko I G , Kutlubaev I M 2014 *Metodika opredelenija signalov upravlenija antropomorfnym robotom Mehatronika,* Avtomatizacija, upravlenie. *Teoreticheskij i prikladnoj nauchno-tehnicheskij zhurnal.* **5** 41-46

[3] Filatov V V 2014 Ispol'zovanie SAPR Solid Works Motion dlja issledovanija plavnosti hoda transportnogo sredstva. *Transportnaja tehnika, jelektronnyj nauchnyj zhurnal.* **1** 1-7

[4] Sagirov Ju G 2013 Prochnostnoj analiz metallokonstrukcii gruzopod-emnyh kranov s ispol'zovaniem Solid Works. *Vestnik Priazovskogo gosudarstvennogo tehnicheskogo universiteta. Serija: Tehnicheskie nauki.* **26** 194-203

[5] Warnakulasooriyaa S., Bagheria A., Sherburnb N., Shanmugavel M. 2012 Bipedal walking robot – a developmental design *Procedia Eng.* **41** 1016–1021

[6] Lima S.C., Yeapa G.H. 2012 The locomotion of bipedal walking robot with six degree of freedom *Procedia Eng.* **41** 8–14

[7] Yoo J.K., Lee B.J., Kim J.H. 2009 Recent progress and development of the humanoid robot. *Hansaram. Rob. Auton. Syst.* **57** 973–981

[8] Buschmann T., Lohmeier S., Ulbrich H. 2009 Humanoid robot Lola: design and walking control. *J. Physiol.* **103** 141–148

[9] Mohameda Z., Capi G. 2012 Development of a new mobile humanoid robot for assisting elderly people. *Procedia Eng.* **41** 345–351

[10] Nakashima M., Tsunoda Y. 2015 Improvement of crawl stroke for the swimming humanoid robot to establish an experimental platform for swimming research. *Procedia Eng.* **112** 517–521

[11] Poppy [Jelektronnyj resurs] https://www.poppy-project.org/ (Data obrashhenija 25.08.16)

[12] DARwin OP [Jelektronnyj resurs] http://www. robotis.com/ (Data obrashhenija 25.08.16)

[13] Koval'chuk A K, Kulakov D B, Semenov S E, Jaroc V V, Verejkin A A, Kulakov B B,

Karginov L A 2014 Metod proektirovanija prostranstvennyh drevovidnyh ispolnitel'nyh mehanizmov shagajushhih robotov *Inzhenernyj vestnik: jelektronnyj nauchno-tehnicheskij zhurnal* **11** 6-10

[14] Verejkin A A, Koval'chuk A K, Kulakov D B, Semenov S E 2014 Analiz i vybor kinematicheskoj struktury ispolnitel'nogo mehanizma jekzoskeleta *Nauka i obrazovanie: jelektronnoe nauchno-tehnicheskoe izdanie* **7** 72-93

[15] Seliverstov G *V* 2013 Opredelenie koncepcii naprjazhenij pri pettinge kranovyh metallokonstrukcij *Izvestija Tul'skogo gosudarstvennogo universiteta. Tehnicheskie nauki.* №7-**1** 221-225

[16] Verejkin A A, Koval'chuk A K, Karginov L A 2014 Issledovanie dinamiki ispolnitel'nogo mehanizma jekzoskeleta nizhnih konechnostej s uchjotom reakcij opornoj poverhnosti *Nauka i Obrazovanie. MGTU im. N.Je. Baumana* **12** 256-278

[17] Pavluk N, Ivin A, Budkov V, Kodyakov A, Ronzhin A. 2016 Mechanical Leg Design of the Anthropomorphic Robot Antares International conference on interactive collaborative robotics. *ICR 2016* (Springer, LNAI 9812) 113–123

[18] Karpov A A, Ronzhin A L. 2009 Information Enquiry Kiosk with Multimodal User Interface *Pattern Recognition and Image Analysis, Moscow: MAIK Nauka/Interperiodica* **19 (3)** 546-558

[19] Karpov A, Ronzhin A, Kipyatkova I. 2011 An Assistive Bi-modal User Interface Integrating Multi-channel Speech Recognition and Computer Vision. *In Proc. 14th International Conference on Human-Computer Interaction HCII 2011*, Springer-Verlag Berlin Heidelberg, LNCS 6762, Orlando, FL, USA. 454-463

[20] Motienko A I, Tarasov A G, Dorozhko I V, Basov O O 2016 Proactive control of Robotic Systems for Rescue Operations. *SPIIRAS Proceedings* **46** 169-189

[21] Yusupov R M, Ronzhin A L, Prishchepa M V, Ronzhin, Al L. 2011 Models and hardware-software solutions for automatic control of intelligent hall. *Automation and Remote Control.* **72 (7)** 1389-1397

[22] Kryuchkov B I, Karpov A A, Usov V M 2014 Promising Approaches for the Use of Service Robots in the Domain of Manned Space Exploration. *SPIIRAS Proceedings* **32** 125-151

IOP Conf. Series: Journal of Physics: Conf. Series **803** (2017) 012075 doi:10.1088/1742-6596/803/1/012075

The processing of information from sensors in intelligent systems

V A Kokovin and A N Sytin

State University "Dubna", Branch "Protvino", 9, Severny pas., Protvino, 142281, Russia

E-mail: kokovin@uni-protvino.ru

Abstract. The article describes the processing of information obtained from sensors in intelligent systems. The paper analyzes the need of advanced treatment for a paralleling operation calculator which reduces the time of response to input events. A realization of a speculative processing algorithm in the FPGA by streaming control is based on a data flow model. This solution can be used in applications related to telecommunications networks of distributed control systems.

1. Introduction

Analysis of solutions in the field of process automation (PA) of large industrial or scientific complexes, where there is a large territorial and algorithmic distribution of the individual subsystems, provides a clear choice of these process control systems – an intelligent distributed control system (IDCS). Intellectualization of the control system is not using a formalized management model, but a model based on the receipt of the information about the real outside world through sensors and the system on the basis of this information, which has the ability to adapt to external changes. In addition, an important feature of the intelligent control system is the ability to change the properties of the control devices in real time.

For example, modern cyclic particle accelerators, integrated in large accelerator structures, being one of the main tools in basic research of high-energy physics, have complex, distributed and computerized control systems. In addition, each accelerator is equipped with some means of timing required to provide a real-time mode of the control system and synchronization of technological processes in the accelerator complex. The distinguishing factor of such complexes is that the processes of the individual subsystems [1, 2] are interrelated in terms of solving the main task - to obtain a stable beam of charged particles with the specified parameters.

A survey of a large array of sensors, subsystems and technological limitations of the time spent for transformation of the information received into the events makes it necessary to use integrated networks with distributed intelligent sensors. However, this solution allows one to attain the gain in the speed only at the stage of preliminary transformations of the physical parameters of the PA into an encoded data sequence. Further, the stage of the conversion of data into the events and messages will be the "bottleneck". A good solution to this problem may be the real-time paralleling of processing of data, obtained from sensors based on Field-Programmable Gate Array (FPGA). Using the element base on the basis of FPGA for solution of management tasks allows scaling of encoded information,

IOP Conf. Series: Journal of Physics: Conf. Series **803** (2017) 012075 doi:10.1088/1742-6596/803/1/012075

parallelization of data streams obtained from sensors to improve performance and a real-time control process for processing data streams in a FPGA.

2. The choice of the model for processing of the information flows outcoming from sensors

The task of raising the performance of control systems is achieved, both by increasing the computing resources and by using more effective methods of information processing.

2.1. Comparison of computational models

The use of the classical model of von Neumann computations [3], when a large amount of input data streams is ineffective, since the "narrow" place becomes a serial data processing. In this case, it is necessary to increase the clock frequency of the calculator, which is not always justified in terms of increased cost of implementing solutions. Another option - the development of architecture of the calculators on the basis of computational model ***data flow***, proposed in the works of D. Dennis, for example [4]. In this case, the implemented algorithm is "sewn up" in the equipment management system in the form of required computational and logical operators, which allows parallelizing various processes and unlimited scale.

One of the main indicators of the functioning of IDCS is the reaction of the system to the change of the state of the input event. Modern technological processes are characterized by a large number of parameters (channel) to be analyzed. However, the information, which is continuously incoming from the sensor, is not always events and requires preprocessing. It may be a situation, when an event is a function of the time intervals of incoming information rather than the information itself, or even the lack of information. Finding solutions is subjected to a predetermined logic processing of input data streams and requires significant computing resources and energy.

2.2. Selection of processing models

There are systems, such as navigation aircraft systems, which have limited capabilities of the enumerated resources. In addition, many embedded systems, process control (for example, technological systems accelerators [1]), should have a very short response time. The experience in the development of real-time control systems for power converters of physical units [1] shows that the provision of the required reaction of the control system to the events (signals) of the fast processes can only be realized by the architecture with maximum parallelized processes. In this case, it is more efficiently to apply the model of stream processing, when it becomes possible to parallelize the processing of individual channels of incoming information and to hardwire the processing algorithm. The most efficient result, taking into accounts these requirements and conditions, gives a solution based on the use of the method of streaming control.

2.3. Streaming control method

Until recently, the implementation of the effective in performance, but highly specialized in terms of function task has remained the disadvantage of the ***data flow*** model. When changing the algorithm of the task, the hardware of control systems has to be also upgraded. With the advent of FPGA technology, the situation has changed. As it was mentioned above, the narrowest bottlenecks in the processing of large amounts of data, arrived from sensors, and formation of the rapid response of control systems are a sequential execution of calculators' instructions, applied in IDCS. The solution to this problem may be the maximum parallelization of data processing and execution of the algorithm, which is the aim of the stream control method. The Stream Control Method (SCM) is based on the classical models of management that solves the problem of providing the required output parameters of the control object (process, installation, etc.) through the formation of control actions based on analysis of the information arrived from the sensors, controlling the state of the control object.

Data streams and control actions underlie the SCM, and the information from the sensors being processed based on the ***data flow*** model and the control algorithm being implemented using hardware, not software. The feature of the SCM consists in parallel processing of data streams and parallel formation of control streams.

Any control system, including based on SCM, characterized an execution model. Since the basis of the SCM is flows, the execution model, representing the SCM, is convenient to consider using DFD (Data Flow Diagramming). The diagram allows one to analyze the information flows, entering the control system input, and the control flows, formed at its output. In addition, one can use the DFD for the development of requirements specification to create control systems. Classical diagram DFD is designed for structural analysis and design of information systems. There are several commonly used notations DFD, different syntax. Diagrams describe the sources and data streams are processes and memory for storing processing results of data flows in the processes. At the same time, functionality diagram does not always correspond to the solved problems, which forces modification of DFD [5]. These diagrams are not directly used in the analysis of flow control systems with the use of the SCM for the following reasons:

- Treatment processes data streams must be strictly determined.
- Every process usually runs on a certain condition; therefore, at the process input, it is necessary to determine the logic of condition, forming the signal for initiation.

A diagram for the analysis of flows is suggested, where the elements of the *event*, the *process*, the *capsule* are used as the key elements with the assumptions described above. The word "*event*" in the system of technological object control will imply a limited set of named signals, the active state of which affects the processes. *Events* do not have the time length (in case of some action), otherwise we can talk about two *events* of the action interval: the start of the *event* and the *event* of its termination. *Events* may be external to the elements of the diagrams, or internal, if they are the results of performing the processes. We define the *process* as a set of P of the $P = (S,\ s0,\ F)$-kind, the components of which have the following meaning:

- S - set whose elements are called a P-state process,
- $s^0 \in S$ - a special state called the initial state of P,
- R - a subset of the $R \subseteq S$-type whose elements are called transitions between states.

Each *process* converts input data streams into output ones according to a specified algorithm, thus forming the control signals (exposure). Conversion of flows in the processes is carried out on the basis of the model of a finite state machine. The transition from one state into another is due to the appearance of the *event* (external or internal). *Capsule* is a structural abstraction that integrates processes implementing the input processing algorithm and control algorithm. We can say that the algorithm is encapsulated in this structure. *Processes* in the *capsules* are carried out consistently and run through the logic conditions. *Capsule* can be nested, i.e. possess a hierarchyare.

In [6], an adaptive control system of the power converter is considered, which is implemented on the basis of the SCM. The flow diagram has been developed to analyze the information and control flows presented in the form of interconnected capsules, in which a control algorithm is encapsulated. Fig.1 shows a flow diagram of the system the power converter control. The information arrived from sensors is processed by ADC and data stream enters capsules P1, P2, P3 and P4 for processing. The objective of the processes in these *capsules* is to implement the algorithm of comparing the incoming data to the set point in the case of performing the specified comparisons conditions to form event *sf* and control signal F to change the oscillator frequency. *Capsules* P6-P10 perform the processes of converting the information from sensor D and reference register RR for generating control signal A at input DC of the converter. Signals *clk* and *str* are external *events* to synchronize all the processes with the control system. The diagram in Fig.1 gives a visual representation of the processed streams and the generated control signals. The diagram can have a hierarchical structure and have a lot of processes. In this case, the diagram can be divided into levels and individual fragments can be visualized.

3. Speculative processing of information obtained from sensors

When solving the problems of automation of interrelated processes, a large number of cross-links appear between distributed subsystems. Telecommunication networks that have a fixed transport delay

IOP Conf. Series: Journal of Physics: Conf. Series **803** (2017) 012075 doi:10.1088/1742-6596/803/1/012075

are used for delivery of the information from sensors in the form of a events and messages. In this case, the increase of the processing speed may significantly affect the IDCS performance.

3.1. Analysis of the processing time of the data obtained from sensors

Fig. 2 shows IDCS structure consisting of three sub-systems, each of which includes the following blocks: *ISN* - intelligent sensor network, *DAS* - data acquisition system, *LCS* - local control system and a power convertor. A telecommunications network, connecting subsystems, has a ring structure. Let us consider

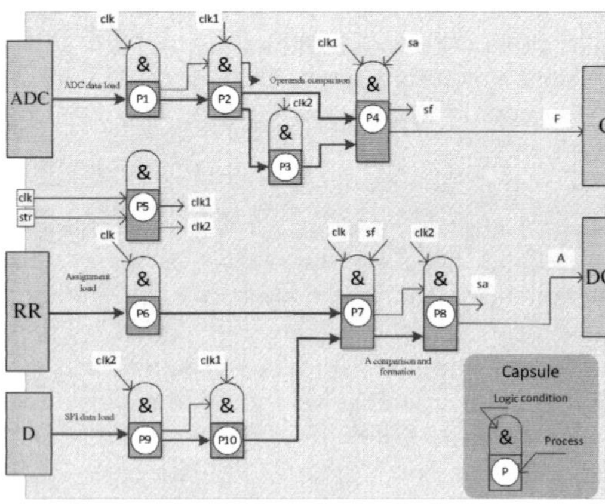

Figure 1. The diagram of information and control flows.

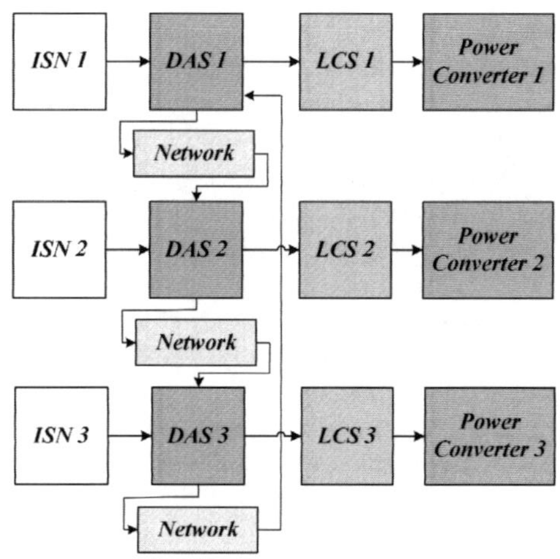

Figure 2. Directions of information dissemination in the IDCS structure.

what constitutes the delivery time of events or messages from one subsystem to another. Let the time of information dissemination $t_{propagation}$ from one subsystem to another is summarized as:

$$t_{propagation} = t_{is} + t_p + t_{tn} \tag{1}$$

when t_{is} - the time of initial information processing (for example, a temperature parameter) by intelligent sensors, t_p – the time of information processing by subsystem *DAS*, obtained from intelligent sensors, and the time of preparation for transmission via a telecommunications network, t_{tn} - the transmission time of messages via a telecommunication network. As in equation (1), time intervals t_{tn} and t_{is} are almost constant, it is possible to optimize $t_{propagation}$ only by reducing t_p.

3.2. Speculative methods of eventing

Time t_p depends on the number of sensors, realization of the calculator in the *DAS* structure, i.e. the model of processing the data obtained from the sensors, as well as sequencing of incoming messages. When using ***data flow*** in model *DAS*, when it is possible to perform scaling processing of information from the sensors, value t_p depends mainly on the speed of realization of the ordering algorithm. Articles [7, 8, 9] propose an adaptive method of eventing (adaptive speculative processing of Out-of-Order Event), arriving from the sensors, using buffering. The method is based on the assumption that processed the events that have come first, even if they violate certain order. To restore the proper order of events entering buffering. The method is based on the assumption that those events are processed that have come first, even if they violate a certain order. To restore the proper order of events entering the control system through the network, a buffer was introduced, in which events are retained for a certain time. This method significantly reduces the processing time and allows more uniform loading of the telecommunications network.

The second area associated with the use of advanced processing of the information obtained from the sensors refers to optimization of the state change of finite automata.

Article [10] presents a hardware method of optimization of transitions FSM (Finite State Machines) between states on the basis of statistical analysis of the states of automaton. This method gives a good saving time result, but cannot be realized in the SCM, as the transitions of the automaton calculator are strictly defined when synthesizing the project in the FPGA.

4. Realization of the method of speculative processing

When receiving information from a large number of sensors, the problem, concerning receiving parallel streams of events (if intelligent sensors themselves convert information into events), conversion of events into a serial stream of messages by means of the *DAS* unit and its transmission to a telecommunications network, arises [1]. The solution to this problem by the *SCM* method [6] allows the simultaneous processing of all events and performing the following steps:

- transformation of events into messages along with encoding of the address of the input port of the host system;
- encapsulation of protocol and service information into messages;
- adding the time stamp of events, entering the *DAS* unit, into the messages, if a network synchronization of the operation of individual subsystems is necessary, the messages are transmitted with a fixed period of time. Article [1] describes the network with the messages transmission period of 100ms. If you solve the problem of reducing time $t_{propagation}$, the messages are transmitted to a telecommunications network as far as *DAS* is processed.

The solution to this problem is well-established, if there are no requirements to preserving the priority of data entering the input ports when transmiting via serial communication.

Figure 3 shows an example of using the method of speculative processing of event *E1, E2* and *E3* in the *DAS* unit. This example corresponds to a model of processing in *DAS*, presented in [11]. Let us assume that events entered the *DAS* unit in sequence *E1* → *E2* → *E3*. The converter of events, incorporated in *DAS*, has a scanning type of formation and preparation of messages for telecommunications network *TS*. That is, when one or more events enter the *DAS* input, scanning of the inputs and the formation of messages begins. Since the processing of events *E1* - *E3* is performed in the order in which position the scanner's pointer is, the event that came later may be treated as if it came first. For example (Fig.3), event *E3* entered *DAS* last, but it was converted into message *M3* and sent to TS like first. This means that according to the speculative processing method, it is necessary to minimize the downtime *DAS* and process the event, which is pointed at by the scanner. Similarly, the rest of the events is processed. To restore the correct order of the events transmitted to the TS messages, the buffer delay of the events is used in [7].

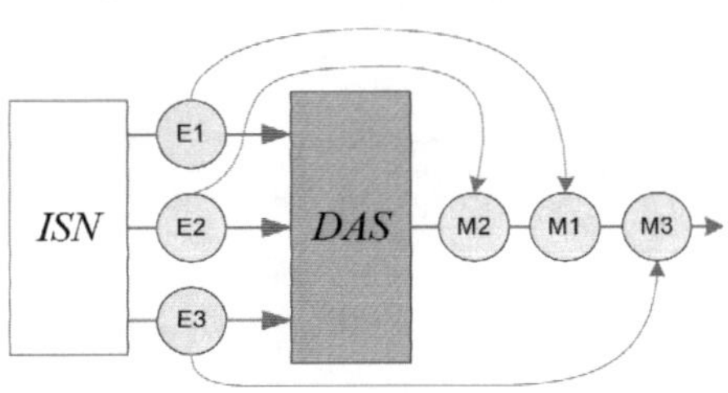

Figure 3. Speculative processing of information from sensors.

We offer to transfer the messages out of order to TS, but to restore order after their reception in the other subsystem if necessary. This solution provides the following advantages:

- in the presence of a large number of events (sensors), time data processing t_p by the *DAS* subsystem reduces significantly;
- the model of the scanning converter is implemented simply enough on FPGA;
- the method of streaming control solution, based on the **data flow** model, can be used for processing.

5. Conclusion

Trial operation of means intended for assembly and transformation of information obtained from sensors, is implemented on the basis of speculative processing in each of the subsystems of the telecommunications system [1], has shown an effective work of *DAS*. The *DAS* unit, equipped with the

64 - channel input, peak traffic > 7 Ge/s (gig events per second) and the rate of transmission of the telecommunications network equal to 100 Mb/s (UDP Protocol), is used in article [11]. Application of the speculative event processing and the buffer memory allowed a significant reduction of the time of information processing by the *DAS* subsystem.

References

[1] Ivanov S 2010 Accelerator Complex U70 of IHEP: Present Status and Recent Upgrades *Proc. of the Conf.* (RuPAC-2010, Protvino) pp. 27-31

[2] Gotsev V, Gres V, Dunaitsev A, Kim L, Komarov V and Sytin A 1993 The control and operation of the beam diagnostics for UNK *Proc. Int. Conf. ICALEPCS'93 on Accelerator and Large Experimental Physics Control Systems* (BERLIN)

[3] Burks A W, Goldstine H H and von Neumann J 1946 Preliminary Discussion of the Logical Design of an Electronic Computing Instrument *Institute for Advanced Study* (Princeton, N. J.)

[4] Dennis J 1984 Data flow ideas for supercomputers *IEEE Society, 28th International Conference* (San Francisco) pp 15-19

[5] Abedjan Z, Schulze P and Naumann F 2014 DFD: Efficient Functional Dependency Discovery *CIKM '14: Proceedings of the 23rd ACM Int. Conf. on Conf. on Information and Knowledge Management* pp 949-58

[6] Diagilev D, Kokovin V and Uvaysov S 2015 Designing a Power Converter with an Adaptive Control System for Ultrasonic Processing Units *Advances in Mechanical Engineering, Lecture Notes in Mechanical Engineering* (Springer Intern. Publ, Switzerland) pp 39-47

[7] Mutschler C and Philippsen M 2013 Reliable Speculative Processing of Out-of-Order Event Streams in Generic Publish/Subscribe Middlewares *DEBS '13 Proceedings of the 7th ACM international conference on Distributed event-based systems* (New York, USA) pp 147-58

[8] Mutschler C and Philippsen M 2014 Adaptive Speculative Processing of Out-of-Order Event Streams, *ACM Transactions on Internet Technology (TOIT)- Special Issue on Event Recognition* (New York, USA) **14** pp 1-24

[9] Hung E. and Wilton S.J.E. 2011 Speculative Debug Insertion for FPGAs *Proc. Int. Conf. FPL - 2011* pp 524-531

[10] Jackson S and Burtscher M. Self Optimizing Finite State Machines for Confidence Estimators. 2006 *Workshop on Introspective Architecture* (Austin, Texas, USA)

[11] Kalinin A Yu, Kokovin V A, Kryshkin V I and Skvortsov V V 2016 An absolute intensity beam monitor *Instruments and Experimental Techniques* **59 (4)** 536–38

Mixed technologies of artistic ceramics processing for the jewelery manufacture

L E Kutsenko, N A Arventyeva

Tomsk Polytechnic University, 30, Lenina ave., Tomsk, 634050, Russia

E-mail: l.e.kutsenko@gmail.com

Abstract. The need for the study of different technologies of the jewellery manufacture is due to the demand of the modern world in a variety of high-quality environmentally friendly products. While working with ceramics, it is possible to get the product, which is unique in its form, a wide range of colors, possibility of harmoniously combining it with a number of other modern materials. Images of ancient birds are used, stages of technology, which allow connecting the ceramics and metal, the technology for working with underglaze paints are represented, application limits associated with a particular operation of the product (fragility) are analyzed in the research. The society need for exclusive products determines the relevance of different materials connection, the characteristics research of their compounds, while working with them. The research objective is a bird image stylized design for a piece of jewellery, different materials compounds, the substantiation of technology of its manufacturing "narikomi" technique. Also, the research objective includes the development of the technology stages, allowing the connection of ceramic and metal to get jewellery that is unique in its form, a wide range of colors, a possibility of harmonious combination of it with a number of other modern materials.

1. Introduction

A magical function ("amulets") was the jewellery primary function in all nations. Amulets provided protection against evil spirits, dangers to those who wore them. The second function was decorative, as an expression of aesthetic needs, self-affirmation, desire for a beauty. The next function could be called as socially distinctive; it means that jewellery was a sign of social distinction, a sing of material well-being and creditworthiness, a rational method of material resources. Women jewellery was kept as arelic, handed down from mother to daughter for generations.

We are going to match these functions all together in this research. We choose the image of future product and fill it by some symbolism (fig. 1). We have a look at the bird image and symbolic meaning. Nowadays a lot of things which were significant early are forgotten, for instance, a bird. The bird image became significant among the eastern Slavs and it reached up to now, changing in conjunction with customs and a culture.

A

b

Figure 1. Design stages: a– the sketch, b – the sample.

A bird is an ancient symbol of imperishable, soul, spirit, divine manifestation, air spirits, spirits of the dead, ascension to the Heaven, ability to communicate with gods or go into a higher state of consciousness, thought, imagination. Great birds as an ancient symbol are often identified with solar deities, gods of thunder and wind. Their tongue is considered as lightning. Birds are a part of the tree symbolism: the deity power descends into the tree or its symbol - the trunk. The samples of relief birds performed by the folk motives are represented in Figure 2.

A

b

Figure 2. Stylization of ancient Russian folk images in ceramics: a – "Sirin" bird, b– "Alkonost" bird

The unique technology that connects various traditions of artistic ceramics and metal was developed for "Bird" ring manufacture.

Different materials are used for this purpose:

– metal- cupronickel – the base
– copper – bird wing
– porcelain - bird figure.

The manufacturing process involves working with bisque (glazed surface) in the "narikomi" technique (Japanese technique of clay colored masses mixing (in this case - white porcelain) with color pigments). The bird body is decorated with colored enamels in Figure 3 (the colors used: blue and lilac). Metal parts, imaging wing and legs, are attached by soldering.

Porcelain in "narikomi" technique is used in figure 3. Manufacturing technology involves working with porcelain bisque (unglazed surface) in "narikomi" technique.

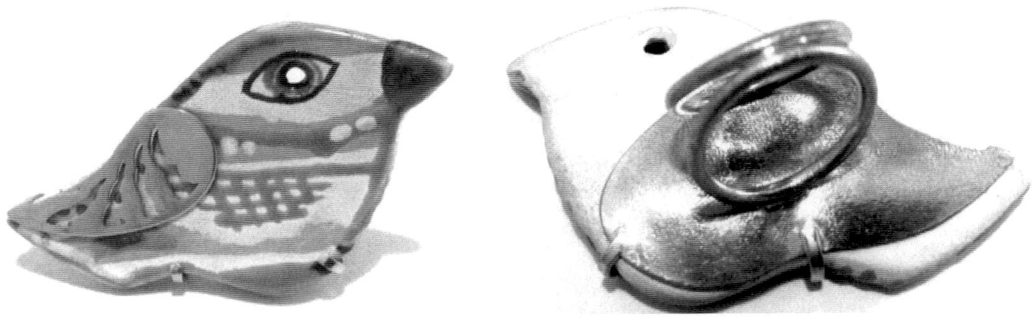

a b

Figure 3. A "Bird" ring: a – front view, b – back view.

2. Technique

Narikomi firstly appeared in ancient Egypt (while ceramic jewellery was created), in Chine and Japan; the English learnt this technique from Romans. Italian artists were one of the first who learnt how to work with underglaze paints in ceramic at the end of the 15th century. They found a majolica painting technique. The technique laid in the fact that ceramic paints and glaze were plotted on crude enamel and were ablated simultaneously with it. Such painting differs by mild colorful coloring, saves tenuous image details, which has pleasant blurred contours after baking. This technique has started arousing interest in Russia only recently. There is a different meaning – agate loam (named after the stone - agate). A distinctive feature of this technique is a combination of different clays in fantasy patterns in one product. Various metal oxides are injected into the white clay, the basis. Then the color clay is cut into layers, stacked on each other in a certain order and pressed. Beautiful pattern is obtained by cutting the formation. It can be chaotic. It also possible to create an elaborate structure and solve different color palette (fig. 4).Clay itself contains an incredible beauty!

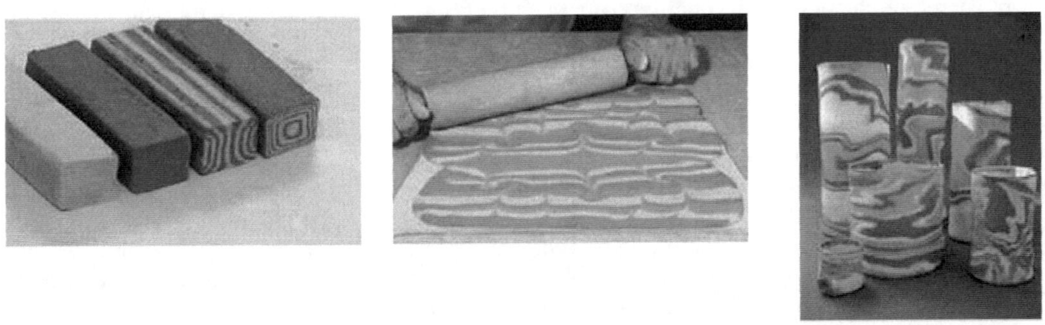

a B c

Figure 4. A working process with agate clay: a –combination of different clay types; b – agate clay layer; c – finished products.

3. General characteristics of underglaze painting

A decor is laid off the unglazed surface of ceramic ware by colors, after it was glazed and fired based on the underglaze painting technique. There are several methods of underglaze painting: a painting on scrap (past primary burning - 700 degrees); a painting on crude material as in our case (fig. 4a). The majolica painting technique is performed on crude enamel. Painting of a "Bird" plaque on crude white enamel is made by salts - liquids (cobalt nitrate Co(NO3)) and underlay paints (pigments). Such painting required accuracy and workmanship, since corrections are not desirable. Painting is made by cobalt salt; after that, underglaze paint is laid on the top of the brushstroke to make painting more expressive and volumetric.

The product is purged from the top by a colorless glaze and fired at 1050 degrees. Paints are fused into the upper layer of enamel during the firing process. Underglaze painting is more durable and strong, as it is preserved under a layer of glassy glaze. Underglaze paints are a mixture of glaze pigments (there are 20-50% of glaze typically) and other materials. These supplements help to increase the mass viscosity and better conformity between paint, shard and glaze. Colors samples for products are made at a given shard at the required temperature with colorless glaze before the work. Paints are diluted with water on a glass, that is, they are pulverized with a spatula with a few drops of glycerin until obtaining smoothness. Water is added to the painting process as needed. Paint must be resistant to high temperatures actions, to the flux and glaze actions during the process of glaze firing. Paints use should be economically beneficial. Defective painting can be corrected by glaze painted at lower temperatures from 720-900 degrees.

4. Conclusion

Ceramic jewellery is widely used in industry, although it is rough at first sight. However, it is possible to create truly elegant and simple decoration out of ceramic due to unique technologies. Despite its fragile appearance, the jewelry obtained is very strong and has chemical resistance to environmental influences. So jewellery does not spoil, ceramics does not fade and does not lose luster over the years.

The jewelry made of ceramics cost is much lower than the cost of the jewellery made of precious, semi-precious metals and stones. Moreover, ceramic jewellery does not yield to other jewellery, as it is based on originality, the unique range of used colours and a huge variety of forms and combinations with different materials. It is possible to reach a unique colour if we use ceramic jewellery with colouring addition. Based on that, ceramic jewellery is a very bright and original product, which underline individuality, beauty and the discriminating taste of its owner.

Methods of ceramic processing allow making it very gentle and smooth to the touch. That is a definite pro, when you make a choice of jewellery resting on tactile perception. Ceramic is light as well, so, even large-sized products have little weight. One more significant advantage of ceramic is that it is a hypoallergenic material, which does not cause any rejection, and its contact with skin is proper. Therefore, this jewellery could be worn even by people who are sensitive to metals. The risk of allergic reactions to ceramic product is minimal.

References

[1] Dajneko V V 2015 *Glass products made based on fusing technology and baking defects. Proceedings of the Technical Aesthetics and Design Academy* **2** 22–25

[2] Akunova L F, Pribluda S Z *Materials Science and technology of artistic ceramic products manufacturing* 25–28

[3] Kukhta M, Kazmina O, Sokolov A, Arventjeva N, Soroka A, Homushku O, Zaitseva S and Sergeyeva M 2014 The influence of glass and metal properties on the peculiarities of an item of art's shaping in ethnostyle. *IOP Conference Series: Materials Science and Engineering* **66(1)** 1-6

[4] Chia K Y, Seehadzi S G and Kuuz S M 1986 Mechanical behavior of PSZ at elevated temperatures. *Ceramic Engineering and Science Proceedinge* **7(8)** 784-794

[5] A study of the application and fabrication of advanced ceramics 1986 *Forschrift-Berichte VDI.* **5** 109

[6] Morrell R 1984 Ceramics in molern engineering *Physics in technology* **15** 252-261

[7] Sokolov A P, Kukhta M S and Pelevin E A 2014 *Modern technologies of decorative surface treatment.* Mechanical Engineering, Automation and Control Systems: Proceedings of the International Conference, Tomsk, October16-18, National Research Tomsk Polytechnic University (TPU). 1-4

[8] Boiko J A 2010 Print as a means of ceramics' decorating *Design. Theory and Practice* **4** 36-52 (in Russian)

[9] Galanin S I Vologdina V O 2010 Decorative etching of coloured glass insertions in the items of

jeweller and art Design. *Theory and Practice* **5** 38-50 (in Russian)

[10] Kunicki-Goldfinger J, Freestone I, McDonald I, Hobot J, Gilderdale-Scott H, Ayers T 2014 Technology, production and chronology of red window glass in the medieval period – rediscovery of a lost technology *J. Arch. Sci.* **41** 89-105

[11] Vilarigues M, Fernandes P, Alves L C, Silva R C 2009 Stained glasses under the nuclear microprobe: A window into history Nuclear Instruments and Methods in Physics Research *Section B: Beam Interactions with Materials and Atoms* **267**, issues 12-13, 15 2260-2264

The intellectual information system for management of geological and technical arrangements during oil field exploitation

N G Markov[1], E E Vasilyeva[1], I V Evsyutkin[1]

[1] Tomsk Polytechnic University, 30, Lenina Ave., Tomsk, 634050, Russia

E-mail: markovng@tpu.ru

Abstract. The intellectual information system for management of geological and technical arrangements during oil fields exploitation is developed. Service-oriented architecture of its software is a distinctive feature of the system. The results of the cluster analysis of real field data received by means of this system are shown.

1. Introduction

Field development and new wells drilling on the exploited fields are expensive projects. That is why, majority of oil-extracting enterprises use the methods for increasing oil output from productive formations on these fields in the form of various geological and technical arrangements (GTA). In addition to that, enterprise specialists have to constantly solve problems of GTA management. It is known that GTA management includes management of the following aggregated business processes (BP): the actualization of monitoring data of a well stock and exploited productive formations that are initial data for the analysis; the selection of candidate wells for GTA as the result of the analysis of initial data; the selection for these wells in the relevant arrangements; the assessment of economic (and technological) efficiency of selected GTA; planning of carrying out GTA and workover crew (WC) schedules; and carrying out GTA by forces of WC.

Today, in most cases, qualified specialists of the enterprises have to analyze pactically "manually" a large amount of diverse geological and field data in order to make reasonable decisions at GTA management. The solution of the problem for the automation level increase for GTA management processes remains still an actual direction of economically effective activity achievement for enterprises due to labor inputs of carrying out BP, the low level of their automation and high dependence on the subjective opinion of experts. The development of fundamentally new information systems is gained special relevance due to the growth of GTA amount.

The intellectual information system (IIS) for GTA management that implements methods of the intellectual analysis of data (IAD) is considered in the article.

2. Methods and algorithms of the intellectual analysis of field and geological data

Nowadays, in oil industry in Russia and abroad there are highly specialized information systems (IS) providing the low level of automation to carry out business processes for GTA management. Usually they provide a "manual" operating mode for the specialist: they can help to extract and transform the initial data from an automated process control system (or other industrial IS of enterprise) to the form required for the analysis. Automation software for decision-making support in highly specialized systems is absent. Results of the work in the "manual" mode of the specialist depend only on his experience, what causes the risk increasing of mistake emergence in the data analysis, first of all, during selection of candidate wells for GTA.

The mode of the automated analysis of initial data is implemented in some Russian GTA management information systems along with the "manual" mode, where the method of the geological potential is used [1] which allows the specialist to calculate an expected oil output of each well and to make the decision whether well is the candidate for carrying out GTA according to results of calculations. Calculations results of this method depend on the scenario conditions created on the basis of experience of the specialist and, therefore, do not give the demanded accuracy at the time of candidate wells selection for GTA. In some information systems, it is possible to implement the methods and algorithms based on decision-making of GTA management by means of conformance results of the analyzed geological and field parameters of wells and productive formations to the set intervals of these parameters [2]. If the considered parameters values of wells and (or) productive formations go beyond the established boundary values, the decision is being made, taking into account the weight of each of the parameters according to which the well becomes the candidate for the wells list for GTA. Obvious shortcomings of these methods and algorithms are complexity of parameters boundary values determination and their weight coefficients, thus the accuracy of the received results for the candidate wells selection depends on the experience of the specialist.

In recent years, the methods of IAD are used for the automatic control mode as an alternative to the simple and inexact methods used today for automated GTA management. So, results of research while solving a problem for candidate wells selection for GTA with the use of fuzzy logic are given in the work [3]. Results have the low accuracy because they are received for the oil well stock in the conditions of inexact values of some initial geological parameters.

Other perspective method is the use of artificial neural networks (ANN). Initial field data are used as input parameters values during ANN training in the research [4] based on ANN application for efficiency forecasting of carrying out one of GTA types for oil wells. However, ANN application assumes its architecture design that is based on ANN developer experience, which is quite difficult to create within the oil-extracting enterprises.

Unsupervised learning can be used in cases when the training selection is absent; the most striking example of these methods are cluster analysis methods. In work [5], the cluster analysis of initial geological parameters is applied to each well for division of the well stock into clusters (groups) depending on priority of carrying out GTA, then into subclusters by GTA types. The authors showed that the advantage of the cluster analysis algorithm applied by them is an opportunity to automatically receive the division of the analyzed well stock into clusters with allocation of the cluster (group) of candidate wells for GTA. It means that separate methods and algorithms of the cluster analysis can be a basis for IS creation, these methods will allow creating software for high-level automation during the solution of problems of GTA management. The main shortcomings of this approach are the use for the analysis of only geological data (trade data are not used) and the application of the elementary algorithm of the clustering developed within a method of the K-means.

Summing up the results of the IAD methods application for GTA management, it is possible to consider that they were received during solution of highly specialized tasks (subtasks) of GTA management. However, the first results of research indicate perspectiveness of cluster analysis methods in relation to these tasks.

Our analysis of four main groups of cluster analysis methods (algorithms) showed that specifics of real field and geological parameters are most fully considered by density and hierarchical methods. The Single-link algorithm (implementing one of the main hierarchical methods) and the DBSCAN algorithm (implementing one of density methods) [6, 7] are chosen as the subject to implementation.

3. Intellectual information system features

The oil-extracting enterprise usually has a large number of specialized software and information systems; among them there can be monolithic and client-server applications, various programming languages and DBMS can be used. Extremely large volume of data is dispersed on all these systems, which complicates the process of their integration and data maintenance in a correct state. All that should be considered in the creation concept of IIS, first of all, during architecture designing of the system.

The main requirement to IIS architecture is its flexibility in many aspects: in relation to data formats, platforms, and also to the methods set implemented in the form of services and the third-party IS set interconnected with IIS.

Flexibility, the cross-platform, the minimum dependence on the data format, loose connections are the principles that are relating to a service-oriented architecture (SOA) approach [8]. SOA represents the perspective model of loosely connected components interaction of the created IIS. This model allows connecting components among themselves by means of exactly defined interfaces. Interfaces do not depend on the hardware platforms used at the enterprise, operating systems and programming languages used for the IIS components development.

The main criterion when choosing a development environment and the IIS programming languages is implementation completeness of the chosen IIS SOA model. The enterprise usually has the formalized structure of BP, however it tends to modification and improvement because of market conditions and technological changes. The flexible adaptive management system of BP (BPMS - Business Process Management System) is necessary to react to these changes.

The most popular BPMS (Bizagi BPM, Bonita BPM, ELMA BPM, etc.) are approximately functionally equal. However, the Russian development ELMA BPM has an advantage, because it combines the BPMS and ESB (Enterprise Service Bus) functions [9]. ELMA BPM is in some way the development environment of IS based on the SOA principles, that is why it was chosen as the basis for IIS development.

Every BP implemented in IIS was originally described in the BPMN of the ELMA BPM development environment. The BPMN contains a very full set of various elements that allows describing any BP.

The detailed architecture of IIS software was developed (figure 1) taking into account the IIS features described above.

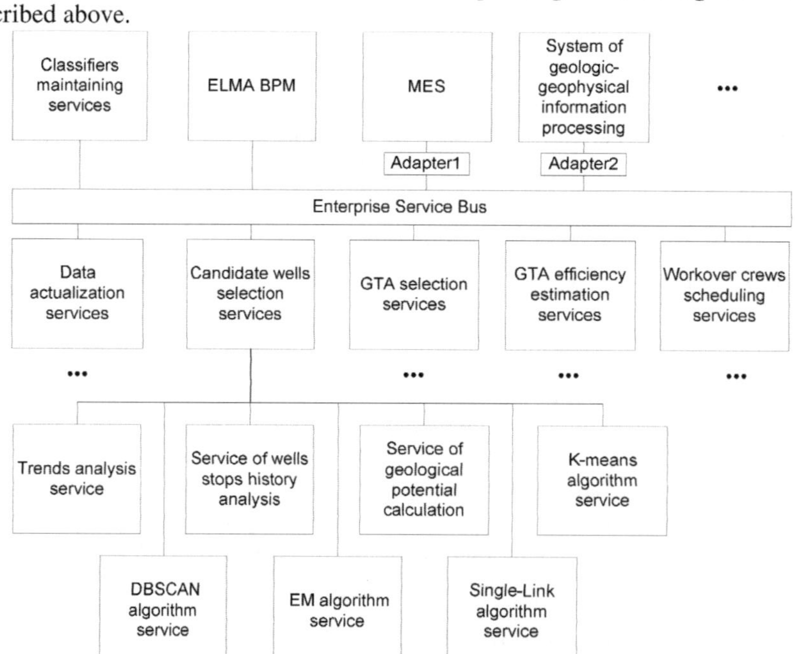

Figure 1. An architecture of IIS software

The main component of IIS architecture is ESB, which functionality is implemented in the ELMA BPM development environment. System services interact among themselves and with enterprise IS (external in relation to the created IIS) through ESB. Two systems are given as the example of external IS. The first of these systems is Manufacturing Execution System (MES) which is intended for operational management of oil-extracting enterprise productions. Instead of it, the enterprise can have a simpler system of production dispatching control. Data actualization services receive basic data from these systems about wells stops, their idle time, etc. Other example of external IS is the system of geologic-geophysical information processing, from which services, for example, geological parameters values of productive formations are obtained. Other external IS can be also connected to ESB if they contain the data used by the IIS services.

External systems must optionally have the standardized interfaces, because they could be created not within the SOA concept, therefore adapters of the ESB must be used for them. It is possible to see the following composite services which are implementing the GTA management BP, and implemented in IIS: data actualization, candidate wells selection, GTA selection, GTA technological and economic efficiency estimation, and workover crew scheduling. In figure 1, services of composite candidate wells selection service (for GTA on them) are displayed in more detail. We see that each method of candidate wells selection is implemented as the separate service. At first, the specialist can choose a desirable methods set of candidate wells selection for GTA, and then run them serially for actually wells selection. If the work result of this or that method does not satisfy specialist, he can choose other methods from the shown set of services.

Let us dwell on features of software IIS realization. The main functionality part of IIS is implemented in the ELMA BPM development environment in the C# language built-in designer and the Razor visualization mechanism. Razor is used for services of interaction with IIS users. ELMA BPM forms the register of services for all functionality realized in it, because ELMA BPM contains possibilities of ESB, that is why internal services in this development environment can be used from the outside by means of other ESB or other system with the documented API.

All descriptions of the main oil-extracting enterprise field objects were created in the specialized FireBird database (DB) under control of ELMA BPM which includes all classifiers and entities of the GTA management subject domain [10].

The external web services implemented by means of Microsoft Visual Studio 2015 WCF (Windows Communication Foundation) in the C# language and also in WebStorm on the Node.JS platform (Express framework) in the JavaScript language were used besides opportunities of the ELMA BPM development environment. These web services are used for the complicated resource-intensive tasks which can demand serious computing capacities. It allows unloading the ESB as much as possible. Standard interfaces of SOA architecture (SOAP 1.1 and WSDL 1.1.) are used for communication with web services in Visual Studio. And the HTTP and WebSocket protocols are used for the Node.JS web services (the AJAX and COMET technologies).

ADO.NET technology is used for initial data loading. The adapter was implemented in the ELMA BPM development environment. This adapter synchronizes the internal DB FireBird condition with the MS SQL DBMS DB which is often used in external IS of the oil-extracting enterprises.

4. Results of IIS development

Developed IIS has been tested on the basis of archival real field data for 10 well pads of the oil field in the Tomsk region. The values of the following parameters of each well were used as initial data for cluster analysis: Q_l – a liquid rate (oil, water, and condensate), Q_g – a gas flow rate, Q_o – an oil flow rate and a combination $Q_l + Q_g$. The available data were picked for each well at the end of the month that was before the month when some wells from analyzed well pads had archival GTA information. The decision to carry out these GTA was based on geological survey expert analysis of specialists that was made in the "manual" operating mode. The specialists divided all wells of well pad into two clusters: "candidate wells for GTA" and "wells without GTA", when they were planning GTA for wells in any month. These data about GTA were used as a reference. For example, well pad № 1 consists of 7 wells: GTAs were carried out for 4 wells in May 2014 and no GTA was applied to other 3 wells. We also suggested that it is possible to divide wells of the well pad into 3 clusters and carried out experiments in this case. In essence, we proposed dividing the cluster "wells without GTA" for each well pad into 2 clusters: "effective" wells (high liquid rate Q_l) and "medium" wells (medium liquid rate Q_l). The wells from the "medium" cluster also can be candidate wells for GTA, while the list of candidate wells is forming, but with a lower priority.

The accuracy of the following cluster analysis algorithms was evaluated: K-means, DBSCAN, EM and Single-link. The accuracy of the algorithm is the ratio of the number of wells correctly classified to clusters that were specified by the enterprise specialists, to the total number of analyzed wells in the well pad.

Prospects of suggested idea about dividing wells of the well pad into 3 clusters were also evaluated during the analysis of field data, along with the division into 2 clusters used by specialists. The results of numerical experiments for well pad № 1 are shown in Table 1, as an example. Clusters that were obtained at the output of the analysis are displayed as two (the case of dividing into two clusters) or three (the case of dividing into three clusters) columns, and the number of candidate wells for GTA is marked with a bold font. For example, Table 1 shows that we have obtained the following clusters as a result of the wells dividing of well pad № 1 into 3 clusters using the DBSCAN algorithm for Q_l parameter: the first cluster (4 candidate wells for GTA and 1 well without GTA, this cluster is indivisible, because it contains wells of different groups), the second cluster (1 "medium" well that has not been chosen for GTA) and the third cluster (1 "effective" well that has not been chosen for GTA).

Table 1. The cluster analysis results of the real field data for well pad № 1

Algorithms	**K-means**					**DBSCAN**					**EM**					**Single-link**				
Clusters	2		3			2		3			2		3			2		3		
Q_l	**4**	0	**2**	**2**	0	**4**	0	**4**	0	0	**4**	0	**4**	0	0	**4**	0	**4**	0	0
	2	1	0	2	1	2	1	1	1	1	2	1	1	1	1	2	1	1	1	1
Q_g	**1**	3	0	**2**	**2**	**4**	0	**3**	0	**1**	**4**	0	**3**	0	**1**	**4**	0	**3**	0	**1**
	2	1	2	0	1	1	2	1	2	0	1	2	1	2	0	1	2	1	2	0
Q_o	**3**	**1**	**1**	**2**	**1**	**4**	0	**3**	**1**	0	0	**4**	**3**	0	**1**	**4**	0	**3**	**1**	0
	1	2	0	1	2	2	1	2	0	1	2	1	1	1	1	2	1	2	0	1
$Q_l + Q_g$	**2**	**2**	**2**	**2**	0	**4**	0	**4**	0	0	**4**	0	**4**	0	0	**4**	0	**4**	0	0
	2	1	0	2	1	2	1	0	2	1	1	2	0	2	1	2	1	0	2	1

We have made the following conclusions based on the results of cluster analysis and the calculation of the accuracy for investigated algorithms on the basis of 10 well pads of the one oil field:

1. In many cases the dividing of a well pad into 3 clusters, where one of the output clusters is a cluster of candidate wells for GTA, allows obtaining separable clusters (clusters that does not contain wells from different groups). The result often is not obtainable in the case of wells division into 2 clusters.

2. Table 1 shows that the K-means algorithm tends to divide candidate wells for GTA from well pad № 1 to several clusters, what leads to low accuracy of the clustering results. A similar result was obtained for other well pads.

3. The cluster analysis of wells from well pad № 1 by two field parameters (Q_l + Q_g) using DBSCAN, EM and Single-link algorithms showed the best result in terms of accuracy, because the set of these parameters allowed obtaining separable clusters: the candidate wells for the GTA cluster (4 wells), the "medium" wells cluster (2 wells) and the "effective" wells cluster (1 well). This set of parameters takes into account all flow rates for extracted carbohydrates by wells. That increases the accuracy of the cluster analysis.

4. The EM algorithm has shown the same accuracy result for the division wells from well pad № 1 into 3 clusters for some parameters as DBSCAN and Single-link algorithms have. But the EM algorithm has shown worse results for some other well pads than DBSCAN and Single-link had.

5. The suggested idea about the prospects of density and hierarchical methods of the cluster analysis and the division of wells from the well pad into 3 clusters were largely confirmed, but the results of the cluster analysis do not always have a high accuracy.

5. Conclusion

The analysis has shown that some researchers prefer to use IAD methods to manage GTA and create IIS based on these methods. Furthermore, hierarchical and density methods of the cluster analysis can be considered as the most correct for the candidate wells selection for GTA and the selection for these wells the relevant arrangements. They have been implemented into IIS along with traditional methods to manage GTA.

The detailed architecture of IIS software was developed by using the SOA-model, relevant services for this IIS were created. During the research of developed IIS capabilities by using the real data from the one of the Tomsk region oil fields, we have found out that the use of hierarchical and density methods (algorithms) of the cluster analysis allows us to solve the problem of candidate wells selection for GTA, but results are not always highly accurate. The further research of these methods applicability depending on the amount and type of initial geological and field parameters of wells and productive formation is required.

References

[1] Kudinov A V, Markov N G 2012 *Problemi avtomatizatsii proizvodstva gazodobyvayuschey kompanii* TPU Pusblisher 247
[2] Sultanov A S, Latifullin F M and Nasybullin A V 2010 *Neftyanoe hozyaystvo* **7** 48-51
[3] Galiullin M M, Zimin P V and Vasilev V V 2011 *Neftyanoe hozyaystvo* **6** 120-3
[4] Keller Yu A 2014 *Izvestiya TPU* **325** 60-5
[5] Perminov D E, Valeev S V 2013 *Nauchno-tehnicheskiy vestnik "NK Rosneft"* **30** 31-5
[6] Guha S, Rastogi E and Shim K 1998 *The Intern. Conf. on Management of Data* **1** 73-84
[7] Kaiking K, Kriegel H P and Kroger P 2004 *The 4th SIAM Intern. Conf. on Data Mining* **2** 240-57
[8] Juric M 2007 SOA approach to integration Packt Publishing Ltd 366
[9] http://www.elma-bpm.ru 2016
[10] Evsyutkin I V, Markov N G 2016 *Atlantis Press* **ITSMSSM** 379-83

REVISITING THE PLANE ELECTROMAGNETIC WAVE TRANSMISSION AND REFLECTION COEFFICIENTS FOR THE LAYER WITH AN ALTERNATING-SIGN DISTURBANCE OF RELATIVE DIELECTRIC PERMITTIVITY

V R Milov [1], L P Kogan [2], P V Gorev [3], P N Kuzmichev [4], P A Egorova [5]

[1] Federal State Educational Institution of Higher Education Nizhny Novgorod State Technical University named after R.E. Alekseev, Nizhny Novgorod, Minin Str., d. 24, Russia
[2] Federal State Educational Institution of Higher Education Nizhny Novgorod State University of Architecture and Civil Engineering, Nizhny Novgorod, Ilyinskaya Str., 65, Russia,
[3] Giprogascenter JSC, Nizhny Novgorod, Alekseevskaya Str., 26, Russia,
[4] Giprogascenter JSC, Nizhny Novgorod, Alekseevskaya Str., 26, Russia,
[5] Federal State Educational Institution of Higher Education Nizhny Novgorod State Technical University named after R.E. Alekseev, Nizhny Novgorod, Minin Str., d. 24, Russia

E-mail: L_kog@list.ru

Abstract. In this paper, we consider the question of the plane electromagnetic wave incidence at the inhomogeneity with an arbitrary profile of the relative permittivity disturbance. Module estimation of Neumann series remainder is carried out for the field of the wave passing through the nonhomogeneous section. Based on that, the number of summands in the series, required to calculate with a given accuracy, the transmission and reflection coefficients have been determined.

1. Introduction. The given article discusses the issue of electromagnetic waves incidence on a disturbed region with the known line of the relative dielectric impermeability. One of the most important problems is calculation of the wave reflection coefficient using the given inhomogeneity and determination of wave transmission coefficient $\varepsilon = \varepsilon(x)$ [1]-[6]. In particular, this issue is relevant when calculating the average characteristics of the field strayed on an accidental layer, consisting of a wide range of such one-dimensional nonhomogeneities with a topologically similar profile of the dielectric permittivity fluctuations [7], [8]. If for given type $\varepsilon(x)$, the solution of the Helmholtz wave equation is known, the determination of these coefficients is simple and is carried out based on the requirements for the fulfilment of the conditions at the boundary [1], [2]. In case of any kind of $\varepsilon(x)$, this problem in general can be solved by successive approximations. However, there are very strict boundaries [5] for the convergence of the corresponding Neumann series [9], [10]. In this paper, we show that for alternating disturbances under certain conditions, there are much less onerous conditions of the specified convergence, which ultimately leads to a significantly lower assessment for the remainder of the series and the minimum required number of iterations.

2. Setting the problem and estimation of the Neumann series remainder

Let us consider a plane electromagnetic wave with electric field intensity of $\boldsymbol{E_0} = E_0(x)\boldsymbol{y_0}$, where $E_0(x) = \exp(-ikx)$, $k = k_0\sqrt{\varepsilon^{(0)}}$, $k_0 = \frac{2\pi}{\lambda}$, $\varepsilon^{(0)}$ are an unperturbed value of the relative dielectric permeability of the medium, $\boldsymbol{y_0}$ - axis unitary vector y (see. Fig. 1) perpendicular to the x axis. The wave propagates along axis x and is normally incident on the disturbed region with a relative dielectric impermeability of the following kind:

$$\varepsilon(x) = \varepsilon^{(0)}[1 + \alpha(x)]. \tag{1}$$

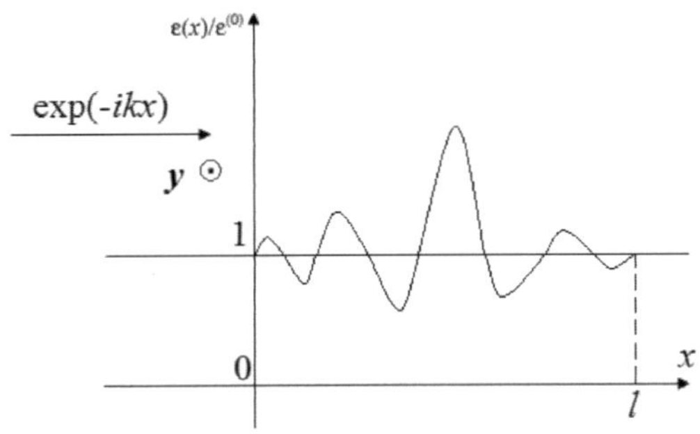

Figure1. A schematic view of the inhomogenuity profile.

In what follows, $\varepsilon(x)$, $\varepsilon^{(0)}$ and $\alpha(x)$ are the real values. The definition of inhomogeneity profile function $\alpha(x)$ can be alternating within the range of $0 \le x \le 1$ and a negligible quantity beyond its limits. We assume that Max $|\alpha(x)| = a$. Time dependence $\sim\exp(i\omega t)$ is omitted for brevity.

For field $\boldsymbol{E}(x) = E(x)\boldsymbol{y_0}$, created by the initial wave in the whole space, we write the following integral equation:

$$E(x) = E_0(x) - k^2 \int_0^l E(x_1)\alpha(x_1)G(x_1,x)\,dx_1, \tag{2}$$

where $G(x_1,x) = \frac{\exp(-ik|x - x_1|)}{-2ik}$ is the Green's function of the problem. For $G(x_1,x)$ here and elsewhere, the first coordinate corresponds to the radiation source, and the second one corresponds to the observation point.

Assuming $x = l$, we have $E(l) = E_0(l) - k^2 \int_0^l E(x_1)\alpha(x_1)G(x_1,l)\,dx_1$. Let us write $E(l)$ as

$$E(l) = \sum_{m=0}^{\infty} E_m(l). \tag{3}$$

Here, $E_m(l) = (-k^2)^m \int_0^l ... \int_0^l E_0(x_m)\,G(x_m,x_{m-1}) ... G(x_2,x_1)G(x_1,l)\,\alpha(x_m)\alpha(x_{m-1}) ... \alpha(x_1) \times$ $\times\, dx_m\, dx_{m-1} ... dx_1 = \sum_{M=0}^{m} \sum_{j_1, j_2, ..., j_M} E_m(l, j_1, j_2, ..., j_M)$.

In the last equation,

$$E_m(l, j_1, j_2, ..., j_M) = (-k^2)^m \int\limits_{\Omega_{j_1, j_2, ..., j_M}} \int\limits_0^l E_0(x_m)\,G(x_m,x_{m-1}) ... G(x_2,x_1)G(x_1,l) \times$$
$$\times\, \alpha(x_m)\alpha(x_{m-1}) ... \alpha(x_1)\, dx_m\, dx_{m-1} ... dx_1 =$$

$$= \exp(-ikl)\left(-\frac{ik}{2}\right)^m \int\limits_{\Omega_{j_1,j_2,\ldots,j_{\mathcal{M}}}} \exp(-2ikx_{j_1})\exp(+2ikx_{j_2})\ldots\exp(-2ikx_{j_{\mathcal{M}-1}}) \times$$
$$\times \exp(+2ikx_{j_{\mathcal{M}}})\ \alpha(x_m)\alpha(x_{m-1})\ldots\alpha(x_1)\ dx_m\ dx_{m-1}\ldots dx_1, \quad (4)$$

where $1 \leq j_1 \leq j_2 \leq \ldots \leq j_{\mathcal{M}} \leq m$. Here, $x_{j_1}, x_{j_2}, \ldots, x_{j_{\mathcal{M}}}$ are reflection points on the corresponding line. The left reflection corresponds to uneven subscripts (for example $j_1, j_3, j_5\ldots$), and the right reflection corresponds to the even ones (for example $j_2, j_4, j_6\ldots$). This fact is associated with argument signs of the corresponding exponential factors in (4) (here, exponents signs $\exp(-2ikx_{j_{M-1}})\exp(+2ikx_{j_M})$ correspond to even value $\mathcal{M}$; at uneven $\mathcal{M}$, there will be $\exp(+2ikx_{j_{\mathcal{M}-1}})\exp(-2ikx_{j_{\mathcal{M}}})$). For any variable x_j, that is included in the list above, integration $(x_j - x_{j-1})(x_j - x_{j+1}) > 0$ is provided (wherein $G(x_{j+1}, x_j)\,G(x_j, x_{j-1}) = \frac{\exp[\pm ik\,(2x_j - x_{j-1} - x_{j+1})]}{(-2ik)^2}$; sign "+" corresponds to the left reflection and "-" – to the right one).

Thus, in (4), variable x_j is included in the list of $x_{j_1}, x_{j_2}, \ldots, x_{j_{\mathcal{M}}}$, integral $L(j) = \int_{x_{min}}^{x_{max}} \exp(\pm 2ikx_j)\ \alpha(x_j)\ dx_j$ (where upper and lower limits $x_{min} = Min\{x_{j-1}, x_{j+1}\}$, $x_{max} = Max\{x_{j-1}, x_{j+1}\}$).

Next, we assume that $\alpha(x) = \Lambda(x)\cos[\varphi(x)]$, where real function $\Lambda(x) \geq 0$, has the only maximum at $0 \leq x \leq l$ and is equal to zero outside this interval, $\varphi(x)$ – the real phase. (These disturbance properties do not lead to the substantial reduction in the domain applicability area of the proposed method, since it is easy to present almost any dependence in form $\alpha(x)$). At first, we consider the module of the real component of L_c of integral $L(j)$:

$$|L_c| = |Re\{L(j)\}| = \left|\int_{x_{min}}^{x_{max}} \cos(2kv)\ \Lambda(v)\cos[\varphi(v)]\,dv\right| =$$
$$= \frac{1}{2}\left|\int_{x_{min}}^{x_{max}} \{\cos[2kv + \varphi(v)] + \cos[2kv - \varphi(v)]\}\ \Lambda(v)dv\right|$$

(for notational convenience here and elsewhere, instead of x_j in integrating, we use designation v).

Let

$$Max|\varphi'(x)| \leq \Delta < 2k. \quad (5)$$

Then, as is easy to understand,

$$|L_c| \leq S = \left|\int\limits_{x_{min}}^{x_{max}} \cos[\zeta\, v + \varphi_0]\ \Lambda(v)dv\right|.$$

Here, $\zeta = 2k - \Delta$, and it remains to be seen as the properties of phase summand φ_0.

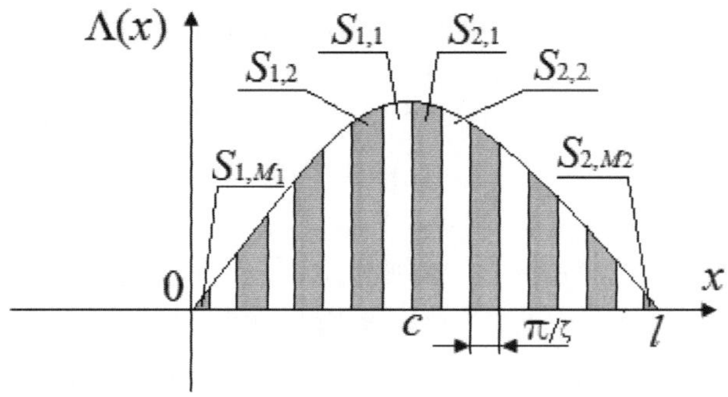

Figure 2. The graph of function $\Lambda(x)$. For the light areas, $\cos[\zeta\, v + \varphi_0] \geq 0$, and for the painted ones, $\cos[\zeta\, v + \varphi_0] \geq 0$. The increase in the indexes for $S_{1,j}$ goes opposite axis *x*, and for $S_{2,j}$ it goes along axis *x*.

Figure 2 shows graph $\Lambda(x)$. The light areas correspond to the areas where $\cos[\zeta v + \varphi_0] \geq 0$, and painted ones - to the intervals, where $\cos[\zeta v + \varphi_0] < 0$. The length of each segment (possibly except for the two outermost) is $\frac{\pi}{\zeta}$. Phase φ_0 corresponds to equation $\cos[\zeta c + \varphi_0] = 0$.

Obviously, $S = |S_1 + S_2|$, where $S_1 = \int_0^c \cos[\zeta v + \varphi_0]\ \Lambda(v)dv$, $S_2 = \int_c^l \cos[\zeta v + \varphi_0]\ \Lambda(v)dv$. As is evident, $S = |S_1 + S_2|$, where $S_1 = \int_0^c \cos[\zeta v + \varphi_0]\ \Lambda(v)dv$, $S_2 = \int_c^l \cos[\zeta v + \varphi_0]\ \Lambda(v)dv$. Let us consider value S_1, having written it in the type of $S_1 = \sum_{j=1}^{M_1} S_{1,j}$. Here, each term under the sign of the sum is equal to the integral over the corresponding part length of no more than $\frac{\pi}{\zeta}$ (see Fig. 2), the number of which is M_1. This sum includes only decreasing elements with the increase of the index and in such case alternating elements. Their number can formally be increased to the infinity, assuming the summands with numbers $j > M_1$ to be arbitrary small and also alternating. Then, by Leibniz's theorem about the convergence of alternating series, we find that $|S_1| \leq |S_{1,1}|$.

By analogy, we get the following for integral $S_2 = \sum_{j=1}^{M_2} S_{2,j}$ (the designated meaning is similar to S_1 and understood from Fig. 2): $|S_2| \leq |S_{2,1}|$. Since the signs of integrals S_1 and S_2 are different (by this theorem, these signs coincide with signs $S_{1,1}$ and $S_{2,1}$, but these values differ in sign, see. Fig. 2), then $S = \big||S_1| - |S_2|\big| \leq Max\{|S_{1,1}|, |S_{2,1}|\} \leq a\frac{\pi}{\zeta}$. Consequently, $|L_c| \leq a\frac{\pi}{2k-\Delta}$, and performed calculations are similar to imaginary part $L_s = Im\{L(\jmath)\}$, we come to the same inequality: $|L_S| \leq a\frac{\pi}{2k-\Delta}$.
As a result, we have:

$$|L(\jmath)| = \sqrt{{L_c}^2 + {L_s}^2} \leq \mathcal{L} = \frac{\sqrt{2}\pi a}{2k - \Delta}. \tag{6}$$

Now let us consider the integrals over variables $x_{\hat{m}} \in \{x_1, x_2, \dots, x_m\}$ which are not included in the list of reflection points $x_{j_1}, x_{j_2}, \dots, x_{j_{\mathcal{M}}}$ on the traces. The given integration is under the condition that $x_{\hat{m}-1} \leq x_{\hat{m}} \leq x_{\hat{m}+1}$, or under the condition that $x_{\hat{m}+1} \leq x_{\hat{m}} \leq x_{\hat{m}-1}$. Each of these two-sided inequalities corresponds to the wave propagation through coordinate $x_{\hat{m}}$ without any reflection; in these conditions, product $G(x_{\hat{m}+1}, x_{\hat{m}})G(x_{\hat{m}}, x_{\hat{m}-1}) = \frac{\exp[\pm ik(x_{\hat{m}-1} + x_{\hat{m}+1})]}{(-2ik)^2}$ does not depend on $x_{\hat{m}}$. Any such coordinate (their total amount is $m - M$) in (4) is the integral type corresponding to $\mathcal{J}(\hat{m}) = \int_{\mathcal{X}_{min}}^{\mathcal{X}_{max}} \alpha(v)\, dv$, where $\mathcal{X}_{min} = Min\{x_{\hat{m}-1}, x_{\hat{m}+1}\}$, $\mathcal{X}_{max} = Max\{x_{\hat{m}-1}, x_{\hat{m}+1}\}$.

In what follows, we assume that

$$|\mathcal{J}(\hat{m})| \leq A = al_0. \tag{7}$$

Also, let us introduce the condition where at any possible values $\mathcal{X}_{min}$ and $\mathcal{X}_{max}$ the relation $|\mathcal{J}(\hat{m})| \ll al$ is valid. Given inequality means that function $\alpha(x)$ is tangibly alternating-sign at $x \in [0, l]$, in other words, in the given interval, the region integral, where $\alpha(x) \geq 0$, is associated with the region integral, where $\alpha(x) < 0$.

In accordance to (4), (6) and (7), we get relation $|E_m(l, j_1, j_2, \dots, j_{\mathcal{M}})| \leq \left(\frac{k}{2}\right)^m \mathcal{L}^{\mathcal{M}} A^{m-\mathcal{M}}$. It is easy to understand that the number of different index sets $j_1, j_2, \dots, j_{\mathcal{M}}$ does not exceed the number of $C_m^{\mathcal{M}}$ combinations of m elements according to $\mathcal{M}$. In this way,

$$|E_m(l)| \leq \sum_{\mathcal{M}=0}^{m} \sum_{j_1, j_2, \dots, j_{\mathcal{M}}} |E_m(l, j_1, j_2, \dots, j_{\mathcal{M}})| \leq \left(\frac{k}{2}\right)^m \sum_{\mathcal{M}=0}^{m} C_m^{\mathcal{M}} \mathcal{L}^{\mathcal{M}} A^{m-\mathcal{M}} = q^m,$$ where

$$q = \frac{k(\mathcal{L} + A)}{2} \leq Q = \frac{ka}{2}\left(\frac{\sqrt{2}\pi}{2k-\Delta} + l_0\right). \tag{8}$$

In what follows we assume that $q < 1$.

Thus, when recording an approximate solution of this kind, we obtain:

$$E(l) \approx E(l,m) = \sum_{\tilde{m}=0}^{m} E_{\tilde{m}}(l). \tag{9}$$

take into account only the first m elements of series (3), for its remainder

$$\Theta_m(l) = \sum_{\tilde{m}=m+1}^{\infty} E_{\tilde{m}}(l)$$

we get to the following estimation:

$$|\Theta_m(l)| \le \sum_{\tilde{m}=m+1}^{\infty} |E_{\tilde{m}}(l)| \le \frac{q^{m+1}}{1-q}. \tag{10}$$

Recorded solutions in the form of (9) are valid under condition

$$|E(l,m)| \gg \frac{q^{m+1}}{1-q}. \tag{11}$$

It should be noted that under the conditions imposed, when the amplitude of the wave incident on the area of the disturbance is taken as one, value $E(l)$ can be regarded as the transmission coefficient of the wave through the nonhomogeneity. Therefore, by the law of the energy conservation, for the reflection coefficient module we have:

$$|V| = \sqrt{1 - |E(l)|^2}. \tag{12}$$

3. Discussion of the results

Inequality (10) is a common form of estimation of series remainder (3) using approximation (9). However, value q calculated as (8) in some cases is substantially less than the known one in scientific literature. Indeed, according to [5], for q there is the following restriction:

$$q \le \frac{kal}{2}. \tag{13}$$

From a comparison of (8) and (13), it follows that the method proposed in this paper allows one to obtain more accurate estimates under certain conditions for specified module remainder $|\Theta_m(l)|$ than it follows from (13). It is possible under condition

$$\mathcal{L} + A \ll al, \tag{14}$$

for the truth of which the following relation is necessary:

$$\frac{2\pi}{2k-\Delta} \ll l. \tag{15}$$

See (6) and (5), so in view of (7) the following inequality holds:

$$l_0 \ll l. \tag{16}$$

As is evident, inequality (15) is meaningful if

$$2k - \Delta \gg \frac{2\pi}{l}, \tag{17}$$

Relation (17), in particular, is under the condition that disturbed wavelength $\lambda \ll l$ and the smallest spatial scale of disturbance $\sim\Delta^{-1}$ are connected with the presence of oscillating factor $\cos[\varphi(x)]$ when disturbance recording $\alpha(x)$ is significant at least one and a half or two times more than λ. Condition (16) means that for disturbance $\alpha(x)$, its medium notation in interval $x \in [0, l]$ is significantly less than medium for $|\alpha(x)|$.

As a result, when the verity of (16) and (17) is proved, transmission coefficient $E(l)$ can be calculated with the given accuracy according to (9), then the reflection coefficient is calculated (12).

As an example of the application of the results, we define the conditions when in (9) we may be limited by case $m = 2$, assuming

$$E(l) \approx E(l,2) = E_0(l) + E_1(l) + E_2(l). \tag{18}$$

Here, $E_0(l) = \exp(-ikl)$, module $|E_0(l)| = 1$. For the second and third summands we have: $E_1(l) = \exp(-ikl)\left(-\frac{ik}{2}\right) \int_0^l \alpha(v)\,dv$, module $|E_1(l)| \le \frac{kA}{2}$ and module

$$E_2(l) = \exp(-ikl)\left(-\frac{ik}{2}\right)^2 \int_0^l \int_0^l \exp(-ik[x_2 - x_1]) \exp(-2ik|x_2 - x_1|)\, \alpha(x_2)\, \alpha(x_1)\, dx_2\, dx_1,$$

$|E_2(l)| \leq Q^2$, see (8). Since integral $\int_0^l \alpha(v)\, dv$ is the real value so it is easy to understand
$|E(l)| \geq \left|1 - i\frac{kA}{2} - Q^2\right| > 1 - Q^2$.
Therefore, provided

$$1 - Q^2 \gg \frac{Q^3}{1-Q}, \quad (19)$$

(see (8), (10) and (11)), we can calculate $E(l)$ in the kind of (18), then calculating the modulus of reflection coefficient V in (12).

4. Conclusion

The main results of this paper are relationships (8) and (19). They allow us to estimate the reminder module of the series of the disturbance theory, by which the field passing through a non-uniform layer has been calculated. When it is required that the disturbance of the dielectric permeability is substantially alternating in sign, it becomes possible to prove the convergence of the Neumann series under much less strict conditions as compared with the known in the scientific literature [5]. This condition makes it possible to limit the number of summands of the series in the calculation of transmission and reflection coefficients corresponding to the plane of incidence on that nonhomogeneity. Awareness of these two quantities makes it possible to simulate the scattering of a plane wave in the initial random layer. The layer consists of a large number of disturbances with a topologically similar profile for a given distribution of their width, amplitude and positions within the layer.

References

[1] Brekhovskih L M 1957 *Waves in the layer environment.* (Moscow, Pub. AN SSSR) 46–50
[2] Ginzburg V L 1960 Propagation of the elementary waves in plasma. (Moscow, Science) 297–312
[3] Krajnov V P, Presnyakov L P 1993 *Phisycal sciences sucess.* **7** 85–92
[4] Samohin A V 2007 *Scientific Journal of the Moscow State Technical Aviation University.* **114** 24–33
[5] Eliseev P G, Osinskii M A 1980 *Soviet Journal of Quantum Electronics* **10(7)** 811–816
[6] Borisov S A 1999 News of the South Federal University. *Technical sciences* **4(14)** 123–127
[7] Abramovich B S, Gurbatov S N, Ryzhov YU A 1979 *News of the Universities. Radiophysics.* **22 (5)** 566–576
[8] Grigor'ev G I, Zaboronkova T M, Kogan L P 2014 Materials *XXIV All-Russia scientific conference*, 29 June – 5 July, 2014. Irkutsk, ISZF SO RAN 156-157
[9] Vladimirov V S, Zharinov V V 2004 Mathematical physical equations. (Moscow, Fizmatlit) 202–212
[10] Gorev P V 1989 Analysis of optical of the temporal series. *University News. Radioelectronics.* **32(1)** 62–65

The dynamic model of enterprise revenue management

A A Mitsel[1,2], M Yu Kataev[1,2], S V Kozlov[1], K V Korepanov[3]

[1] National Research Tomsk Polytechnic University, 30, Lenina Ave., 634050, Tomsk, Russia
[2] Tomsk State University of Control Systems and Radioelectronics, 40, Lenina Ave., Tomsk, 634050, Russia
[3] National Research Tomsk State University, 36, Lenina Ave., 634050, Tomsk, Russia

E-mail: Maa@asu.tusur.ru

Abstract. The article presents the dynamic model of enterprise revenue management. This model is based on the quadratic criterion and linear control law. The model is founded on multiple regression that links revenues with the financial performance of the enterprise. As a result, optimal management is obtained so as to provide the given enterprise revenue, namely, the values of financial indicators that ensure the planned profit of the organization are acquired.

1. Introduction

The financial stability of the organization from the long-term perspective standpoint is normally assessed by the system of indicators that includes several main sections: assessment of property position; liquidity assessment; assessment of financial stability; assessment of business activity; and profitability assessment. In each group there are 6 to 12 different coefficients (with a total of 41) [1-12].

Apart from the coefficient analysis, a number of classification models that separate bankrupt companies from stable borrowers and predict the possible bankruptcy of the borrower company are used [6-16].

There are numerous proprietary methodologies for assessing the probability of bankruptcy, which operate on a wide range of indicators. Such factor models have been developed on the basis of a multi-dimensional (multiplicative) discriminant analysis.

The best-known models for assessing the probability of bankruptcy are:

- the Altman models [5-7];
- the four-factor Lis model which is suitable for Russian enterprises with such legal organization forms as CJSC and OJSC [8];
- the Fulmer model – a nine-factor model for bankruptcy risk assessment [8];
- the Springate model for predicting enterprise bankruptcy [10,11];
- the four-factor Taffler's bankruptcy model [12];
- the Sayfullin-Kadykov enterprise bankruptcy model – a medium-term rating model for bankruptcy risk prediction that has been developed by Russian scientists, which can be applied to any industry and businesses of various sizes [13, 14].
- in addition to these, the following models must also be mentioned: by G. V. Davydova and A. Yu. Belikov [15], O. P. Zaytseva [16], A. N. Makaryeva and L. V. Andreeva [17] and others.

Analysis of models has shown that there exists no universal model for assessing the bankruptcy risk of an enterprise. The limits of model applicability are related with the economic environment in which the models were obtained. [18] displays that foreign models are unsuitable for Russian conditions. Moreover, the models of financial sustainability for enterprises of different industries may also substantially differ from each other. For example, the performed analysis of financial stability of 30 oil and gas industry companies with the legal form of an open joint stock company according to the balance sheet for 2010, 2011 and 2012 has shown that the five-factor Altman model poorly reflected the real situations at the Russian enterprises and did not determine further development of the organizations. Thus, according to the Z-index outcome, 17 per cent of enterprises had a very high probability of bankruptcy, 30 percent – a high probability of bankruptcy, and 53 percent – a very low probability of bankruptcy. In fact, 100 percent of all businesses continued to operate in 2012; therefore, it can be assumed that 100 percent of enterprises had to have a very low probability of bankruptcy.

Nevertheless, the Russian models may not claim to be universal and suitable for businesses of any industry. For instance, though calculations according to the Sayfullin-Kadykov probability of bankruptcy model for the same 30 oil and gas companies showed high performance in the modern economic environment, this model is not deprived of disadvantages – there is room for some companies to vary between the R model key indicators, the degree of bankruptcy probability and the real state of affairs at the enterprises. This can be explained by the fact that both the model and its coefficients were calculated at the end of the 1990s, when Russia had other economic conditions and other development strategies for enterprises, as well as another tax climate. Therefore, we come to the following conclusion: for the oil and gas industry companies under consideration, the Altman model is pessimistic, whereas the Sayfullin-Kadykov model, on the one hand, allows assessing the current financial condition of the company; it still does not allow accurate assessment of the probability for the beginning of a crisis situation.

For enterprises of various branches, we have constructed proprietary models for assessing the enterprise revenues, which is the most significant indicator related to financial stability. [19] has already analyzed the issue of selecting factors that influence the main source of the enterprise's financial receipts – the revenue from product sales. The selection of factors is conducted by the principal component method. Based on the factors selected, an equation of regression that indicates the dependence of the revenues on those factors is formed. In this work, on the basis of the financial statements of 33 machine-building enterprises, there were selected 37 indicators. With the principal component method, 11 indicators were distinguished that had a significant impact on the enterprise revenues.

2. Model Description

Let us select n indicators as most critical for assessing the risk of bankruptcy that also affect revenues from product sales as the main source to form the financial resources of the enterprise. Let the values of these parameters exceed the lower boundaries of the permissible and let it be necessary to increase their values to avoid bankruptcy. This requires investments in the company.

As $x_i(t)$, $i=1,...,n$, let us denote indices at time moment t, $t=0,...,T-1$, where T is the planned time moment for the enterprise to leave the critical state; $V(t)$ – revenues from product sales that the company must have to lead itself out of crisis (it is in fact due to an additional investment into the company necessary to overcome crisis); $V^0(t)$ – planned revenues needed for sustainable functioning of the enterprise. It is assumed that part of the revenues is invested in production proceeds.

The dependence of the revenues on time is presented as a multiple regression:

$$V(t)=\sum_{i=1}^{n} a_i x_i(t), \tag{2}$$

where a_i denotes the regression coefficients.

The values of $y_i(t)=a_i x_i(t)$ may be considered as revenues conditioned by the i indicator of the industrial activity. Then

$$V(t)=\sum_{i=1}^{n} y_i(t) \quad (3)$$

The dependence of the planned revenues is presented in the form of:

$$V^0(t+1)=\left(1+\mu_0(t)\right)V^0(t) , \quad (4)$$

where $\mu_0(t)$ – the desired rate of the company's revenue growth. The revenue growth rate must be such that the planned revenues, at least, cover the inflation.

Let us present the dynamics of the $V(t)$ revenue in the form of:

$$V_i(t+1)=\left[1+\mu_i(t)\right]\left(y_i(t)+u_i(t)\right),\ i=1,...,n \quad (5)$$

where $\mu_i(t)$ is the parameter of the required revenue increase conditioned by the i indicator of the industrial activity; $u_i(t)$ is the management connected with the change of the $x_i(t)$ indicator (while $u_i(t)>0$ denotes an increase in the indicator, and $u_i(t)<0$ – its decrease). Let us introduce vector $z(t)=\left(y(t),\ V^0(t)\right)^T$. Then equations (4) and (5) can be rewritten as:

$$z(t+1)=A_0(t)z(t)+B_0(t)u(t) , \quad (6)$$

where $A_0(t)$ is the diagonal matrix of dimension $(n+1)\times(n+1)$

$$A_0(t)=diag\left(1+\mu_1(t),...,1+\mu_n(t);\ 1+\mu^0(t)\right);$$

and $B_0(t)$ is the diagonal matrix of dimension $(n+1)\times n$

$$B_0(t)=\begin{pmatrix} 1+\mu_1(t) & ... & 0 \\ ... & ... & ... \\ 0 & ... & 1+\mu_n(t) \\ 0 & ... & 0 \end{pmatrix};$$

Let us take the quadratic functional as management quality criterion

$$J=\sum_{t=0}^{T-1}\left[[V(t)-V^0(t)]^2+(u(t))^T R(t)(u(t))\right]+[V(T)-V^0(T)]^2, \quad (7)$$

which characterizes the quality of the process, tracking the planned revenues of enterprise $V^0(t)$. By minimizing this functional, we hereby provide the company's access to the planned mode. Here, $R(t)$ is the diagonal matrix of weight coefficients.

Let us introduce vector $z(t)$ to expression (7). Value $\left(V(t)-V^0(t)\right)$ takes the form of $\left(V(t)-V^0(t)\right)=Cz(t)$, where $C=(1,1,...,1,-1)\in R^{n+1}$, and quality criterion J will take the shape of:

$$J=\begin{Bmatrix} \sum_{t=1}^{T-1} z^T(t)C^T Cz(t)+\sum_{t=0}^{T-1}\left(u^T(t)R(t)u(t)\right)+ \\ +z^T(T)C^T Cz(T) \end{Bmatrix}\to \min_{u(t)} \quad (8)$$

Thus, we have an optimal management problem in which the equation of the state is described by multi-step process (6), and the quality functional – by expression (8). Management is defined by vector $u(t)$.

It is necessary to find an optimal solution of $\left(\bar{z}(t),\bar{u}(t)\right)$, in which functional (8) takes the minimum value. The method of solving problems (6) and (8) may be found, for example, in [20, 21].

3. Simulation results

Based on the multiple regression, an analysis of the "KuibyshevAzot" financial condition was conducted, as well as an economic and mathematical model of the firm's profit was built. For the analysis of the "KuibyshevAzot" financial activity, which is specialized in the chemical industry [22], 33 coefficients (factors) were selected, including net income [23]. The values of the selected coefficients were determined by the financial statements according to the corresponding formulas [19]. Upon further selection of coefficients, one of the most common methods for reducing the dimension of the feature space under study was used, i.e. factor analysis [24]. This method has made it possible to reduce the number of the indicators under analysis to 15. As a result, the following model was obtained:

$$V = 1{,}16x_1 + 146{,}4x_2 - 300{,}68x_3 + 42{,}76x_4 + 226{,}73x_5 - 41{,}04x_6 - \\ 7{,}86x_7 + 3401{,}59x_8 + 53{,}85x_9 - 3373{,}32x_{10} - 26{,}9x_{11} + 1{,}94x_{12} + \\ 2{,}59x_{13} - 202{,}45x_{14} - 8{,}21x_{15} - 264{,}74. \qquad (9)$$

Here the following notations are used:

- V – revenues from product sales,
- x_1 – net profit in rubles,
- x_2 – the index of the permanent (capital) asset,
- x_3 – the equity-assets ratio,
- x_4 – the acid ratio,
- x_5 – the concentration ratio of the debt capital,
- x_6 – marginal liquidity ratio (of the intermediate coating),
- x_7 – the current assets to equity ratio,
- x_8 – the ratio of borrowed and own funds,
- x_9 – the debt capital structure ratio,
- x_{10} – the financial activity ratio (the capital leverage),
- x_{11} – the leverage ratio,
- x_{12} – the turnover of the equity capital,
- x_{13} – the turnover of funds in calculations (in turns),
- x_{14} – the profitability of the total capital,
- x_{15} – the return on assets.

The "KuibyshevAzot" enterprise under study does not belong to the category of bankrupt enterprises and was taken only as an example for building a business management model.

Graphs of the planned and actual revenues of the enterprise obtained as a result of solving optimization problems (6) and (8) are shown in Figure 1.

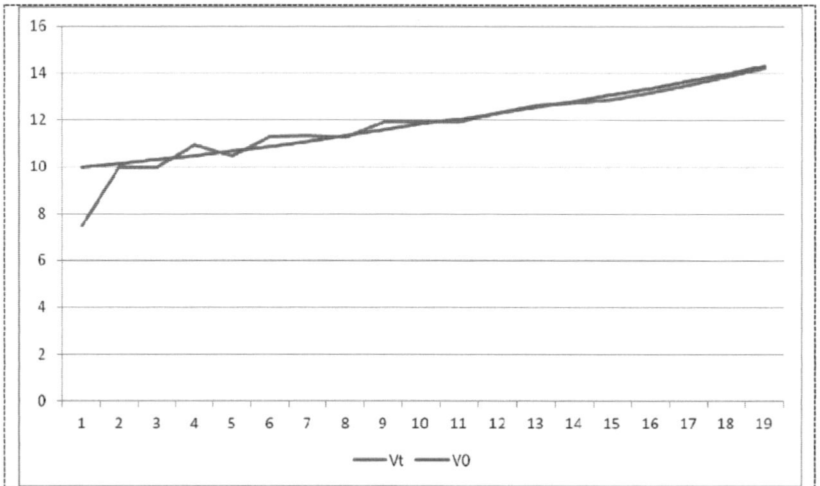

Figure 1. Planned $V0_t$ and actual V_t revenues of the enterprise.

The graphs show that over time the real value of the revenues is approaching the planned one. The values of indicators providing a given level of revenues are given in Figure 2.

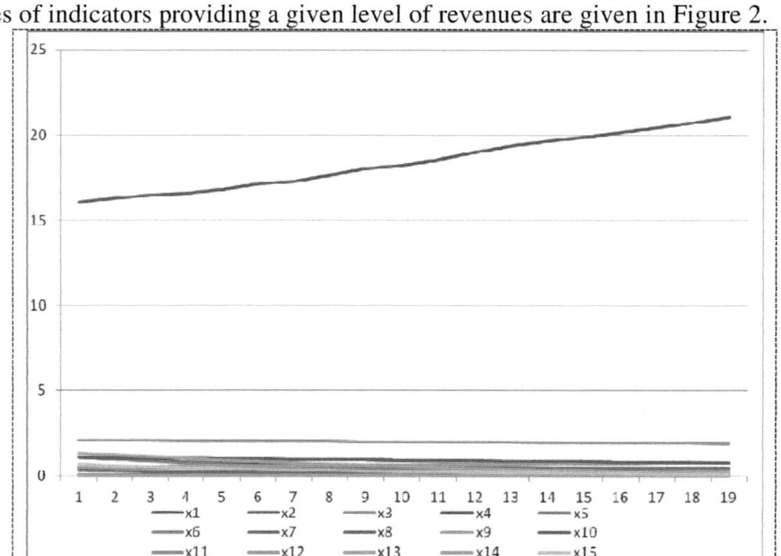

Figure 2. The optimal values of indicators for which the revenue reaches the predetermined level.

Analyzing the values of the indicators, we can conclude that in order to achieve the planned level of revenues it is necessary to reduce the following indicators: the first, second, ninth, thirteenth, fourteenth and fifteenth (i.e. the total amount of economic resources at the organization's disposal, the net fixed (capital) assets index, the debt capital structure ratio, the turnover of funds in the calculations (in turns), the profitability of the total capital, and the return on assets, respectively. At the same time, one of the indicators has adopted a negative value (the turnover of the equity capital), which conflicts with the economic sense. This is due to the fact that the optimal management problem was being solved without any restrictions on the permissible values of the controlled variables. Therefore, the value of this indicator should be set to zero, as shown in Figure 2.

4. Conclusion

On the basis of the regression model that relates to the dependency of the key indicator on the enterprise's financial performance, there was built a dynamic model of the enterprise revenue management. It is based on the quadratic criterion and the linear control law. As a result, optimal management was obtained to achieve the predetermined revenue, i.e. the values of financial indicators that provide access to the planned profit of the enterprise.

References

[1] Telipenko, E. V., Zakharova A. A. 2014 Bankruptcy risk management of a machine builder. In: *Applied Mechanics and Materials*. **682** 617-622

[2] Bankruptcy models (diagnosis and assessment of the probability of bankruptcy) [electronic resource]. In: *Financial finance-m.info site manager*. Access mode: http://finance-m.info/bankruptcy_models.html (reference date 05/05/2014)

[3] Telipenko, E. V., Zakharova A. A. , Sopova S. P. 2015 Forecasting risk of bankruptcy for machine-building plants. In: *IOP Conference Series: Materials Science and Engineering* **91** 012066

[4] Zakharova, A. A., Telipenko E. V. 2012 Information system of bankruptcy risk management of an enterprise. In: *7th International Forum on Strategic Technology (IFOST – 2012): Proceedings: in 2 vol., Tomsk, September 18-21, 2012*. (Tomsk: TPU Press) **1** 539-543

[5] Altman, E. 1983 *Corporate Financial Distress* (New York, Wiley)

[6] Altman, E. I., Haldeman R. G. , Narayanan P. 1977 Zeta analysis: a new model to identify bankruptcy risk of corporation. In: *Journal of Banking and Finance* **June 1977**

[7] Altman, E. I. 1968 Financial ratios. Discriminant analysis and the prediction of corporate bankruptcy. In: *Journal of Finance* **September 1968**

[8] The R. Lis's four-factor model of bankruptcy risk assessment [electronic resource]. In: *Site Financial Analysis and Investment Analysis of a Company*. Access mode: http://www.beintrend.ru/2011-12-05-17-20-28 (reference date 07/09/2013).

[9] Fulmer, J. G. et al. 1984 A bankruptcy classification model for small firms. In: *Journal of Commercial Bank Lending* **July 1984** 25-37

[10] The Springate's forecast model of solvency [electronic resource] In: *Analysis of the Financial Condition of the Company*. Access mode: http://afdanalyse.ru/publ/finansovyj_analiz/1/prognoznaja_model_platezhesposobnosti_springejta/13-1-0-39 (reference date 07/09/2013).

[11] The Springate's bankruptcy prediction model of an enterprise [electronic resource]. In: *Financial Analysis and Investment Analysis of a Company*. Access mode: http://beintrend.ru/springate (reference date 07/09/2013). Original Source: Springate, G. L. V. 1978 *Predicting the Possibility of Failure in a Canadian Firm*. (Unpublished M.B.A. Research Project, Simon Fraser University) January 1978

[12] Taffler R., Tishaw H. 1977 Going, going, gone ñ four factors which predict. In: *Accountancy*. **March 1977** 50-54

[13] The Sayfullin-Kadykov enterprise bankruptcy model [electronic resource]. In: *Site Financial Analysis and Investment Analysis of a Company*. Access mode: http://www.beintrend.ru/2011-06-20-17-05-06 (reference date 07/09/2013).

[14] Sheremet, A. D., Saifulin R. S. 1998 *Methods of Financial Analysis Companies*. (Moscow: Case) 320 p.

[15] Davydov, G. V.,.Belikov A. Yu. 1999 Methods of quantitative risk assessment bankruptcy. In: *Risk Management* **3** 13-20

[16] Zaitseva, O. P. 1998 Crisis management in the Russian company. In: *Siberian Financial School*. **11-12** 66-73

[17] Makarieva V. I., Andreeva L. V. 2004 *Analysis of Financial and Economic Activity of the Enterprise* (Moscow: Finance and Statistics) 264 p

[18] Telipenko, E. V., Zakharova A. A. 2015 Problems of forecasting bankruptcy risk engineering plants. In: *Innovative Technology and Economics in Engineering: Proceedings of the VI International Scientific and Practical Conference. Yurga Institute of Technology*. Tomsk: Publishing house of Tomsk Polytechnic University. Pp. 262-266.

[19] Mitsel, A. A., Telipenko E. V. 2011 Assessing the impact of financial and economic activity of enterprise revenue from sales of products. In: *The Economic Analysis: Theory and Practice* **27 (234)** 57-64

[20] Krotov, M. A. et al. 1990 *Fundamentals of Optimal Control Theory* (Moscow: Higher School) 430 p.

[21] Athans, M. 1967 The matrix minimum principle. In: *Information and Control*. **11** 592-606

[22] JSC "KuibyshevAzot" [Electronic resource]. Access mode: http://www.kuazot.ru/ (reference date 14/03/2016).

[23] *The Content of the Financial Statements: Balance Sheet and Profit and Loss* [electronic resource]. Access mode: http://www.profiz.ru/peo/12_2011/Sod_fin_otch_balans_i_otc/ (reference date 14/03/2016).

[24] *StatSoft, Electronic Textbook on Statistics. The Main Components and Factor Analysis* [electronic resource]. Access mode: http://www.statsoft.ru/home/textbook/modules/stfacan.html/ (reference date 14/03/2016).

Mathematical simulation of the optical system of a fiber-optic measuring micro motion converter with a cylindrical lens modulation element

T I Murashkina, A.V. Motin, E A Badeeva

Penza state university, Penza, Russia

e-mail: timurashkina@mail.ru

Abstract. The paper presents the results of mathematical modeling to determine the physical, structural and technological parameters of differential fiber-optic micro motion converters with a cylindrical lens, which are basic elements of the technical solutions for fiber-optic sensors of various physical quantities used in the industry in automated control systems.

Introduction. The widespread adoption of fiber-optic technologies in the industry requires the development of constructive and technological solutions of fiber-optic measuring instruments based on the application of modern technologies, materials, ensuring their high metrological and operational characteristics, low cost, high technology design, operation in harsh environments, the minimum weight-and-dimensional characteristics.

The basis of the new technical solutions of fiber-optic sensors of various physical quantities is the differential fiber optic micro motion converter with a cylindrical lens, simultaneously fulfilling the functions of optical modulation and a focusing element, and even the function of the inertial mass – in acceleration sensors [1]. The cylindrical lens is simple to manufacture and can be easily attached to the sensor housing. Besides, it significantly reduces the loss of light in the measuring zone.

Fiber optic micro motion converters are characterized by a lack of influence on the measurement result by electromagnetic interference radiation, mechanical factors, high temperature (increased to 200 °C (or higher) and a physical quantity measured on the optical fiber, which improves the reliability of the design.

Mathematic simulation of the optical system for fiber-optic accelerometers with a cylindrical lens modulation element. As a result of a plurality of reflections within the optical fiber, there is symmetrization of a ray beam with respect to its optical axis and averaging of brightness over light radiating end of the fiber [2]. The narrow conical bundle of rays, incident at an angle at the input end of the optical fiber end faces with straight lines, fills the outlet zone bounded by two coaxial surfaces close, so the cross section is perpendicular to the optical axis, which forms a ring zone. This makes it possible with further reforms of the optical signal to divide it in a constructive way in the transverse direction of the beam into two separate two-measurement channels of a micro motion differential converter with a cylindrical lens and implement a differential conversion circuit [3].

Via the supply optical fiber, the luminous flux is sent from the radiation source to a cylindrical lens arrangement area (Figure 1). Modulation of the optical signal in this case is carried out by changing the curvature of the "gas-glass" media boundaries in upper and lower portions of the cylindrical lens as it moves under the influence of the physical quantity measured.

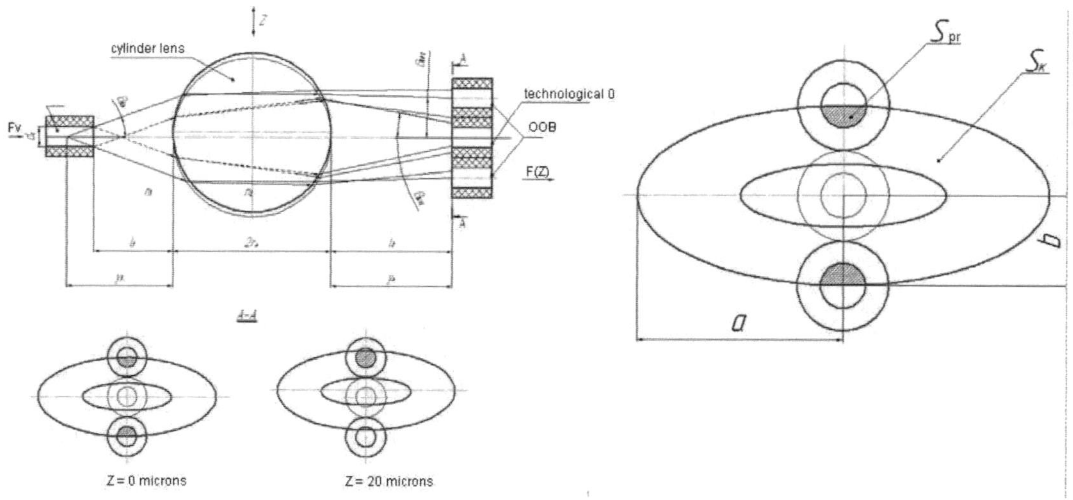

Figure 1. A settlement and a structural scheme of the differential fiber-optic measuring micro motion converter with a cylindrical lens

It undergoes a spatial redistribution in the body of the lens.

To determine the progress of optical beams in the developed mathematical model of the optical system of a differential fiber-optic measuring micro motion converter, we introduced the Cartesian coordinate system: 1) for the surface of the light flux output from supply optical fiber $\{x, y, z\}$,2) for optical modulating element – cylindrical lens $\{x', y', z'\}$ (Figure 2).

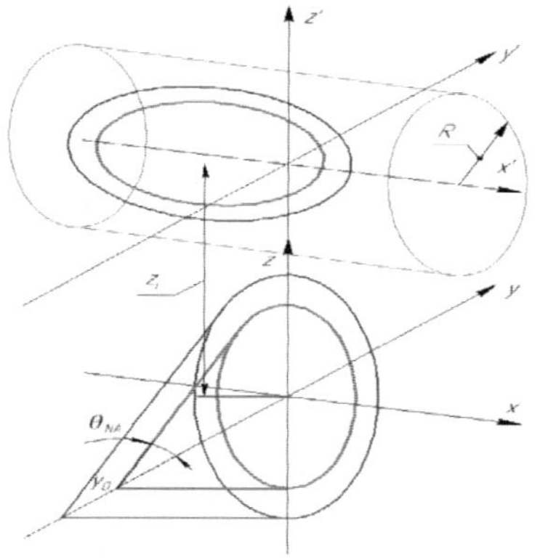

Figure 2. Calculation of optical beams in the fiber-optic measuring micro motion converter with a cylindrical lens

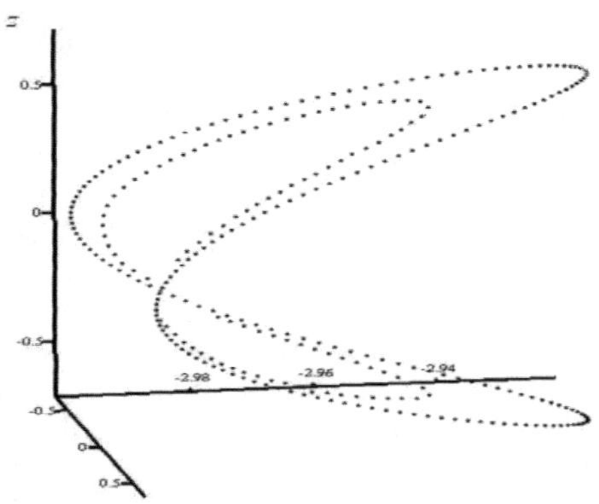

Figure 3. A spatial intersection curve of the luminous flux generated on the supply optical fiber with a cylindrical lens

IOP Conf. Series: Journal of Physics: Conf. Series **803** (2017) 012101 doi:10.1088/1742-6596/803/1/012101

Then the spatial intersection curve of the luminous flux, which is generated at the output of the supply optic fiber (light cone) with cylindrical lens, can be expressed by a one-parameter system in the polar coordinate system (figure 3):

$$\begin{cases} x(\eta) = r(\eta)\cos(\eta), \\ y(\eta) = (-1)^k \sqrt{R^2 - (x(\eta) - x_i)^2}, k = 1,2, \\ z(\eta) = r(\eta)\sin(\eta) \end{cases} \quad (1)$$

where $\eta \in [0, \pi]$, k, it makes sense to order a spatial intersection of the curve of the light cone with a cylindrical lens ($k = 1$ – intersection curve cone on the cylinder inlet, $k = 2$ – on the cylinder outlet);

$$r(\eta) = \frac{ay_0 + z_i \sin(\eta) + (-1)^k \sqrt{(ay_0 + z_i \sin(\eta))^2 - \alpha^2 + \sin^2(\eta)\left((z_i)^2 + y_0 - R^2\right)}}{\alpha^2 + \sin^2(\eta)} \quad (2)$$

where $\alpha = \tan(\theta_{NA})$, θ_{NA} – half angle at the vertex of the cone (in radians), y_0 – coordinate of the top of the light cone, z_i – displacement along the axis of cylindrical lens z, R – radius of the cylindrical lens.

On the basis of Snell's law and vector product $\vec{e}_\tau = [\vec{e}_n \times \vec{r}_n]$, the vector of normal to cylinder surface ($\vec{e}_n$), and the vector, directed from the point of intersection of the light cone with the cylinder to the point with coordinates $(0; -y_0; 0)$ – ($\vec{r}_n$), determined the angle of the refracted beam after the light beam from the outside into the cylindrical lens:

$$\beta = \arcsin\left(\frac{n_1}{n_2}\left|\vec{e}_\tau\right|\right) = \arcsin\left(\frac{n_1}{n_2}\sin\left(\vec{e}_n\, \vec{r}_n\right)\right) = \arcsin\left(\frac{n_1}{n_2}\sin\left(\left[\left|\vec{e}_\tau\right|\right]\right)\right), \quad (3)$$

where n_1 - working environment refractive index; n_2 – refractive index of the lens body.

Mathematical modeling was performed in the MatLab software. It was revealed that in the plane of output optical fibers of first and second measurement channels the light spot has the form of a solid ellipse, if the radiation source is laser (figure 4*a*), and a ring shape, formed by two ellipses if the light source is a light-emitting diode (figure 4*b*).

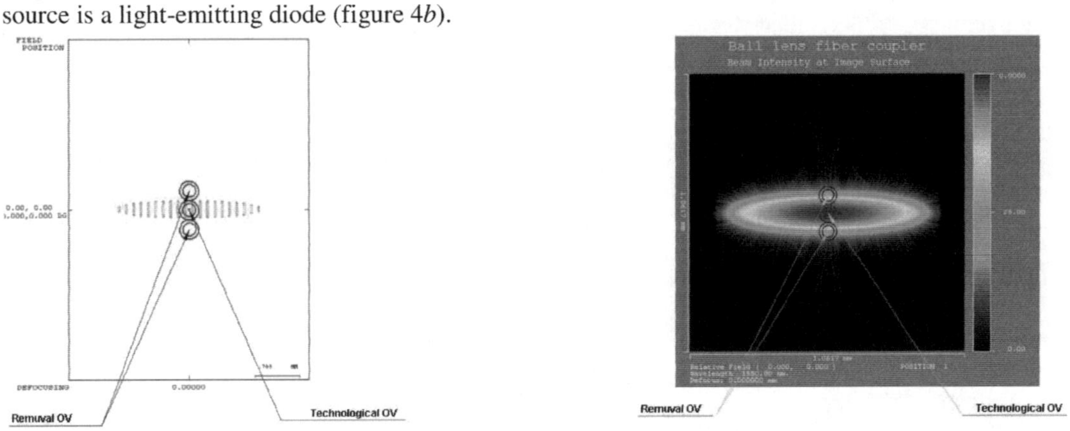

a) The diagram of the light spot when the beam of light is in the form of a solid cone (laser)

b) The diagram of the light spot when the beam of light is in the form of a hollow cone (LED)

Figure 4. The results of mathematical modeling of the differential fiber-optic measuring micro motion converter of the optical system

Light from the LED spot allows splitting it into two parts of two measurement channels, which are arranged vertically one above the other output optical fiber of the first and second measuring channels, and implementing a differential conversion of the luminous flux in the measurement area. In the center of the spot is the cut fiber, which does not perform any conversion of the luminous flux.

The transform functions of optical channels of the first and second differential measurement channels of the fiber optic accelerometer with a cylindrical lens are defined:

$$\Phi(z_i)_{\substack{1chanel\\2chanel}} = \Phi_0 \frac{2\int_0^{x_0} a_B{}^2 \sqrt{1-\frac{x_0^2}{b_B{}^2}} - \left(\sqrt{r_{cent}^2 - x_0^2} \mp p_1 z_i \pm p_2\right) dz_i}{\pi a_B b_B - \pi a_M b_M}, \tag{4}$$

where

$$a = tg\Theta_{BX2}\left[\frac{r_{cent}\sin\gamma_2}{\sin\Theta_{BX2}} - \frac{r_{cent}\sin\gamma_1}{\sin\Theta_{BX1}}\right], \qquad b = \Theta_{NA}\left[\frac{\cos\Theta_{NA}(2d_c + r_{cent}tg\Theta_{NA} \pm Z)}{\sin\Theta_{BX1}} - r_{cent}\right],$$

$$\Theta_{BX1} = 2\left[\arcsin\left(\frac{\cos\Theta_{NA}(2d_c + r_{cent}tg\Theta_{NA} \pm Z)}{r_{cent}}\right) - \arcsin\left(\frac{\cos\Theta_{NA}(2d_c + r_{cent}tg\Theta_{NA} \pm Z)}{r_{cent}}\frac{n_C}{n_{cent}}\right)\right],$$

$$\Theta_{BX2} = 2\left[\arcsin\left(\frac{\cos\Theta_{NA}(d_c + r_{cent}tg\Theta_{NA} \pm Z)}{r_{cent}}\right) - \arcsin\left(\frac{\cos\Theta_{NA}(d_c + r_{cent}tg\Theta_{NA} \pm Z)}{r_{cent}}\frac{n_C}{n_{cent}}\right)\right],$$

$$\gamma_1 = \arcsin\left(\frac{\cos\Theta_{NA}(2d_c + r_{cent}tg\Theta_{NA} \pm Z)}{r_{cent}}\right), \qquad \gamma_2 = \arcsin\left(\frac{\cos\Theta_{NA}(d_c + r_{cent}tg\Theta_{NA} \pm Z)}{r_{cent}}\right),$$

$$\gamma_1 = \arcsin\left(\frac{\cos\Theta_{NA}(2d_c + r_{cent}tg\Theta_{NA} \pm Z)}{r_{cent}}\right), \qquad \gamma_2 = \arcsin\left(\frac{\cos\Theta_{NA}(d_c + r_{cent}tg\Theta_{NA} \pm Z)}{r_{cent}}\right),$$

where a_B, a_M – a major semi-axe of the external and internal ellipse, respectively; b_B, b_M – a minor semi-axe of the external and internal ellipse, respectively; Θ_{BX1}, Θ_{BX2} - angles of incidence of the outside and inside boundary of the luminous flux at the entrance of the output optical fiber, respectively; γ_1 , γ_2 - beam angles to the normal on the outer and inner boundary after the output of the light flux from the cylindrical lens, respectively; Θ_{NA} - the aperture angle at the output from the input optic fiber; d_c - the optical fiber core diameter; r_{cent} - the cylindrical lens radius; Z - cylindrical lens displacement along the Z axis.

A symbol «+» in the double system of signs refers to the first measuring channel, and a «-» sign – to the second measuring channel when moving the cylindrical lens upward relative to the equilibrium position.

Mathematical modeling determines the shape of the light spot in the plane output optic fiber and the intensity of changes in the function output at the optic modulation element at various distances l_1 from the input optic fiber end to the cylindrical surface of the lens and l_2 - from the cylindrical lens surface to the receiving ends of the output optic fiber, and cylindrical lens radius r_{cent} changing within some range, the movement along the Z axis was also simulated [4]. Distances l_1 and l_2 are determined in the following formulas:

$$\frac{1{,}5d_c}{tg\Theta_{NA}} \le l_1 \le \frac{r_{cent}}{tg\Theta_{NA}} - \frac{0{,}5d_C}{tg\Theta_{NA}} - r_{cent}; \qquad l_2 = \frac{\cos\Theta_{NA}(2d_c + r_{cent}tg\Theta_{NA})}{\sin\Theta_{BX1}} - r_{cent}. \tag{5}$$

It was determined that distance l_2 is associated with the core diameter of optical fiber d_C by ratio: $l_2=7.5dc$.

Figure 5 illustrates the results of calculation of refracted light cones at displacement of the cylindrical axis of the lens up to 0.01 and 0.02 mm (z_i).

Analysis of the simulation results allows selecting optimal parameters for the fiber-optic measuring micro motion converter, based on the conditions of maximum sensitivity (1...2 mW/micron, and 0.5…0.6 milliwatt/micron,– of counterparts' sensitivity) and minimum linearity error (no more than 0.07 %, and 0.1 % – of counterparts' sensitivity).

For the optical fiber with parameters d_c=0.2 mm, Θ_{NA}=12, degrees transfer of the maximum possible output power of an optical signal conversion zone is achieved with parameters r_{cent}=1.5 mm, l_l=0.5 mm, l_2=1.5 mm (figure 2c). With any other critical parameters, the loss of the luminous flux or luminous power distribution unevenness is observed.

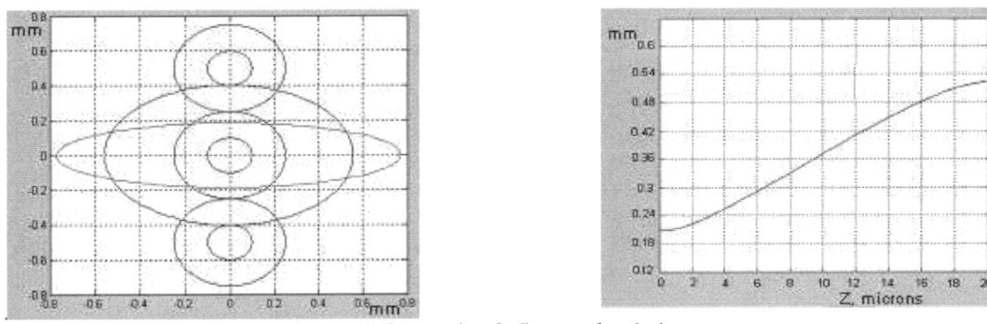

a) r_{cent}=1.5 mm; l_1=0.5mm; l_2=1.1mm

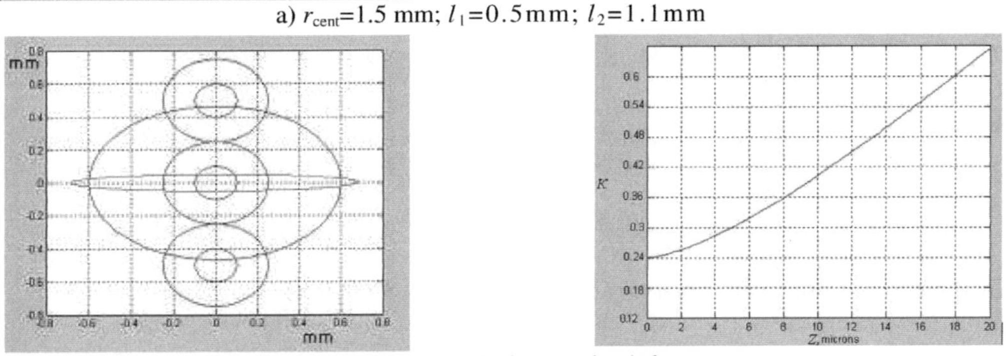

b) r_{cent}=1.5 mm; l_1=0.5 mm; l_2=1.3 mm

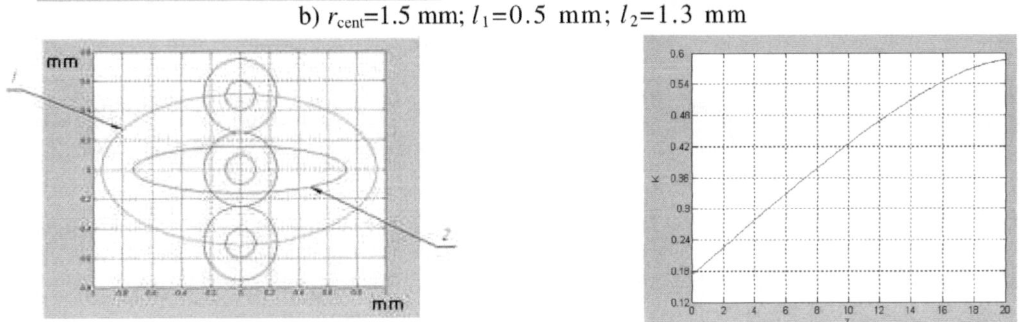

1 - projection of an internal cone
2 - projection of an external cone

c) r_{cent}=1.5 mm; l_1=0.5 mm; l_2=1.5 mm

Figure 5. The results of mathematical modeling of the optical system in the MatLab software (fragment)

Conclusion

Analysis of the dependency has shown that the maximum possible transfer of the light flux into the optical signal radiation conversion zone of the fiber-optic measuring micro motion converter with the cylindrical lens is achieved for the following parameters of the optical system (figure 5, *c*): cylindrical lens radius r_{cent}=1,5 mm, the distance between the end of the input optic fiber and the cylindrical lens surface is l_1=0,5 mm, the distance between the rear surface of the lens and the end surface of output optic fiber is l_2=1,5 mm. There can be other parameters of critical loss of the luminous flux, or uneven distribution of the light output. To achieve the minimum linearity error, it is necessary to limit the range of lens movement from 0 to 16 microns.

Acknowledgement
The study was financially supported by RFBR in the form of grant №15-08-02675

References

[1] Shhevelev A. S., Bazykin S. N., Murashkina T. I. 2010 *Reliability and quality.* (Penza : Izd-vo PGU) **1** 498–500

[2] Badeeva E.A., Kolomiec L.N., Krivulin N.P., Murashkina T.I., Pivkin A.G. 2008 *Development of the light stream theory* (Penza : Izd-vo PGU) p. 85

[3] A.V. Gorish, E.A. Gorish, O.S. Grayevsky, D.A. Kryuchkov, Yu.N. Makarov 2010 The problems of sensors construction for physical parameters measurement are considered. The description of sensors sensitive elements transformation is presented *Metrology* **10** 3-14

[4] Kondratov V. T. and Redko V. V. 2009 Fiber-optic pressure sensors with amplitude modulation of the optical signal: basic definitions and classification *Journal of Sensors and systems* **8** 78-85

[5] Busarin V., Nosov Yu. R. Fiber 1990 *Optic sensors: Physical bases, calculation and application.* (M.: Energoatomizdat) p. 256

[6] Brosilow S. A. 2011 Differential fiber-optic sensor of acceleration. Design and technology *Reliability and quality: Tr. Intern. Symp.* (Penza: Izd-vo PGU) **1** 235-237

[7] Brosilow, S. A. 2013 Embodiments of fiber-optical converters of micromovements reflective type *Proceedings of the international Symposium "Reliability and quality* (Penza: Izd-vo PGU) **2** 13

[8] Shevelev A. S., Krivulin N. P., Volkov V. S. 2009 Modeling of the distribution of light intensity in the fiber-optic displacement transducer with a control in the form of spherical lens *Devices and systems. Management. Control. Diagnosis.* **12** 16-18

An optoelectronic fuel level sensor

T I Murashkina, E A Badeeva, A V Badeev, M M Savochkina

Penza state university, Penza, Russia

e-mail: timurashkina@mail.ru

Abstract. The block and schematic construction diagrams of a new optoelectronic fuel level sensor are considered. The operating principle of the sensor is based on registering the intensity value of the optical path reflected from the mirror, located on the reservoir bottom.

Introduction. Optical techniques are widely used to measure fluid flow parameters and less frequently used to measure the liquids level, the fuel level, in particular. The reason is that the impact on measurement accuracy is caused by the change of radiation source (RS) power, as well as total sensitivity and internal resistance of radiation detectors (RD) due to temperature [1, 2].

Materials and methods

To solve the assigned task, 1) the optical path, which converges at the calculated angle relatively the optical axis of the optoelectronic fuel level sensor (OEFLS), forms; the optical path proceeds a part of the track to the mirror through the medium with refraction coefficient n0, and the second part of the track – in the medium, the level of which is measured, with refraction coefficient n1>n0, thus the optical path reflected by the mirror is registered; 2) the gain-phase method of signal transformation is used [4, 5].

A block diagram and design of the sensor. Paper [3] introduces a new technique to measure the fuel level by recording the intensity value of the optical path reflected from the mirror on the reservoir bottom or another level of the reservoir.

The optical path, converging by the calculated optic axial angle, is formed, one part of the optical path towards the mirror is going through the medium with refraction index n0, and the other part – through the medium with refraction index n1>n0, its level being measured and the optical path being reflected from the mirror being registered. In order to implement the given algorithm of optical signals conversion, the block diagram (Figure 1) and construction design for the optoelectronic fuel level sensor (OEFLS) (Figure 2) have been developed.

Body 1 contains tightened holder 2 with four rolled in lenses 6. Light emitted diodes (LEDs) 3 are mounted in these lenses focus. Operating photodiode 5 is set up in the centre of holder 2, and lens 7 is rolled in closer than the focus distance. Compensating photodiodes 4 are mounted in close proximity to LEDs 3. The axes of LEDs 3 and photodiodes 4 are mutually perpendicular. Photodiodes 4 are supposed to compensate the change of LEDs 3 radiation power under the ambient temperature shift.

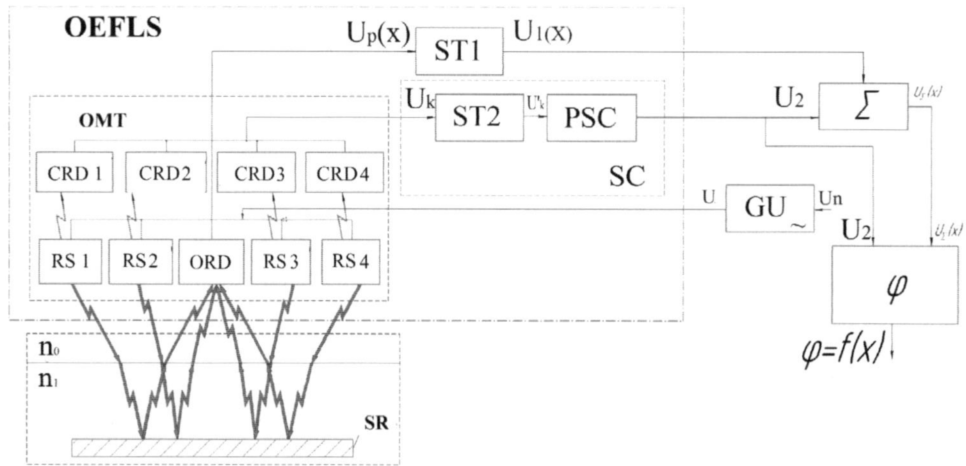

ORD – operating radiation detector; RS – radiation sources; CRD – compensating radiation detector; ST – scaling transducer; GU – generating unit; PSC – phase-shifting circuit; PI – phase indicator; Σ – adder; SC – scaling contour; OMT – optoelectronic measuring transducer; SR – specular reflector.

Figure 1. The block diagram of the optoelectronic fuel level sensor

Holder 2 is mounted in body 1 and tightened by first nut 9. PC-board 8 is tightened inside body 1 with the help of second nut 9. PC-board 8 contains integrated circuit 544УД1А, acting as a scaling transducer and a phase-shifting circuit. The sensor is connected to an intermediate transducer through the cable and plug 15. Plug 15 is inserted in sleeve 14, which, in turn, is sealed into body 1. Prior to mounting collar 14 with plug 15, the sensor's body cavity is first treated under vacuum and then filled up with premium grade argon (National State Standard GOST 10157-79). Glass 12 is mounted at the butt end of body 1 and is sealed against O-ring 11 to guarantee the structural integrity. O-ring 11 is tightened up with nut 13 to stay fixed in body 1. Nuts 10 are used to mount the sensor in a reservoir. Mirror 16 is a polished metal plate.

In order to ensure the sensor's high reliability while in operation, structural and assembly parts are to be fastened by welding and soldering.

OEFLS operates as follows: luminous fluxes out of LEDs 3 go through lenses 6 and at an angle of α, they first go through air to the sensor optical axis, and then - through the liquid, which level is being measured. When the luminous flux reaches the interface of two media with different refraction indexes n0 and n1, they refract and then expand at an angle of β in the direction of mirror 16. Being reflected from the mirror, luminous fluxes make a return path and, having gone through lens 7, are focused upon the photosensitive surface of photodiode 5.

Transformation function. Luminous fluxes F_{rsi} of radiation sources RS1…RS4 are modulated by sine-curve voltage of low frequency generator unit GU~, and then, by means of the optical system they are transferred into a measurement area onto specular reflector SR. Luminous fluxes cover one part of path h_i in the air, and the other part x_i - in the liquid, which level is being measured. Luminous fluxes, reflected from the mirror, reach the operating radiation detector (ORD), where they are converted into electric signal $U_P(X)$. Some parts of luminous fluxes of RS*i* reach compensating radiation detectors CRD1, …, CRD*i* directly, where they are converted into constant electric signals $U_{CRD1}…U_{CRDi}$, which are summarized in the join point. Signal $U_P(X)$ and combined compensating signal U_C are scaled to levels $U_1(X)$ and U_2 at scaling transducers ST1 and ST2, respectively, and then signals are shifted at an angle of φ_{12} relative to each other with the help of the phase-shifting circuit (PSC). Initially signal U_2 is shifted at an angle of 180 degrees relative to signal $U_1(X)$ (Figure 3). Signals $U_1(X)$ and U_2 reach adder inputs, where they are summarized geometrically. Signal $U_\Sigma(X)$ goes from the adder output onto one of the inputs of phase indicator PI, the other input of which receives a reference signal: either $U_{\sim}$, or $U_1(X)$, or U_2.

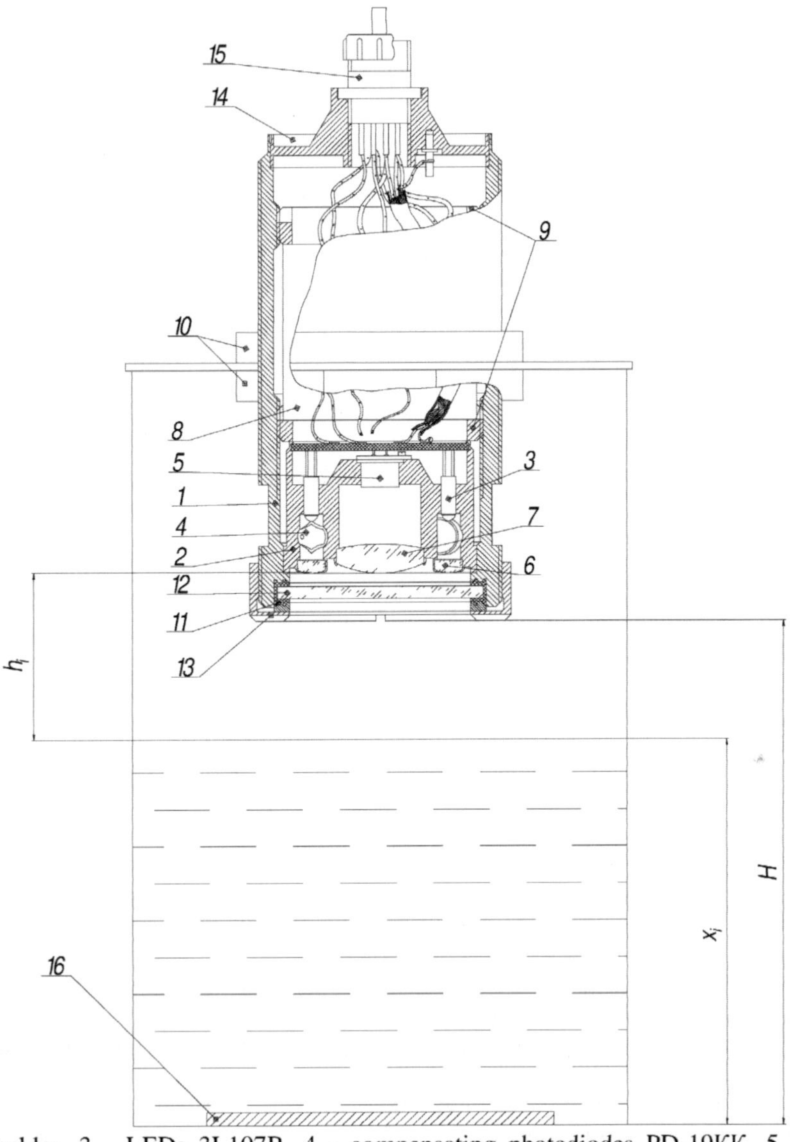

1- body, 2 - holder, 3 - LEDs 3L107B, 4 - compensating photodiodes PD-19KK, 5 - operating photodiode PD20-32K, 6, 7– lenses of a smaller and bigger diameter, 8 - printed-circuit board, 9 – inner nuts, 10 – mounting nuts, 11 - O-ring seal, 12 - safety glass, 13 - nut, 14 - collar, 15 - plug ST1-10-5-V, 16 – mirror.

Figure 2. The Construction Design for the Optoelectronic Fuel Level Sensor

The conversion function of OEFLS is written as:

$$\varphi = arctg\frac{\sin\ _{12}}{\cos\ _{12} + \frac{R_{ORD}(XX)_{V1}R_{ST1}}{R_{ST2}R_{PSC}\sum_{i=1}^{4}R_{CRDi}S_{V2}}}, \qquad (1)$$

where $R_{ORD}(X)$, R_{CRD} - conversion ratios of luminous flux for ORD and CRD; R_{ST1} and R_{ST2} – shrinkage ratios of ST1 and ST2, respectively; R_{PSC} – transfer ratio of PSC; S_{V1}, S_{V2} - voltage sensitivities of ORD and CRD, respectively.

Basic elements responsible for sensor temperature error *t* are RS, RD and measuring circuit:

$$\Delta\ _t = \sqrt{(\Delta\ ^t_P)^2 + (\Delta\ ^t_S)^2 + (\Delta\ ^t_{ИС})^2}, \qquad (2)$$

where $\Delta\varphi^t_P, \Delta\varphi^t_S, \Delta\varphi^t_{ИС}$ - temperature errors, caused by the change in the power of radiation source (RS), total sensitivity of radiation detectors (RD), and parameters of the measuring circuit (MC).

When measuring the ambient temperature, the radiation power of RS and the total sensitivity of RD change.

Then:

$$\Delta\varphi_t = \sqrt{(\Delta\varphi^t_P)^2 + (\Delta\varphi^t_S)^2 + (\Delta\varphi^t_{ИС})^2}, \qquad (3)$$

where $\Delta\varphi^t_P, \Delta\varphi^t_S$ - measurement errors conditioned by the change in radiation power of RS and total sensitivity of RD, respectively.

In order to reduce the temperature error, it is necessary that $\Delta\varphi^t_P = \Delta\varphi^t_S$, that is, the changes of phase φ of sum signal UΣ(X), caused by the change in the radiation power of RS and by the change in photosensitivity of RD, mutually compensated. This is possible in case of the following equality:

$$arctg\frac{\sin\varphi_{12}}{\cos\varphi_{12} + R\frac{S_{V1}}{S_{V2}}f(\Delta F_0)R_{ORD}(X).} = arctg\frac{\sin\varphi_{12}}{\cos\varphi_{12} + R\frac{R_{1S}}{K_{2S}}\frac{S_{V1}}{S_{V2}}R_{ORD}(X).} \qquad (4)$$

where f(ΔF) – some nonlinear function depending on the luminous flux of RS, where ΔF is additional luminous flux of RS conditioned by the change in the radiation power of RS. Hence, the following equality should be fulfilled for full temperature compensation:

$$R1S/R2S = -f(\Delta F).$$

The stated equality is possible in case the radiation power of RS and the ratio of RD photosensitivities change will vary in opposite directions along with the ambient temperature change. For instance, when the ambient temperature rises, the RS power will increase, while the ratio of RD will decrease. It is possible to obtain a similar equality by changing the ratio of the number of ORD and CRD, as well as by changing the CRD position relatively RS (for example, bringing it closer to RS or, otherwise, moving away).

If the level indicator is built according to the scheme, shown in Figure 1, to obtain full temperature compensation, it is necessary to fulfill condition 5:

$$arctg\frac{\sin\varphi_{12}}{\cos\varphi_{12} + R\frac{S_{V1}}{\sum_{i=1}^{4}R_{CRDi}S_{V2}}f(\Delta F_0)R_{ORD}(X)} = arctg\frac{\sin\varphi_{12}}{\cos\varphi_{12} + R\frac{R_{1S}}{R_{2S}}\frac{S_{V1}}{\sum_{i=1}^{4}R_{CRDi}S_{V2}}R_{ORD}(X)}. \qquad (5)$$

The studies have demonstrated that in order to reduce the value of the error caused by ratios RST1, RST2, RPSC change, it is necessary to arrange parts ST2 and PSC3 out of the disturbing factors range;

to arrange ST1 in the OEFLS housing, as it serves as amplifier UP(X) of a small-signal, the transmission of which via the electrical cable is impossible without preamplification.

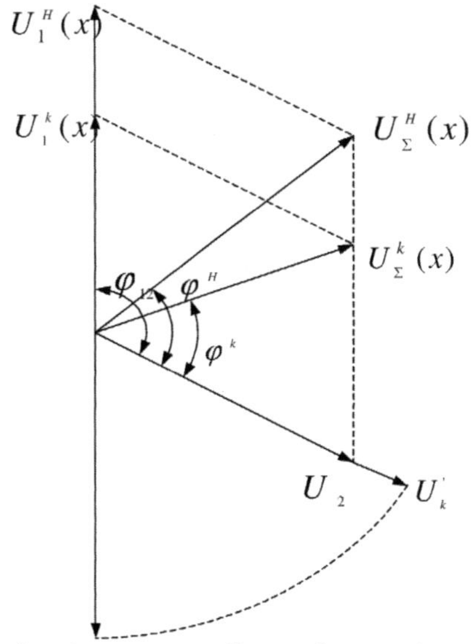

index «H» refers to the signals corresponding to the opening of the measurement range, index «K» refers to the signals corresponding to the ending of the measurement range.

Figure 3. A vector diagram of OEFLS signals.

The study has defined that to reduce the temperature error caused by S_{V1}, S_{V2} nonproportional changes under the ambient temperature shift, it is necessary to use single-type radiation detectors with the same point of sensitivity deviation, which leads to partial compensation of the temperature error.

4. Conclusion

The reduction of temperature error of OEFLS can be achieved by:

- application of amplitude-phase transformation of signals;
- introduction of additional compensating channel;
- using ORD and CRD of a single type;
- changing of the quantitative ratio of RS in RD;
- changing of the ORD position relatively RS.

New OEFLS will have the temperature error which is no more than 1…2 %.

Acknowledgments

This research was supported by the Russian Fund of Basic Research (grant No. 15-08-02675).

References

[1] Badeev A.V., Meshcheryakov V.A, Murashkina T.I. 2003 New method of liquid level measurement *Sensors and systems* **7** 50-52

[2] Grayevsky O. S. 2010 Technological ways of improvement of metrological characteristics of optoelectronic level gages of liquid environments: *thesis... Candidate of Technical Sciences* (Penza : Izd-vo PGU) p.197

[3] Murashkina T. I., Badeev A. V. 2003 Optoelectronic sensor of movements *Sensors and systems* **5** 30 – 32

[4] Chigvintsev S. V. 2005 Phase-shifting sensors of parameters of the movement for control systems of shtangovy deep pump installations *The Oil and gas case* **3** 139-152

[5] Chigvintsev S. V., Chigvintsev I. S. 2012 Amplitude-phase measuring converters of movement for sensors of parameters of the movement *The Oil and gas case* **3** 96-101

Implementation of the Algorithm for Congestion control in the Dynamic Circuit Network (DCN)

H S Nalamwar[1], M A Ivanov[1] and G U Buddhawar [2]

[1] Tomsk Polytechnic University, 30, Lenina Avenue, Tomsk, 634050, Russia
[2] NCET, Mouza Bodli, Dhanora Road, Gadchiroli, MS, 442605, India

E-mail: hitesh@tpu.ru

Abstract. Transport Control Protocol (TCP) incast congestion happens when a number of senders work in parallel with the same server where the high bandwidth and low latency network problem occurs. For many data center network applications such as a search engine, heavy traffic is present on such a server. Incast congestion degrades the entire performance as packets are lost at a server side due to buffer overflow, and as a result, the response time becomes longer. In this work, we focus on TCP throughput, round-trip time (RTT), receive window and retransmission. Our method is based on the proactive adjust of the TCP receive window before the packet loss occurs. We aim to avoid the wastage of the bandwidth by adjusting its size as per the number of packets. To avoid the packet loss, the ICTCP algorithm has been implemented in the data center network (ToR).

1. Introduction

In this work, our discussion is mainly focused on the congestion problem that usually occurs in incast, a condition which is the opposite of broadcast. In broadcast, one node sends messages to multiple nodes; in incast - on the contrary - multiple nodes send messages to one and the same node. The TCP (Transport Control Protocol) is a reliable protocol and has been widely used in the Internet. The congestion happens when many synchronized servers work under the same switch and TCP is not very reliable for high-bandwidth and low latency networks. Previous solutions focused on either reducing the wait time for packet loss recovery with faster retransmissions [1, 2, 3], or controlling switch buffer occupation to avoid overflow by using ECN and modified TCP on both the sender and receiver sides [4, 5]. This report focuses on avoiding the packet loss before incast congestion, which seems more appealing. Our idea is to perform the incast congestion avoidance at the receiver side by preventing incast congestion. The receiver side is a natural choice since it knows the throughput of all TCP connections and the available bandwidth. The receiver side can adjust the receive window size of each TCP connection so that the aggregate bursts of all the synchronized senders are kept under control. We suggest calling this design Incast congestion Control for TCP (ICCTCP). However, adequately controlling the receiver window is challenging. The receiver window should be small enough to avoid incast congestion, but also large enough for good performance and other non-incast cases. The technical novelties of this work are as follows:

1) To perform congestion control on the receiver side, we use the available bandwidth on the network interface as a quota to coordinate the receiver window increases of all incoming connections.

2) Our per flow congestion control is performed independently of the slotted time of RTT of each

connection, which is also the control latency in its feedback loop.

3) Our receiver window adjustment is based on the ratio of the difference between the measured and expected throughput over the expected.

The live RTT is necessary for throughput estimation as we have observed that TCP RTT in a high bandwidth of low latency network increases with throughput, even if the link capacity is not reached. TCP incast has been identified and described in a distributed storage cluster [5]. In distributed file systems, the files are deliberately stored in multiple servers. However, TCP incast congestion occurs when multiple blocks of a file are fetched from multiple servers at the same time. Several applications of specific solutions have been proposed in the context of parallel file systems.

2. TCP Incast Congestion

Fig. 1 shows a typical data-center network structure. There are three layers of switches/routers, in particular the ToR switch, the Aggregate switch, and the Aggregate router. The figure also illustrates a detailed case for a ToR switch connected to dozens of servers.

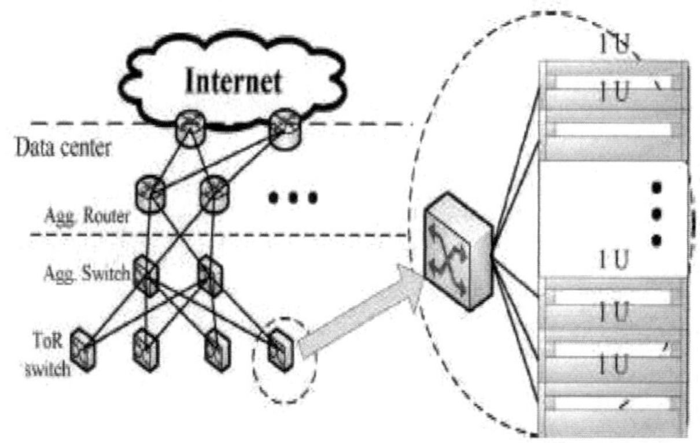

Figure 1. Data-center networks and a detailed illustration of a ToR switch connected to multiple rack-mounted servers.

Incast congestion happens when multiple sending servers under the same ToR switch send data to one receiver server simultaneously. This process is demonstrated in Fig. 2.

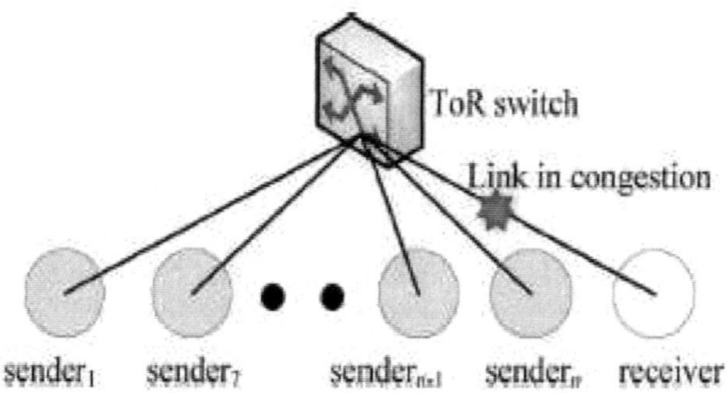

Figure 2. The scenario of incast congestion in data-center networks, where multiple TCP senders transmit data to the same receiver under the same ToR switch.

The amount of data transmitted by each connection is relatively small, e.g. 64kb. The term goodput is effective throughput obtained and observed at the application layer. The multiple TCP connections are barrier-synchronized. First, these connections are established between all senders and the receiver, respectively. Then, the receiver sends out a (very small) request packet to ask each sender to transmit data, that is, multiple requests packets are sent using multiple threads. The TCP connections are issued round by round, and one round ends when all connections on that round have finished their data transfer to the receiver. Here, we can observe similar goodput trends for three different traffic amounts per server, but with slightly different transition points.

3. Motivation

As we have mentioned, in broadcast, one node sends messages to multiple nodes, while in incast the process is quite opposite, that is multiple nodes send messages to one and the same node. When a number of servers work in parallel with a single server, network congestion occurs. Network congestion is the situation in which an increase in data transmission results in proportionately smaller increase, or even a reduction, in throughput. Congestion degrades the performance of the network. Packet loss is the major reason for the congestion.

Packets are having two numbers that are associated with each packet.

1. A sequence number indicates the ID of a packet.

2. An acknowledgement number (ACK) indicates the expected acknowledge (ACK) number of the packet that is associated with it.

4. TCP Timeout

Timeout happens when the sender does not receive an ACK for a long period (retransmission or RTO). There are several reasons for the timeout due to congestion given below:

a) In distributed file systems, as the files are deliberately stored in multiple servers at the time of fetching, congestion occurs.

b) With the recent progress in data center networking, TCP incast problems in the data center network have become a practical issue.

c) TCP throughput is severely degraded by incast congestion since one or more TCP connections can experience timeout caused by packet drops.

d) TCP variants sometimes improve performance, but fail to prevent incast congestion collapse since most of the timeouts are caused by buffer overflow.

Based on the analysis of the characteristics of the data center network, the common pattern and the TCP congestion control algorithm, the results of TCP incast are as follows:

1. Since the top of rack switches is shallow buffered, highly bursty, fast and simultaneous, data transmissions overfill the switch buffer to make packet losses.

2. Mass Packet losses will lead to TCP congestion control, reducing the sending window by half and decreasing the sending rate.

3. The intense packet loss results in TCP timeouts. The TCP timeout lasts for 100 milliseconds, but the round trip time of the data center network is around 100 microseconds. Coarse grained RTO reduces the application throughput by 90%.

5. Problem Analysis

The root cause of TCP incast collapse is that the highly bursty traffic of multiple TCP connections overflows the Ethernet switch buffer in a short period of time, causing the intense packet loss resulting in the TCP retransmission and timeouts. In distributed file systems, TCP incast congestion occurs when multiple blocks of a file are fetched from multiple servers, as files are stored at multiple servers. To resolve issue, several application specific solutions have been proposed in the context of the parallel file system. Due to recent progresses on DCN, the TCP incast problem is a common problem. Since there are various applications running in DCN, the best solution is to keep the application free and keep responsibility at a transport layer for building effective solutions. It is a preferred way to avoid congestion and timeout for TCP incast. Our idea is to perform incast congestion avoidance at the receiver side by preventing incast congestion. The receiver side is a natural choice since it knows the throughput of all TCP connections and the available bandwidth. The receiver side can adjust the receiver window size of each TCP connection.

IOP Conf. Series: Journal of Physics: Conf. Series **803** (2017) 012106 doi:10.1088/1742-6596/803/1/012106

6. Working Modules

Here, we have the TOR switch which is a datacenter network.

a. The ToR switch is acting as a data center network.

b. All traffic is managed at the ToR Switch.

c. All the data/Packets from a number of servers are collected at ToR and then transferred to a client.

d. ToR manages all the traffic from different servers and transfers it to a client to avoid the packet loss.

7. Experimental Results and Discussions

Figures 3 and 4 show the results obtained after the implementation of the ICTCP algorithm in the platform.

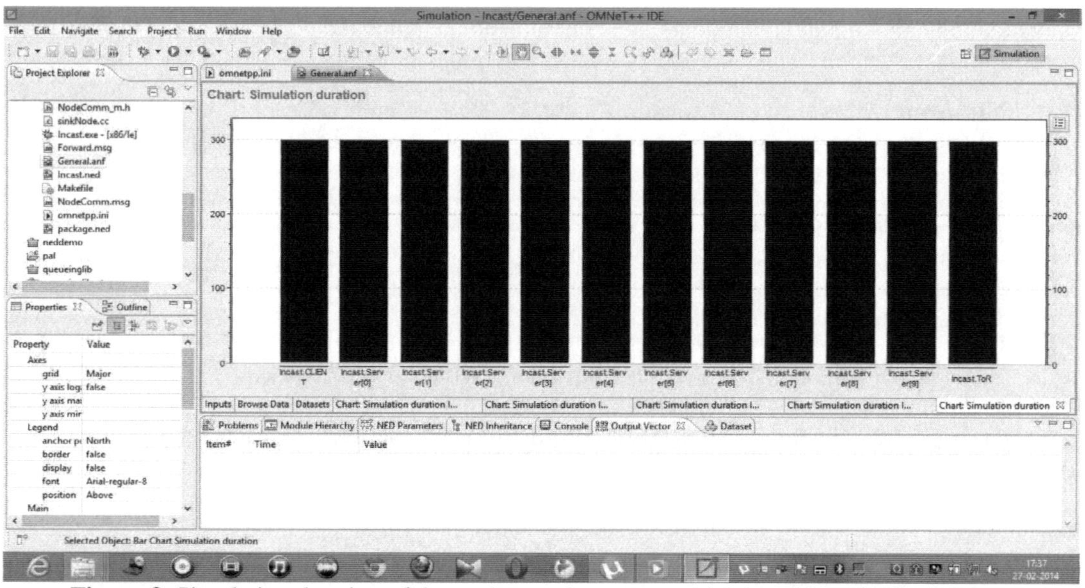

Figure 3. Simulation duration of the packet on the server, which is uniform in the system.

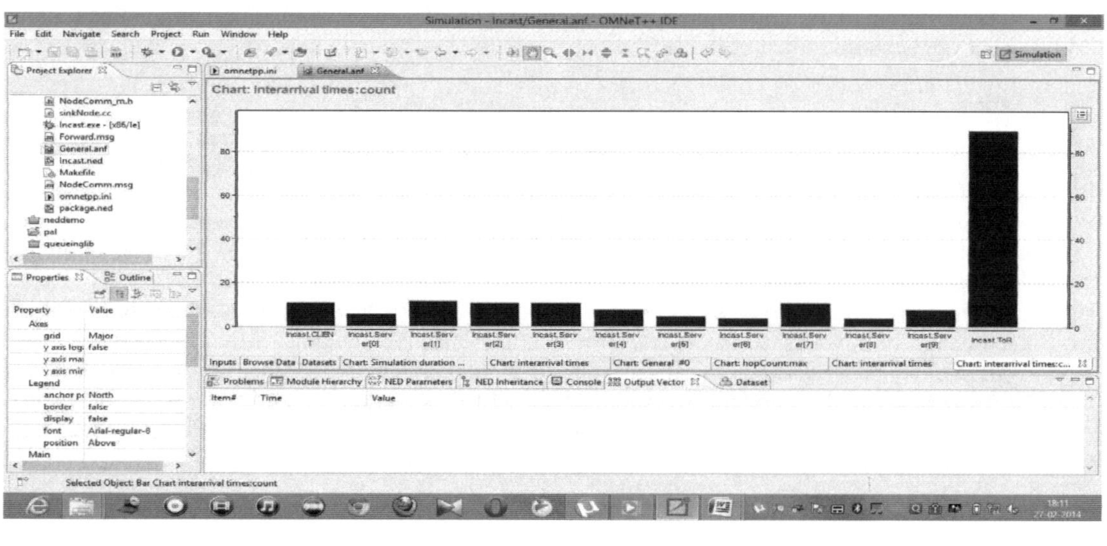

Figure 4. The inter-arrival time representing time between each arrival of the packet on the receiver side

8. Advantages
Table 1 compares the efficiency of the retransmission algorithm to the efficiency of the ICTCP algorithm, with the latter demonstrating obvious advantages

No	Retransmission algorithm	ICTCP algorithm
1	Buffer Overflow Problem	Buffer overflow is impossible
2	Timeout problem	Time out problem is less possible
3	Load increase on the receiver side	Load does not increase on the receiver side
4	Both windows need to be adjusted	Only receiver windows need to be adjusted
5	Wait time reduced by retransmission	Not in the ICTCP algorithm
6	Bandwidth is not preserved	Bandwidth is preserved to a large extent as a means of retransmission avoidance

Table 1. Efficiency of the ICTCP algorithm over the retransmission algorithm

9. Conclusion
The paper describes the design and implementation of the ICTCP algorithm to improve TCP performance for TCP incast. The previous approaches based on a fine tuned timer for faster retransmission or controlling of a switch buffer are different from ICTCP. The paper focuses on the receiver based congestion control algorithm where ICTCP proactively adjusts the TCP receiver window based on the ratio of difference of achieved and expected throughputs per connection over expected throughputs, as well as the available bandwidth at the receiver end. Our experimental results demonstrate that ICTCP is effective to avoid congestion by achieving almost zero timeout for TCP incast, and packet losses are reduced to a large extent, and thus the bandwidth is considerably preserved. The ICTCP method is also used for the preservation of congestion in the network.

References
[1] Vasudevan V, Phanishayee A, Shah H, Krevat E, Andersen D, Ganger G, Gibson G, and Mueller B 2009 Safe and effective fine-grained TCP retransmissions for datacenter communication *Proc. ACM SIGCOMM* 303–314

[2] Kandula S, Sengupta S, Greenberg A, Patel P, and Chaiken R 2009 The nature of data center traffic: Measurements & analysis *Proc. IMC* 202–208

[3] Alizadeh M, Greenberg A, Maltz D, Padhye J, Patel P, Prabhakar B,Sengupta S and Sridharan M 2010 Data center TCP (DCTCP) *Proc. SIGCOMM* 63–74

[4] Krevat E, Vasudevan V, Phanishayee A, Andersen D, Ganger G, Gibson G and Seshan S 2007 On application-level approaches to avoiding TCP throughput collapse in cluster-based storage systems *Proc. Supercomput.* 1–4

[5] Al-Fares M, Loukissas A and Vahdat A 2008 A scalable, commodity data center network architecture *Proc. ACMSIGCOMM* 63–74

[6] Guo C, Lu G, Li D, Wu H, Zhang X, Shi Y, Tian C, Zhang Y and Lu S 2009 BCube: A high performance, server-centric network architecture for modular data centers *Proc. ACM SIGCOMM* 63–74

[7] Brakmo B and Peterson L 1995 TCP Vegas: End to end congestion avoidance on a global internet *IEEE J. Sel. Areas Commun.* **13** 1465–1480

Engineering applications of metaheuristics: an introduction

Diego Oliva [1,3,4], Salvador Hinojosa [2,3], M V Demeshko[4]

[1] Departamento de Ciencias Computacionales, Tecnológico de Monterrey, Campus Guadalajara, Av. Gral. Ramón Corona 2514, Zapopan, Jal, México
[2] Dpto. Ingeniería del Software e Inteligencia Artificial, Facultad Informática, Universidad Complutense de Madrid, 28040 Madrid, Spain
[3] Departamento de Electrónica, Universidad de Guadalajara, CUCEI, Av. Revolución 1500, Guadalajara, Jal, México
[4] Tomsk Polytechnic University, Lenin Avenue 30, Tomsk, Russian Federation

E-mail: diego.oliva@itesm.mx

Abstract. Metaheuristic algorithms are important tools that in recent years have been used extensively in several fields. In engineering, there is a big amount of problems that can be solved from an optimization point of view. This paper is an introduction of how metaheuristics can be used to solve complex problems of engineering. Their use produces accurate results in problems that are computationally expensive. Experimental results support the performance obtained by the selected algorithms in such specific problems as digital filter design, image processing and solar cells design.

1. Introduction

In engineering, practically everything can be optimized. Optimization can be included in different subfields, for example, quality control, design, surveillance, etc. The aim is to find the best values that solve the problem. For example, less energy consumption. The optimization methods can be divided into two big groups: deterministic and stochastic. The deterministic techniques include classical approaches and most of them use gradient operations. On the other hand, the stochastic approaches include the use of heuristics and metaheuristics that imitates the behavior of different processes in nature. Recently, several metaheuristic methods have been proposed with interesting results. Such approaches use our scientific understanding of biological, natural or social systems, which at some level of abstraction can be represented as optimization processes, as inspiration. These methods include the social behavior of bird flocking and fish schooling such as the Particle Swarm Optimization (PSO) algorithm [1], the cooperative behavior of bee colonies such as the Artificial Bee Colony (ABC) technique [2], the improvisation process that occurs when a musician searches for a better state of harmony such as the Harmony Search (HS) [3], the emulation of the bat behavior such as the Bat Algorithm (BA) method [4], the mating behavior of firefly insects such as the Firefly (FF) method [5], the social-spider behavior such as the Social Spider Optimization (SSO) [6], the simulation of the animal behavior in a group such as the Collective Animal Behavior [6], the emulation of immunological systems as the clonal selection algorithm (CSA) [7], the simulation of the electromagnetism phenomenon as the electromagnetism-Like algorithm [8], and the emulation of the differential and conventional evolution in species such as the Differential Evolution (DE) [9] and Genetic Algorithms (GA) [10], respectively. They have been growing very fast. For that reason, there is the necessity to test them in real world problems. The drawback of the use of metaheuristics is the necessity to adapt the problem to be solved from an optimization point of view. In other words, it is necessary to define the objective function.

This paper presents a study of a selected group of problems solved using metaheuristics. The methods have been selected considering their importance in their own areas. All their areas are adapted and defined as an optimization problem. Different and popular algorithms have been used to search the best solutions. These techniques are Electromagnetism-Like Optimization (EMO) [8], Harmony Search (HS) [3] and Artificial Bee Colony (ABC) [2].

The rest of the paper is organized as follows. Section 2 introduces the basics of metaheuristics and explains the concepts of EMO, HS and ABC. In section 3, the optimization problems are formulated. Section 4 presents the experimental results and, in section 5, there are conclusions presented.

2. Metaheuristic optimization algorithms

From a conventional point of view, a metaheuristic method is an algorithm that simulates a biological, natural or social system at some level of abstraction. To be more specific, a standard EC algorithm includes:

1. One or more populations of candidate solutions are considered.
2. These populations change dynamically due to the production of new solutions.
3. A fitness function reflects the ability of a solution to survive and reproduce.
4. Several operators are employed in order to explore an exploit appropriately the space of solutions.

The methodology of metaheuristics suggest that, on average, candidate solutions improve their fitness over generations (i. e., their capability of solving the optimization problem). A simulation of the evolution process, based on a set of candidate solutions whose fitness is properly correlated to the objective function to optimize, will, on average, lead to an improvement of their fitness and thus steer the simulated population towards the global solution.

Most of the optimization methods have been designed to solve the problem of finding a global solution of a nonlinear optimization problem with box constraints in the following form:

$$\begin{aligned} &\text{maximize} \quad f(\mathbf{x}), \quad \mathbf{x} = (x_1,\ldots,x_d) \in \mathbb{R}^d \\ &\text{subject to} \quad \mathbf{x} \in \mathbf{X} \end{aligned} \tag{1}$$

where $f:\mathbb{R}^d \to \mathbb{R}$ is a nonlinear function, whereas $\mathbf{X} = \{\mathbf{x} \in \mathbb{R}^d \,|\, l_i \le x_i \le u_i, i = 1,\ldots,d\}$ is a bounded feasible search space, constrained by the lower (l_i) and upper (u_i) limits.

In its initial point, the algorithm begins by initializing the set of N candidate solutions with values that are randomly and uniformly distributed between the pre-specified lower (l_i) and upper (u_i) limits. In each iteration, a set of operators is applied over population $\mathbf{P}^k$ to build new population $\mathbf{P}^{k+1}$. Each candidate solution $\mathbf{p}_i^k$ ($i \in [1,\ldots,N]$) represents d-dimensional vector $\{p_{i,1}^k, p_{i,2}^k, \ldots, p_{i,d}^k\}$ where each dimension corresponds to a decision variable of the optimization problem at hand. The quality of each candidate solution $\mathbf{p}_i^k$ is evaluated by using objective function $f(\mathbf{p}_i^k)$ whose final result represents the fitness value of $\mathbf{p}_i^k$. During the iterative optimization process, best candidate solution $\mathbf{g}$ ($g_1, g_2, \ldots g_d$) seen so-far is preserved considering that it represents the best available solution. Fig. 1 presents a graphical representation of a basic cycle of a metaheuristic method.

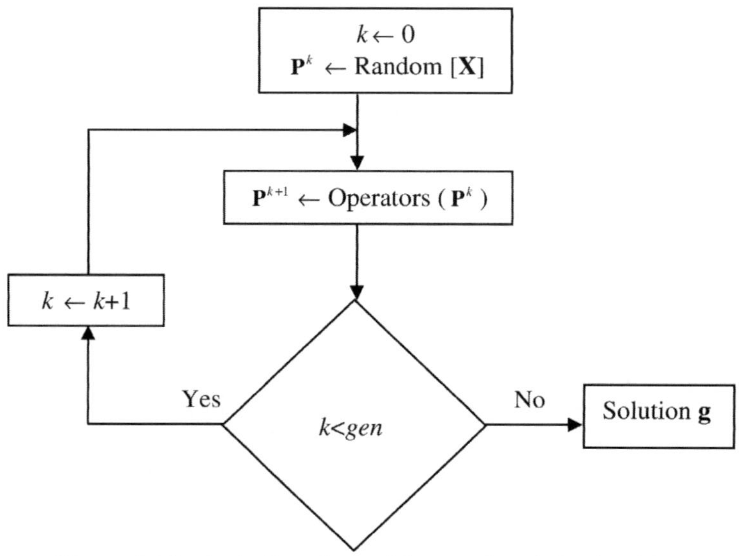

Figure 1. The basic flowchart of the metaheuristic algorithm.

2.1. Electromagnetism-Like Optimization (EMO)

The EMO method, initially designed for bound constrained optimization problems, [8] utilizes N, n-dimensional points $x_{i,t}$, $i=1,2,...,n$, as a population for searching feasible set $\mathbf{X}=\{x\in\mathbb{R}^n | l_i \le x \le u_i\}$, where t denotes the iteration (or generation) number in the algorithm. Initial population $\mathbf{S}_t=\{x_{1,t},x_{2,t},...,x_{N,t}\}$ (being $t=1$) is taken of uniformly distributed samples of search region, $\mathbf{X}$. We denote the population set at the t-th iteration by $\mathbf{S}_t$, as the members of $\mathbf{S}_t$ changes with t. After the initialization of $\mathbf{S}_t$, EMO continues its iterative process until a stopping condition (e.g. the maximum number of iterations) is met. An iteration of EMO consists of two steps. In the first step, each point in $\mathbf{S}_t$ moves to a different location by using the attraction-repulsion mechanism of the electromagnetism theory [11]. In the second step, points, moved by the electromagnetism principle, are further moved locally by a local search and then become members of $\mathbf{S}_{t+1}$ in the $(t+1)$-th iteration. Both the attraction-repulsion mechanism and the local search in EMO are responsible for driving the members, $x_{i,t}$, of $\mathbf{S}_t$ to the proximity of the global optimizer.

2.2. Harmony Search (HS) Optimization

The Harmony Search Algorithm (HSA) introduced by Geem, Kim, and Loganathan [12] is an evolutionary optimization algorithm which is based on the metaphor of the improvisation process that occurs when a musician searches for a better state of harmony. The standard HSA, considers each solution as a "harmony" and is represented by an n-dimension real vector. An initial population of harmony vectors are randomly generated in the search space and stored within a Harmony Memory (HM). A new candidate harmony is thus generated from the elements in the HM by using a memory consideration operation either by a random re-initialization or a pitch adjustment operation. Finally, the HM is updated by comparing the new candidate harmony and the worst harmony vector in the HM. The worst harmony vector is replaced by the new candidate vector in case it is better than the worst harmony vector in the HM. The above process is repeated until a certain termination criterion is met. The basic HS algorithm consists of three main phases: HM initialization, improvisation of new harmony vectors and updating of the HM.

2.3. Artificial Bee Colony (ABC) Optimization

The ABC is an evolutionary algorithm inspired by the intelligent behavior of honey bees [2]. ABC consists of three essential components: food source positions, nectar amount, and several honey-bee classes. Each food source position represents a feasible solution for the problem under consideration. The nectar amount for a food source represents the quality of such a solution (represented by a fitness value). Each bee class symbolizes one particular operation for generating new candidate food source positions (i.e., candidate solutions). The ABC algorithm starts by producing a randomly distributed initial population (food source locations). After initialization, an objective function evaluates whether such candidates represent an acceptable solution (nectar amount) or not. Guided by the values of such an objective function, candidate solutions are evolved through different ABC operations (honey-bee types) until a termination criterion is met.

3. Problems formulation

Metaheuristics can be applied to several problems not only in engineering. However both optimization algorithms and problems, must be adapted according to the problem to solve. In this context, three problems are selected and are solved using one of the methods presented in Section 2. Each problem has been selected from different areas considering their importance. The selected problems are: Digital filters design, Image thresholding and Solar cells design.

3.1 Digital Filters Design

In the filter design process nowadays, the most efficient way to perform is by using metaheuristic algorithms which find the parameters that shape the filter. The main objective of filter design is to approximate the response of the filter to a desired signal by minimizing the objective function while adjusting the parameters of the filter coefficients (see Fig. 2). This paper studies the Infinite Impulse Response (IIR) filters; they are defined in Eq. (2).

$$\frac{Y(z)}{X(z)}=\frac{b_0+b_1z^{-1}+b_2z^{-2}+\ldots+b_rz^{-r}}{a_0+a_1z^{-1}+b_2z^{-2}+\ldots+a_cz^{-c}} \tag{2}$$

Here Y(z) and X(z) are the output and input of the filter. The parameters to be estimates are $\mathbf{A}\cup\mathbf{B}=\{b_0,b_1,b_2,\ldots,b_r,a_0,a_1,a_2,\ldots,a_c\}$. The objective function used to measure the quality of the solutions is the Mean Squared Error (MSE) and is presented in Eq. (3).

$$J=\frac{1}{T}\sum_{t=1}^{T}(Y_p(t)-Y_{fil}(t))^2 , \tag{3}$$

where Y_{fil} is the output generated by the filter using the values estimated by the optimization algorithm. Meanwhile Y_p is the desired output.

3.2 Multilevel Thresholding

One important process in image processing is segmentation. This tasks consist in classifying the pixels of the image. Thresholding is the mostly used process in segmentation, in which the pixels of a gray scale image are divided in sets or classes depending on their intensity level (*L*). For this classification, it is necessary to select a threshold value (*th*) and follow the simple rule of Eq. (4).

$$\begin{aligned} C_1 &\leftarrow p \quad \text{if} \quad 0\le p<th \\ C_2 &\leftarrow p \quad \text{if} \quad th\le p<L-1 \end{aligned} , \tag{4}$$

where *p* is one of the $m\times n$ pixels of gray scale image I_g that can be represented in *L* gray scale levels $L=\{0,1,2,\ldots,L-1\}$. C_1 and C_2 are the classes in which pixel *p* can be located, finally *th* is the threshold. The rule in Eq. (5) corresponds to bi-level thresholding and can be easily extended for multiple sets. The objective function used for this problem is the between class variance proposed by

Otsu [13] that employs the maximum variance value of the different classes as a criterion to segment the image.The objective function is defined as follows:

$$J(\mathbf{TH}) = \max(\sigma^2(\mathbf{TH})), \quad 0 \le th_i \le L-1, \quad i = 1,2,...,k \tag{5}$$

where $\mathbf{TH} = [th_1, th_2, ..., th_{k-1}]$, is a vector that contains multiple thresholds, and the variances are computed using $\sigma^{2^c} = \sum_{i=1}^{k} \sigma_i = \sum_{i=1}^{k} \omega_i (\mu_i - \mu_T)^2$. Here, i represents a specific class. ω_i and μ_j are, respectively, the probability of occurrence and the mean of a class. For MT, such values are obtained as:

$$\omega_0^c(th) = \sum_{i=1}^{th_1} Ph_i^c, \quad \omega_1^c(th) = \sum_{i=th_1+1}^{th_2} Ph_i^c, \ldots, \omega_{k-1}^c(th) = \sum_{i=th_k+1}^{L} Ph_i^c \tag{6}$$

and for the mean values,

$$\mu_0^c = \sum_{i=1}^{th_1} \frac{iPh_i^c}{\omega_0^c(th_1)}, \quad \mu_1^c = \sum_{i=th_1+1}^{th_2} \frac{iPh_i^c}{\omega_0^c(th_2)}, \ldots, \mu_{k-1}^c = \sum_{i=th_k+1}^{L} \frac{iPh_i^c}{\omega_1^c(th_k)} \tag{7}$$

3.3 Solar Cells Design

Solar cell (SC) accurate modeling has received significant attention in recent years [14]. The modeling of PV cells consists of two steps: the mathematical model formulation and the accurate estimation of their parameter values. For the mathematical model, the Current vs. Voltage (I-V) characteristics that rule the behavior of a solar cell are considered. One of the most used methods for parameter estimation is the use of an equivalent circuit. The Single Diode model presented in Fig. 2 is an easy approximation widely used to obtain the best design of an SC.

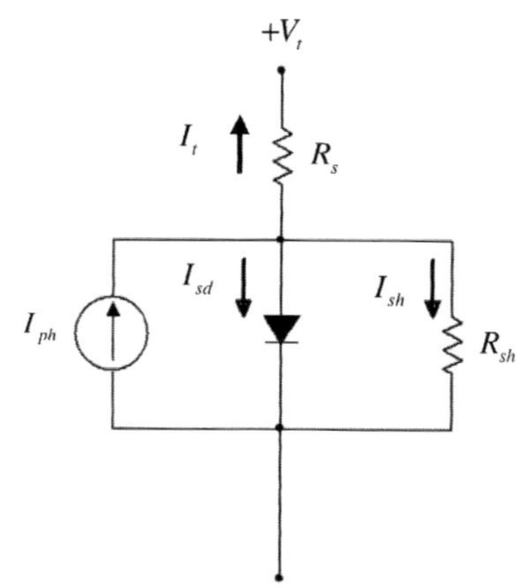

Figure 2. The single diode model of Solar Cells.

From Fig. 2, the mathematical model for the circuit is defined as:

$$I_t = I_{ph} - I_{sd}\left[\exp\left(\frac{q(V_t + R_s \cdot I_t)}{n \cdot k \cdot T}\right) - 1\right] - \frac{V_t + R_s \cdot I_t}{R_{sh}} \tag{8}$$

In Table 1, the range of values for each parameter is presented.

Table 1. The upper and lower range of the solar cell parameters

Parameter	Lower value	Upper value
$R_s\,(\Omega)$	0	0.5
$R_{sh}\,(\Omega)$	0	100
$I_{ph}\,(A)$	0	1
$I_{sd}\,(\mu A)$	0	1
n	1	2

For solar cell design, the objective function is the Root Mean Square Error (*RMSE*) that exists between measurements and the estimated parameters (Eq. (8) and Eq. (9)).

$$f_{\mathrm{SD}}(V_t, I_t, \mathbf{x}) = I_t - I_{ph} + I_{sd}\left[\exp\left(\frac{q(V_t + R_s \cdot I_t)}{n \cdot k \cdot T}\right) - 1\right] + \frac{V_t + R_s \cdot I_t}{R_{sh}} \tag{8}$$

$$RMSE(\mathbf{x}) = \sqrt{\frac{1}{N}\sum_{c=1}^{N}\left(f^c\left(V_t^c, I_t^c, \mathbf{x}\right)\right)^2}, \tag{9}$$

4. Experimental results

4.1. Digital Filter Design Using EMO

The problem of designing IIR filters has been solved using the Electromagnetism-Like Optimization algorithm. A second order system using a first order model has been selected as an example. For this problem the transfer function of the system is defined as:

$$H_s(z^{-1}) = \frac{y_p(z^{-1})}{u(z^{-1})} = \frac{0.05 - 0.4z^{-1}}{1 - 1.1314z^{-1} + 0.25z^{-2}} \tag{10}$$

An IIR filter can model the system of Eq. (10). In this case, it has the following transfer function:

$$H_m(z^{-1}) = \frac{y_m(z^{-1})}{u(z^{-1})} = \frac{b}{1 - az^{-1}} \tag{11}$$

Here $H_s(z^{-1})$ and $H_m(z^{-1})$ correspond to the input-output relation of the system and the proposed model, respectively. A noisy signal is applied to the system and the filter and defined by $u(z^{-1})$. From Eq. (11), the EMO algorithm estimates the values of a and b using the objective function defined in Eq. (3). The results obtained by EMO have been compared with PSO and are described in Table 2 and Fig. 3.

Table 2. Comparison between the values obtained by EMO and PSO for IIR filter design.

Values	a	b	J
EMO	0.9034	-0.3030	0.0107
PSO	0.8950	-0.3142	0.0121
Real	0.906	-0.311	-

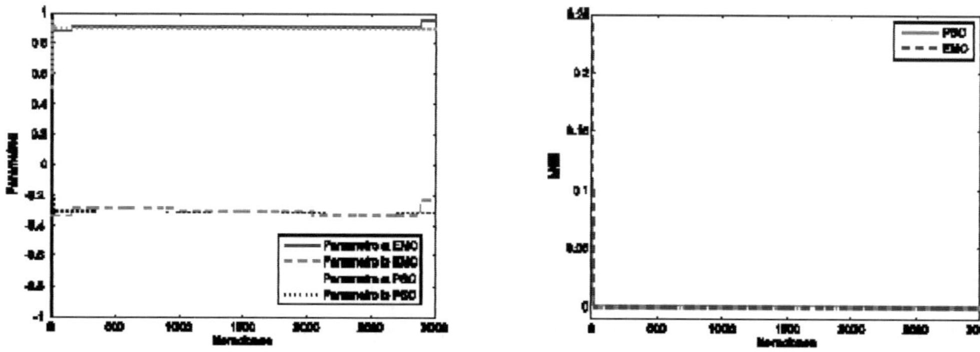

Figure 3. Comparison of parameters for EMO and PSO for IIR filter design.

4.2. Multilevel Thresholding Using HS

A Harmony Search algorithm has been applied to find the best thresholds that can segment an image. The approach is applied over the complete set of benchmark images whereas the results of a selected images are registered in Table 3. Such results present the best threshold values obtained after testing the proposed method, considering four different threshold points $th = 2,3,4,5$. In the Table, there are also the *PSNR* and the *STD* values. According to the results, it is evident that the *PSNR* and *STD* values increment their magnitude as the number of threshold points increases.

Table 3. Results after applying the HSMA using Otsu's function to the set of benchmark images.

Image	k	Thresholds x_{best}	*PSNR*	*STD*
Camera man	2	70, 144	17.247	2.30 E-12
	3	59, 119, 156	20.211	1.55 E-02
	4	42, 95, 140, 170	21.533	2.76 E-12
	5	36, 82, 122, 149, 173	23.282	5.30 E-03
Lena	2	91, 150	15.401	9.22 E-13
	3	79, 125, 170	17.427	2.99 E-02
	4	73, 112, 144, 179	18.763	2.77 E-01
	5	71, 107, 134, 158, 186	19.443	3.04 E-01
Baboon	2	97, 149	15.422	6.92 E-13
	3	85, 125, 161	17.709	1.92 E-02
	4	71, 105, 136, 167	20.289	5.82 E-02
	5	66, 97, 123, 147, 173	21.713	4.40 E-01

The results obtained after applying the HS to each of the images from Table 3 are presented in Fig. 4 as an example. The results show the segmented image, the histogram with the thresholds (best solutions) and the objective function evolution.

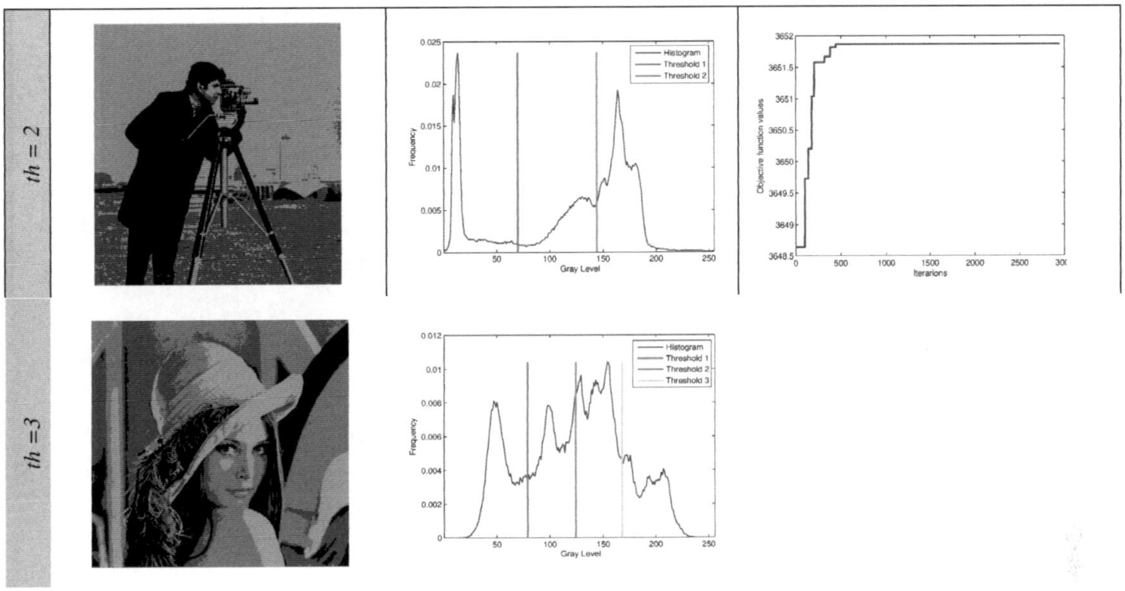

Figure 4. Results of different thresholds over obtained with HS in the selected test images.

4.3. Solar Cells Design Using ABC

For the solar cells design, the Artificial Bee Colony method has been used to estimate the best parameters of the single diode mode. The experiments have been performed using a commercial silicon solar cell (from the R.T.C. Company of France) under the standard test conditions (STC), with a diameter of 57mm. During the data collection process, it is considered that the solar cell operates under the following operating conditions: 1 sun ($1000W/m^2$) at $T=33°C$. In order to test the performance of the ABC method, four more temperatures have been included: $T=25°C$, $T=50°C$, $T=75°C$. The experiments were performed using 26 measurements form the physical SC. Such samples are shown in Table 5. Table 4 shows the values of the SC estimated by ABC; they are also compared with PSO.

Table 4. Comparative results of ABC and PSO applied over the SD model.

Parameter	ABC	PSO
$R_s(\Omega)$	0.0364	0.0354
$R_{sh}(\Omega)$	53.6433	59.0120
$I_{ph}(A)$	0.7608	0.7607
$I_{sd}(\mu A)$	0.3251	0.4000

The reliability model of the fault-tolerant computing system with triple-modular redundancy based on the independent nodes

P A Rahman[1] and E Yu Bobkova[2]
[1] Department of Automated Technological and Informational Systems, Ufa State Petroleum Technological University, Sterlitamak 453118, Russian Federation
[2] Department of Design and Humanities, Samara Cossack Institute of Food Industry and Business, Samara 443031, Russian Federation

E-mail: pavelar@yandex.ru

Abstract. This paper deals with a reliability model of the restorable non-stop computing system with triple-modular redundancy based on independent computing nodes, taking into consideration the finite time for node activation and different node failure rates in the active and passive states. The obtained by authors generalized reliability model and calculation formulas for reliability indices for the system based on identical and independent computing nodes with the given threshold for quantity of active nodes, at which system is considered as operable, are also discussed. Finally, the application of the generalized model to the particular case of the non-stop restorable computing system with triple-modular redundancy based on independent nodes and calculation examples for reliability indices are also provided.

1. Introduction

In the modern world, a rapid development of information technologies and their application in different spheres of human activities is observed. A person has a deal with a large amount of information almost every day. He creates, stores, processes and transmits it by using personal computers and mobile devices. Moreover, the enterprises use specialized data storage and processing systems [1, 2], which are used as the base for information systems to provide enterprise business processes.

A special place in the modern world is occupied by the distributed data processing systems [3, 4], such as high-availability clusters and fault-tolerant computing systems, which provides fault-tolerant data processing. For such systems, it is important to know the reliability indices to assess the risks for the business processes and the degree of reduction of such risks due to the application of fault-tolerant technologies. In this situation, the development of reliability models and analysis of reliability indices for the fault-tolerant computing systems is quite an urgent need of the day.

Currently, a set of fundamental reliability models and calculation methods for reliability indices are discussed in Russian textbooks [5, 6] and several specialized models for the computer systems and networks are available in international literature [7, 8]. However, such models are based on the simplified two-state model of the repairable element and do not take into consideration the specific technical issues of modern data processing systems, based on a set of computing nodes: the finite time for node activation and different failure rates of the node in active and passive states.

Within the research work of the authors, in the field of fault-tolerant data storage, transmission and processing systems [9, 10], the reliability model for restorable non-stop computing systems with triple-modular redundancy, based on independent nodes, taking into consideration the finite time for activation of computing nodes and different node failure rates in active and passive states, was developed. A base three-state reliability model of the computing node and calculation formulas for reliability indices were derived by the authors. Later, this model and formulas were generalized for the specific case of the system, based on independent and identical nodes, with the given bottom threshold for the quantity of active nodes, at which the computing system is considered as operable. At last, the generalized model and formulas were applied for the case of a computing system with triple-modular redundancy.

2. The base three-state reliability model of the computing node

At first, let us introduce a reliability model of the computing node based on Markov chain with the next state-space and conditions of transitions between states:

- State P (passive) – the node is up, but passive, and does not provide computing operations because of software initialization. From this state, the node can either pass to state A with rate γ_N (activation rate), or pass to state F with rate λ_P (failure rate in the passive state).
- State A (active) – the node is up and active, and provides the computing operations. From this state, the node can pass to state F with rate λ_A (failure rate in the active state).
- State F (fault) – the node is down. From this state, the node can pass to state P with rate μ_N (rate of the repair operations).

Figure 1 shows Markov chain, which graphically represents the state-space and transition conditions according to the discussed reliability model of the computing node:

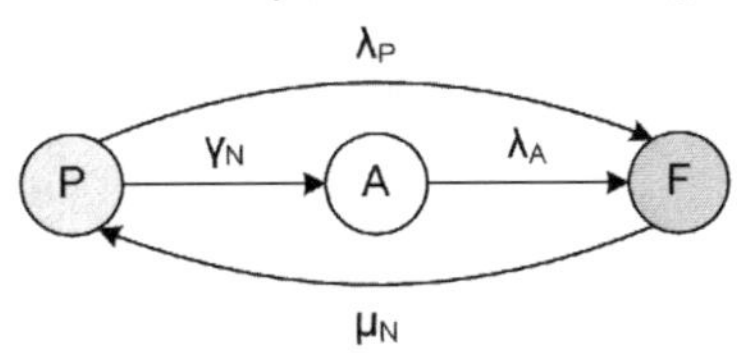

Figure 1. The base three-state reliability model of the computing node.

Accordingly, the differential equations of Kolmogorov-Chapman for this Markov chain is as follows:

$$\begin{cases} P_P(0)=1; \quad P_A(0)=0; \quad P_F(0)=0; \\ P_0(t)+P_1(t)+P_2(t)=1; \\ \dfrac{dP_P(t)}{dt}=-(\lambda_P+\gamma_N)P_P(t)+\mu_N P_F(t); \\ \dfrac{dP_A(t)}{dt}=\gamma_N P_P(t)-\lambda_A P_A(t); \\ \dfrac{dP_F(t)}{dt}=\lambda_P P_P(t)+\lambda_A P_A(t)-\mu_N P_F(t). \end{cases} \tag{1}$$

Where λ_A - the failure rate of the computing node in the active state, λ_P - the failure rate of the computing node in the passive state, μ_N - the repair rate of the computing node, and γ_N - the activation rate of the computing node (the rate of passing from the passive state to the activate state).

Considering $t \to \infty$, when Markov process reaches the steady state and derivatives of probability functions tending to zero, we obtain the following formulas for stationary probabilities of the states:

$$\begin{cases} P_{\mathrm{P}}(\infty) = \dfrac{\mu_{\mathrm{N}}\lambda_{\mathrm{A}}}{\mu_{\mathrm{N}}\gamma_{\mathrm{N}} + \lambda_{\mathrm{A}}(\mu_{\mathrm{N}} + \gamma_{\mathrm{N}} + \lambda_{\mathrm{P}})}; \\ P_{\mathrm{A}}(\infty) = \dfrac{\mu_{\mathrm{N}}\gamma_{\mathrm{N}}}{\mu_{\mathrm{N}}\gamma_{\mathrm{N}} + \lambda_{\mathrm{A}}(\mu_{\mathrm{N}} + \gamma_{\mathrm{N}} + \lambda_{\mathrm{P}})}; \\ P_{\mathrm{F}}(\infty) = \dfrac{\lambda_{\mathrm{A}}(\gamma_{\mathrm{N}} + \lambda_{\mathrm{P}})}{\mu_{\mathrm{N}}\gamma_{\mathrm{N}} + \lambda_{\mathrm{A}}(\mu_{\mathrm{N}} + \gamma_{\mathrm{N}} + \lambda_{\mathrm{P}})}. \end{cases} \tag{2}$$

Considering state A as the only operable state for the computing node, we also obtain the formula for the stationary availability factor of the computing node:

$$K = P_{\mathrm{A}}(\infty) = \frac{\mu_{\mathrm{N}}\gamma_{\mathrm{N}}}{\mu_{\mathrm{N}}\gamma_{\mathrm{N}} + \lambda_{\mathrm{A}}(\mu_{\mathrm{N}} + \gamma_{\mathrm{N}} + \lambda_{\mathrm{P}})}. \tag{3}$$

Moreover, the formula for the mean time to failure of the computing node can be easily obtained by using the topological method for the reliability models based on the Markov chains. It can be calculated as a ratio of the availability factor to the weighted sum of probabilities of all operable states, each of which is multiplied by the sum of all rates of transitions to all inoperable states. In our case, we have only one operable state, A, with only one transition to state F with rate λ_{A}, thus we obtain the following simple formula for the mean time to failure:

$$T_{\mathrm{F}} = \frac{P_{\mathrm{A}}(\infty)}{\lambda_{\mathrm{A}} P_{\mathrm{A}}(\infty)} = \frac{1}{\lambda_{\mathrm{A}}}. \tag{4}$$

At last, we can easily obtain the formula for the mean time to the recovery of the computing node, which takes into consideration both repair and activation rates, by using the fundamental relation between the availability factor, mean time to failure and mean time to recovery:

$$T_{\mathrm{R}} = \frac{1-K}{K} T_{\mathrm{F}} = \frac{\mu_{\mathrm{N}} + \gamma_{\mathrm{N}} + \lambda_{\mathrm{P}}}{\mu_{\mathrm{N}}\gamma_{\mathrm{N}}}. \tag{5}$$

3. A generalized reliability model of the system based on independent computing nodes

Let us now introduce a computing system with n independent and identical computing nodes. Each of the nodes can independently be in one of the three states and reliability parameters of all nodes are identical – it is our main assumption within this scientific research.

Next, let us introduce a state-space for the computing system with the two-dimensional numbering scheme, in which each states represent some quantity of i active nodes, j faulty nodes, $0 \le i \le n$, $0 \le j \le n$, $i + j \le n$, and $n - i - j$ passive nodes. The total number of states is equal to $(n+1)(n+2)/2$. From state (i, j), the system can pass to one of the neighbor states:

- State $(i-1, j+1)$ with rate $i\lambda_{\mathrm{A}}$ (failure of one of active nodes).
- State $(i+1, j)$ with rate $(n-i-j)\gamma_{\mathrm{N}}$ (activation of one of passive nodes).
- State $(i, j+1)$ with rate $(n-i-j)\lambda_{\mathrm{P}}$ (failure of one of passive nodes).
- State $(i, j-1)$ with rate $j\mu_{\mathrm{N}}$: (completion of repair of one of faulty nodes).

Moreover, let us consider the computing system as operable only if at least s nodes are active (provide computing operations), $1 \le s \le n$. Accordingly, states $(i \ge s, j)$ of the system are considered as operable, and states $(i < s, j)$ - as inoperable.

Figure 2 shows Markov chain for the reliability model of the computing system with independent and identical computing nodes, and it graphically represents the state-space and transition conditions according to the discussed reliability model of the computing system.

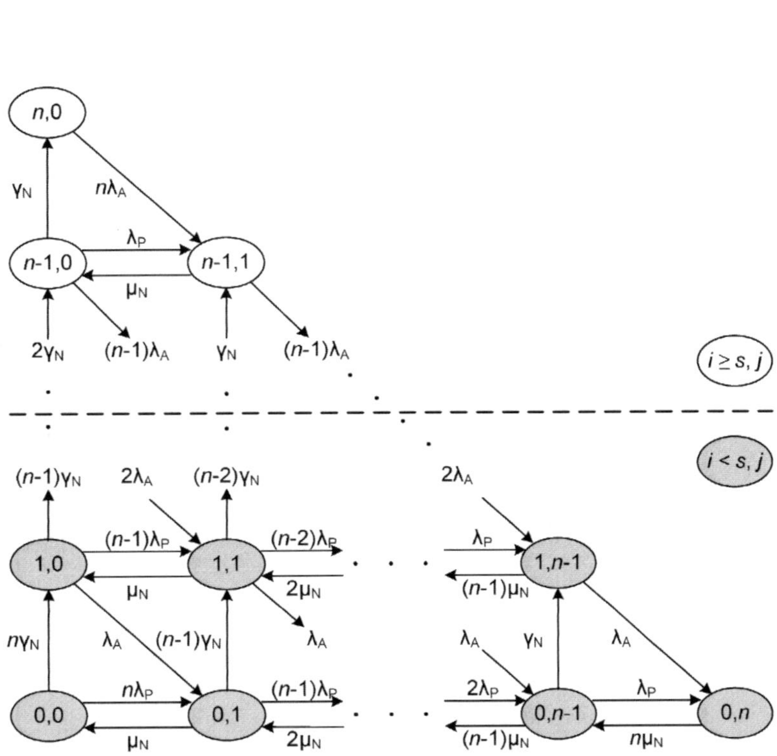

Figure 2. The reliability model of the system based on n identical and independent computing nodes.

Mention must be made, due to node independency and identity we do not need to construct and solve the Kolmogorov-Chapman system for the large Markov chain. The formula for stationary probability of state (i, j) can be obtained as a product of stationary probabilities of i nodes in the active state, stationary probabilities of j nodes in the faulty state, stationary probabilities of $n-i-j$ nodes in passive states, multiplied by the number of all possible combinations $n!/(i!\,j!(n-i-j)!)$ of active, faulty and passive nodes. Finally, we obtain the following formula for probability of state (i, j):

$$P_{i,j}(\infty)=\frac{n!\,P_{\mathrm{A}}^{i}(\infty)P_{\mathrm{F}}^{j}(\infty)P_{\mathrm{P}}^{n-i-j}(\infty)}{i!\,j!(n-i-j)!}=\frac{n!\,\mu_{\mathrm{N}}^{i}\gamma_{\mathrm{N}}^{i}\lambda_{\mathrm{A}}^{j}(\gamma_{\mathrm{N}}+\lambda_{\mathrm{P}})^{j}\mu_{\mathrm{N}}^{n-i-j}\lambda_{\mathrm{A}}^{n-i-j}}{i!\,j!(n-i-j)!(\mu_{\mathrm{N}}\gamma_{\mathrm{N}}+\lambda_{\mathrm{A}}(\mu_{\mathrm{N}}+\gamma_{\mathrm{N}}+\lambda_{\mathrm{P}}))^{n}};\qquad(6)$$

$$i=0\ldots n;\quad j=0\ldots n;\quad i+j\le n.$$

To obtain the formulas for reliability indices of the system, we also need an additional formula for the sum of probabilities of all states for the given i active nodes:

$$\sum_{j=0}^{n-i}P_{i,j}(\infty)=\frac{n!\,P_{\mathrm{A}}^{i}(\infty)(P_{\mathrm{F}}(\infty)+P_{\mathrm{P}}(\infty))^{n-i}}{i!(n-i)!}=\frac{C_{n}^{i}\mu_{\mathrm{N}}^{i}\gamma_{\mathrm{N}}^{i}\lambda_{\mathrm{A}}^{n-i}(\mu_{\mathrm{N}}+\gamma_{\mathrm{N}}+\lambda_{\mathrm{P}})^{n-i}}{(\mu_{\mathrm{N}}\gamma_{\mathrm{N}}+\lambda_{\mathrm{A}}(\mu_{\mathrm{N}}+\gamma_{\mathrm{N}}+\lambda_{\mathrm{P}}))^{n}}.\qquad(7)$$

For given bottom threshold s for the quantity of active nodes, at which system is considered as operable, we obtain the following formula for the stationary availability factor of the system as a sum of the probabilities of all operable states $(i\ge s, 0\le j\le n-i)$:

$$K=\sum_{i=s}^{n}\sum_{j=0}^{n-i}P_{i,j}(\infty)=\sum_{i=s}^{n}\frac{C_{n}^{i}\mu_{\mathrm{N}}^{i}\gamma_{\mathrm{N}}^{i}\lambda_{\mathrm{A}}^{n-i}(\mu_{\mathrm{N}}+\gamma_{\mathrm{N}}+\lambda_{\mathrm{P}})^{n-i}}{(\mu_{\mathrm{N}}\gamma_{\mathrm{N}}+\lambda_{\mathrm{A}}(\mu_{\mathrm{N}}+\gamma_{\mathrm{N}}+\lambda_{\mathrm{P}}))^{n}}.\qquad(8)$$

Accordingly, the mean time to failure can be calculated as a ratio of the stationary availability factor to the weighted sum of probabilities of the «border» operable states $(i=s, 0\le j\le n-i)$, each of which is multiplied by according rate $i\lambda_{\mathrm{A}}$ of transition to the inoperable state. It may be noted that for all «border» states, the transition rate to the inoperable state is the same and is equal to $s\lambda_{\mathrm{A}}$. So, we obtain the following formula for the mean time to failure of the discussed computing system:

$$T_{\mathrm{F}} = \frac{\sum_{i=s}^{n}\sum_{j=0}^{n-i} P_{i,j}(\infty)}{s\lambda_{\mathrm{A}}\sum_{j=0}^{n-s} P_{s,j}(\infty)} = \frac{\sum_{i=s}^{n} C_n^i \mu_{\mathrm{N}}^{i-s}\gamma_{\mathrm{N}}^{i-s}\lambda_{\mathrm{A}}^{n-i}(\mu_{\mathrm{N}}+\gamma_{\mathrm{N}}+\lambda_{\mathrm{P}})^{n-i}}{sC_n^s\lambda_{\mathrm{A}}^{n-s+1}(\mu_{\mathrm{N}}+\gamma_{\mathrm{N}}+\lambda_{\mathrm{P}})^{n-s}}. \quad (9)$$

Finally, the formula for the mean time to recovery can be easily obtained by using the fundamental relation between the availability factor, the mean time to failure and the mean time to recovery:

$$T_{\mathrm{R}} = \frac{1-K}{K}T_{\mathrm{F}} = \frac{\sum_{i=0}^{s-1} C_n^i \mu_{\mathrm{N}}^i\gamma_{\mathrm{N}}^i\lambda_{\mathrm{A}}^{s-1-i}(\mu_{\mathrm{N}}+\gamma_{\mathrm{N}}+\lambda_{\mathrm{P}})^{s-i}}{sC_n^s\mu_{\mathrm{N}}^s\gamma_{\mathrm{N}}^s}. \quad (10)$$

4. The reliability model of the restorable non-stop computing system with triple-modular redundancy

Lastly, let us introduce a computing system with three identical and independent nodes and triple-modular redundancy. All three nodes provide the same computing operations for the same data, and the final result is selected by using some majority voting circuit, so the system provides a valid result only if at least two of three nodes are active (provide computing operations).

It should be noted that for our reliability model we consider the computing system as a non-stop system. The system will not be halted in case of failure of two nodes, when computing results become invalid. Moreover, the system will continue functioning even if the third node fails. So, the system is «on-line» in any case, but is considered as operable (able for providing valid computing results) only if at least two of three nodes are active. This is quite an important assumption for the discussed fault-tolerant computing system with three nodes and the majority voting circuit, which is slightly different from the generic triple-modular redundancy system. In addition, we consider that the majority voting circuit is ideal and its failure rate is negligible within the scope of our reliability model.

Accordingly, the reliability model of the computing system with triple-modular redundancy is a particular case of the generalized model described above with the number of nodes equal to $n = 3$ and bottom threshold $s = 2$ for the quantity of active nodes.

Accordingly, the formula for the stationary availability factor of the computing system can be obtained by substituting $n = 3$ and $s = 2$ in formula 8:

$$K = \frac{\mu_{\mathrm{N}}^2\gamma_{\mathrm{N}}^2(\mu_{\mathrm{N}}\gamma_{\mathrm{N}} + 3\lambda_{\mathrm{A}}(\mu_{\mathrm{N}}+\gamma_{\mathrm{N}}+\lambda_{\mathrm{P}}))}{(\mu_{\mathrm{N}}\gamma_{\mathrm{N}} + \lambda_{\mathrm{A}}(\mu_{\mathrm{N}}+\gamma_{\mathrm{N}}+\lambda_{\mathrm{P}}))^3}. \quad (11)$$

Similarly, we can obtain the formula for the mean time to failure from formula 9:

$$T_{\mathrm{F}} = \frac{\mu_{\mathrm{N}}\gamma_{\mathrm{N}} + 3\lambda_{\mathrm{A}}(\mu_{\mathrm{N}}+\gamma_{\mathrm{N}}+\lambda_{\mathrm{P}})}{6\lambda_{\mathrm{A}}^2(\mu_{\mathrm{N}}+\gamma_{\mathrm{N}}+\lambda_{\mathrm{P}})}. \quad (12)$$

Finally, we can obtain the formula for the mean time to recovery from formula 10:

$$T_{\mathrm{R}} = \frac{(\mu_{\mathrm{N}}+\gamma_{\mathrm{N}}+\lambda_{\mathrm{P}})(3\mu_{\mathrm{N}}\gamma_{\mathrm{N}} + \lambda_{\mathrm{A}}(\mu_{\mathrm{N}}+\gamma_{\mathrm{N}}+\lambda_{\mathrm{P}}))}{6\mu_{\mathrm{N}}^2\gamma_{\mathrm{N}}^2}. \quad (13)$$

5. The reliability indices calculation example

A restorable non-stop computing system with triple-modular redundancy is given. The failure rate of the computing node in the active state is $\lambda_{\mathrm{A}} = 1/2920$ hour^{-1} (on average, one failure per 4 months), the failure rate in the passive state is $\lambda_{\mathrm{P}} = 1/8760$ hour^{-1} (on average, one failure per year), node activation rate $\gamma_{\mathrm{N}} = 20$ hour^{-1} (on average, one activation per 3 minutes) and node repair rate

$\mu_N = 1/24$ hour^{-1} (on average, one repair per 24 hours). The calculations of the reliability indices for the given computing system with triple-modular redundancy by using the obtained above formulas give the following values:

- Stationary availability factor by formula 11 is equal to: $K_F \approx 0.999801$.
- Mean time to failure by formula 12 is equal to: $T_F \approx 60548$ hours.
- Mean time to recovery by formula 13 is equal to: $T_R \approx 12$ hours.

6. Conclusion

Within the scope of this article, a reliability model of the restorable non-stop computing system with triple-modular redundancy based on independent computing nodes, taking into consideration the finite time for node activation and different node failure rates in the active and passive states, is discussed. The generalized reliability model obtained by authors and calculation formulas for reliability indices for the system based on identical and independent computing nodes with the given threshold for the quantity of active nodes, at which the system is considered as operable, are also overviewed. Finally, calculation examples for reliability indices are also provided.

Scientific results, obtained by the authors, were used for designing fault-tolerant computing systems of Moscow Power Engineering Institute, Nuclear Power Plant "Balakovo" and several other medium and large-scale enterprises.

7. Acknowledgements

The authors are grateful to professor I. I. Ladygin, Moscow Power Engineering Institute, for scientific support, and to V. A. Tushavin, an associate professor, Saint-Petersburg State University of Aerospace Instrumentation, for technical support.

8. References

[1] Tanenbaum A S 2006 *Structured Computer Organization* (New Jersey: Prentice Hall PTR)
[2] Broydo V L and Ilyina O P 2011 *Computer systems, networks and telecommunications* (Saint-Petersburg: Piter)
[3] Tanenbaum A S and Steen M 2007 *Distributed Systems* (New Jersey: Prentice Hall PTR)
[4] Coulouris G, Dollimore J, Kindberg T and Blair G 2011 *Distributed Systems: Concepts and design* (London: Addison Wesley)
[5] Cherkesov G N 2005 *Reliability of Hardware and Software Systems* (Saint-Petersburg: Piter)
[6] Polovko A M and Gurov S V 2006 *Basis of Reliability Theory* (Saint-Petersburg: BHV-Petersburg)
[7] Shooman M L 2002 *Reliability of computer systems and networks* (New York: John Wiley &Sons, Inc.)
[8] Koren I and Krishna C M 2007 *Fault-Tolerant Systems* (San Francisco: Morgan Kaufmann Publishers)
[9] Rahman P A, Muraveva E A and Sharipov M I 2016 Reliability Model of Fault-Tolerant Dual Disk Redundant Array *Key Engineering Materials* **685** 805-810
[10] Rahman P A and Bobkova E Yu 2016 Reliability model of fault-tolerant data processing system with primary and backup nodes *IOP Conference Series: Materials Science and Engineering* **124** 012023

Application of the model of the printed circuit board with regard to the topology of external conductive layers for calculation of the thermal conditions of the printed circuit board

I M Rybakov, N V Goryachev, I I Kochegarov, A K Grishko, S A Brostilov, N K Yurkov

Penza State University, 40, Krasnaya st., Penza, 440026, Russia

E-mail: kipra@mail.ru

Abstract. The paper proves the necessity of taking into account external conductive layers of the printed circuit board with the thermal physical designing radio-electronic means. For example, a single printed circuit board shows the level of influence of the external conductive layer on the thermal conditions of the printed circuit board. It proved the influence of Joule heat in the thermal conditions of a single conductor. Developed geometrical and thermal printed circuit board models take into account the topological layer and can improve the accuracy of determining the thermal conditions of the printed circuit board.

1. Introduction

The problem of ensuring proper thermal conditions of the printed circuit board with a high density of conductors' topology

A modern electronic product can not be imagined without the use of a variety of semiconductor devices and integrated circuits. The characteristics of these electric articles (EA) are often dependent on the deviation of the temperature that adversely affects the reliability and quality of radio-electronic means (REM) and thus has a negative impact on their performance.

The increasing distribution of printed circuit boards (PCB) in the REM both with a dense arrangement of EA and with dense topology of conductors deteriorates the situation (Figure 1).

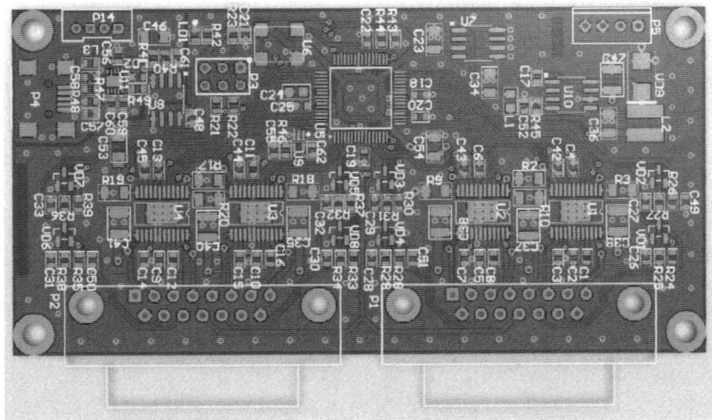

Figure 1. The PCB with high-density topology

All the heat produced by radio-electronic components is released in the form of losses, and if in active elements such as resistors element overheating can be regarded as a directly proportional to the dissipation power, the passive elements, such as condensers, have low power of dissipation, which amounts to 10 - 15%. Most passive elements has a greater sensitivity to changes in temperature, so putting them beside the intense heat radiating elements or unit placement in low temperatures, can lead to failure of the element. Because of this, the arrangement of the EA PCB surface and placing the latter in the housing of REM are also an important factor affecting the thermal treatment of a device (Figure 2).

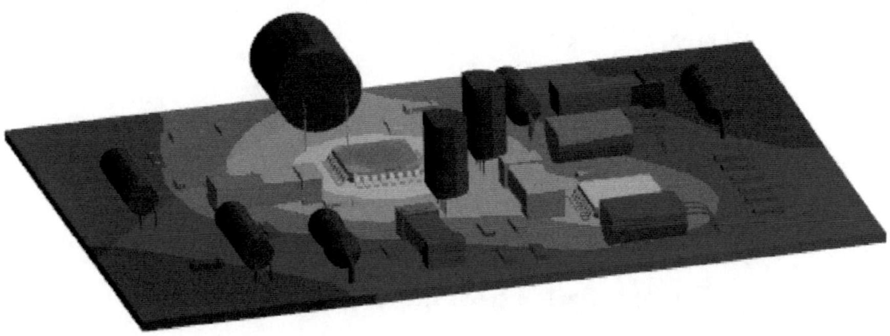

Figure 2. Thermal conditions of the PCB

For the selection of the optimal thermal conditions of operation of the electronic equipment, it is necessary to carry out thermal calculations at the early stages of design to predict the possible outcome of the thermal-physical design.

Existing methods of thermal calculation and analysis of the PCB and the REM have different accuracy and are time consuming. The most important factor in obtaining reliable results is to carry out calculations with a sufficient amount of raw data. Initial data for calculations are the thermal characteristics of radioelectronic components. Thermal characteristics represent the dependence of EA temperature on the power, scattered by EA and by surrounding bodies, at constant temperature of the environment, taking into account the structure and physical properties of the body. For the thermal calculation of PCB and REM, it is necessary to know the thermal characteristics of the EA used in the device, the operating temperature range, etc. Most of the thermal characteristics, used in calculating of the thermal conditions, are taken from the documentation for the EA. External factors, including the ambient temperature, are usually stipulated in the terms of reference for the design of the product.

Modern REM is constructed of a plurality of EA, which have a complex shape and can emit large quantities of heat per unit area. This structure, in thermal-physical respect, is an extremely complex system, and it is often impossible for actual REM to make a complete system of equations and solve it analytically. Therefore, the processes occurring in the REM are in most cases schematized, taking a number of simplifications that helps to get a thermal model of PCB REM, which is convenient to carry out thermal calculations and analysis.

The authors of a number of studies suggest an approach and model of the printed board assembly, taking into account the topology of the outer conductive layers. Here is the experience of the application of this model in the evaluation of the thermal conditions of PU.

2. The use of a model that takes into account the topology of the conducting layers to calculate the thermal conditions of the printed circuit board

In each system, which have an electric current conductivity, and wherein the specific conductivity of the material is finite, the heat losses are expected. Heat losses, also called ohm heating, are in many cases an undesirable side effect of current conductivity of electric current.

The model simulates the system consisting of a small portion of the printed circuit board assembly comprising a chip and copper path (trace), associated with the transistor. This model has a specific performance that later will assess the adequacy of the data (Figure 3). The derived equation to determine the thermal characteristics, allows estimating the operating temperature of the chip and the printed conductors, which is higher than the ambient temperature due to undesirable thermal radiation.

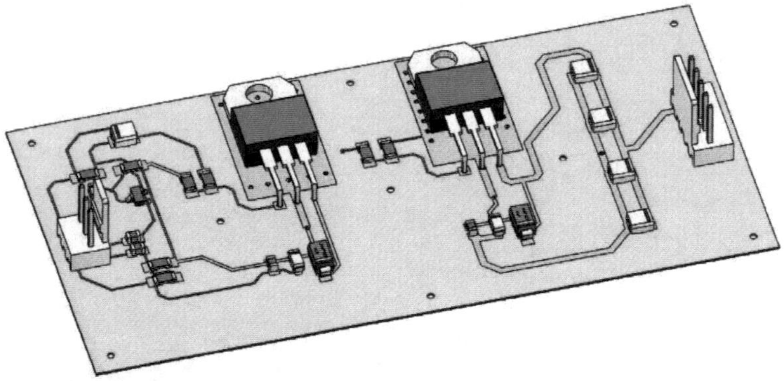

Figure 3. The full geometric model of the PCB

A chip, in this case - a standard product, can be mounted on the radiator, but during determination of thermal conditions it does not have additional cooling elements in the structure. However, for this simulation, the influence of Joule heat and the heat flow, radiated from the chip, on the thermal PCB conditions will be detected and evaluated. A major positive effect, in most cases is to determine the need to install additional cooling elements in the PCB structure. Therefore, it will be defined whether the installation of the radiator is required, or if the operating temperature can be sufficiently low not to cause damage to the system in the absence of the radiator.

Figure 4 shows the geometry of the model to be used in the simulation. The chip is mounted on the circuit board using a through-hole technology. The solder in the holes provides mechanical support and electronic contact between the copper conductors and the legs of the chip.

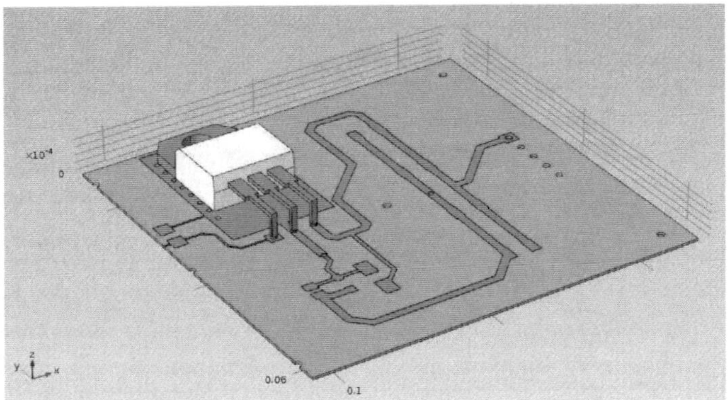

Figure 4. The model of the geometry used in simulation

The core of the chip itself is located in the interior of the housing, shown in Figure 4. It will radiate the heat flow; the core is connected with legs, which transfer the heating from the core owing to heat transmission in a solid body, which further extends both to the printed conductors and to the area of the PCB itself. These compounds, as expected, have significant effects on the thermal conditions of the PCB.

The chip front part is made of ceramics, while the rear portion, which can be clamped to the heat sink, is made of copper. At the feet of the transistor and the front of the pack, there is a correspondence to thermalphysical properties. Brass legs are soldered to the circuit board with the solder material consisting of 60Sn-40Pb (60% tin and 40% lead). The mounting plate is made of fiberglass.

Current conductivity and ohm heating occur in copper routes of solders and in legs. In these parts, the physics of heat transfer and heat release due to ohm heating is fully connected to the electric current conductivity. All other parts of the microcircuit can exchange only heat transfer and heat release with the environment.

The core of the transistor is represented by the inner edge with internal generation of high temperature, corresponding to 1.0 W. The cooling via convection takes place in all external borders having a heat transfer coefficient of 2.5 W / (m^2 * K). This value of the heat transfer coefficient corresponds to the normal case scenario, when the printed circuit board is located in the housing. The ambient temperature is 293.15 K.

The flow enters the circuit board through the vertical borders of copper traces connected with the base, the emitter and the collector, shown in Figure 5. The value of the flow in the path boundary connected with the transmitter, is 0.099 A. The value of the flow in the path boundary connected with the collector is 0.10 A. The difference in the absolute flow between the emitter and the flows of the collector corresponds to the flow in the path boundary, connected with the base, which amounts to 1.0 mA.

The maximum temperature is approximately 340 K. This is appropriate within the acceptable operating temperature range for the chip, which implies that in this case its attachment to the radiator is not required.

Another interesting observation is that the effect of ohm heating (Joule heat) is substantial in the copper tracks in the interval between the chip and the input. Figure 6 represents a graph of temperature along the copper conductors connected to the PCB base.

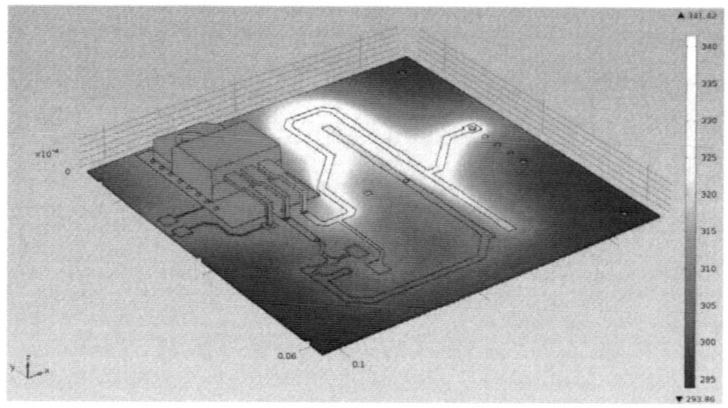

Figure 5. Temperature distribution in the printed conductor

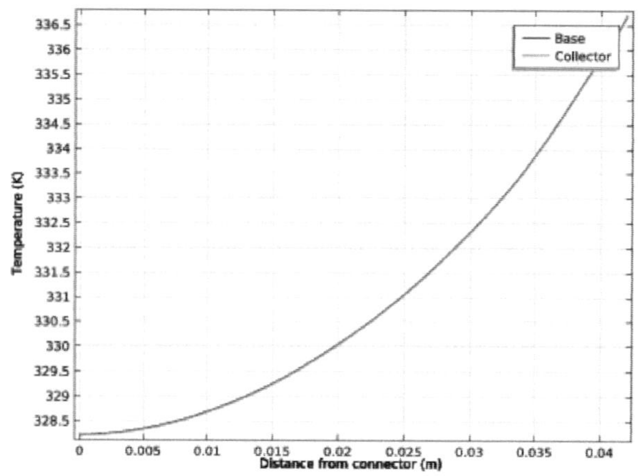

Figure 6. The temperature along the topology of the copper conductors

The fact that the effect of ohm heating increases the temperature in the copper conductors, which leads to the conclusion that a higher temperature in these routes is due to high thermal conductivity of copper. The copper conductors have a high temperature, which is the same as that produced in the chip, but the wiring board has insufficient heat conductivity and therefore is not heated to the same extent as copper conductors.

Further calculations were carried out using a complete geometric model of the PCB.

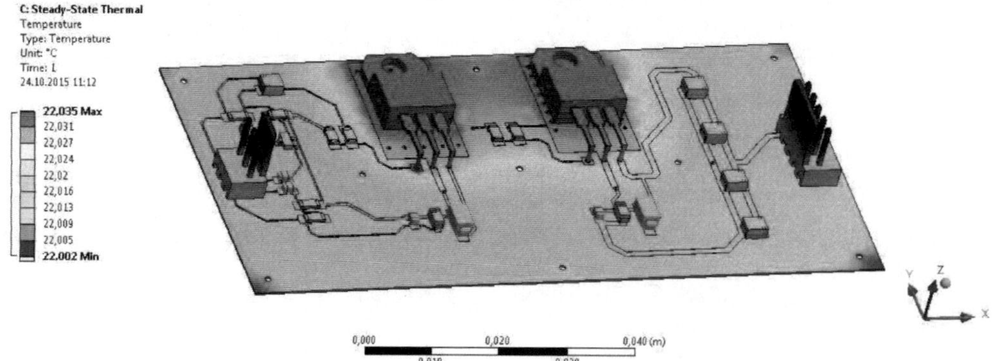

Figure 7. Visualization analysis of the complete geometric model

These studies have helped to create and organize the whole sequence of actions related to the ordering method of thermal calculation. Taking into consideration the heating factors, such as Joule heat emitted from the printed conductors, has a significant effect.

3. Conclusion

The authors conducting the study clearly confirmed the importance of taking into account the topology of the conductive layer in the thermal analysis of the printed board assembly. Thermal conditions of a single conductor are determined not only by a common temperature background of the printed circuit board, but also by the effect of Joule heat. This fact should be considered especially when designing PCB for power devices, since the current density of the conductive layer in such devices is much higher than in the printed board assembly of small-signal devices.

4. Acknowledgments

This article was written as part of the of the state government work "Conducting research projects, fundamental research, applied research and experimental development» «№8.389.2014 / K» on the theme "Information Technology Research designs of radio electronic means under external factors".

References

[1] Kochegarov I. I., Tynda A. and Goryachev N. 2016 The method of obtaining the thermal model of the printed circuit assembly *XIX IEEE International Conference on Soft Computing and Measurements (SCM)* (St. Petersburg, Russia) pp. 183-185

[2] Samarskii A A; Vabishchevich P N 2007 *Numerical Methods for Solving Inverse Problems of Mathematical Physics*

[3] Zhang Y., Pennatini A. and Bagnoli P. E. 2012 A thermal model for the PCB structure including thermal contribution of traces and vias *Thermal Investigations of ICs and Systems (THERMINIC)* (Budapest) pp. 1-6.

[4] Grishko A. K., Goryachev N. V., Kochegarov I. I., and Yurkov N. K. 2016 Dynamic Analysis and Optimization of Parameter Control of Radio Systems in Conditions of Interference *Proceedings of XXI International Siberian Conference on Control and Communications (SIBCON-2016)* (Moscow, Russia)

[5] Dogruoz M. B., Abarhama M. and Shankaranb G. V. 2016 Transient thermal behavior of SOIC packages — an optimization study *15th IEEE Intersociety Conference on Thermal and Thermomechanical Phenomena in Electronic Systems (ITherm)* (Las Vegas, NV) pp. 1083-1092

[6] Zhang Y., Xu Z. and Hu C. 2016 Computing environmental life of electronic products based on failure physics *Journal of Systems Engineering and Electronics* **27** (2) 493-500

[7] Ge Z., Wang K. and Duan K. 2015 Numerical simulation and thermal analysis of PoP packaging Electronic Packaging Technology (ICEPT) (Changsha) pp. 1438-1443

IOP Conf. Series: Journal of Physics: Conf. Series **803** (2017) 012132 doi:10.1088/1742-6596/803/1/012132

The Model and Control Methods of Access to Information and Technology Resources of Automated Control Systems in Water Supply Industry

M Yu Rytov[1], S A Spichyack[1], V P Fedorov[2] and D I Petreshin[3]

[1] Department of Information Security Systems, Bryansk State Technical University, Bryansk, Russia
[2] Department of Mechanical Engineering and Technology, Bryansk State Technical University, Bryansk, Russia
[3] Academic and Scientific Technological Institute, Bryansk State Technical University, Bryansk, Russia

E-mail: rmy@tu-bryansk.ru

Abstract. The paper describes a formalized control model of access to information and technological resources of automated control systems at water supply enterprises. The given model considers the availability of various communication links with information systems and technological equipment. There are also studied control methods of access to information and technological resources of automated control systems at water supply enterprises. On the basis of the formalized control model and appropriate methods there was developed a software-hardware complex for rapid access to information and technological resources of automated control systems, which contains an administrator's automated workplace and ultimate users.

1. Introduction

The object of research is the control procedure of access to information and technological resources in the automatic control system of a water supply enterprise. Modern water supply enterprises belong to critical objects of housing and communal services and have a great number of technological objects and information resources. These resources in practice are integrated under the control of an automated system including such components as ERP-systems, DBMS (database management system), ACS technological equipment (ACS - automatic control system), communication systems with mobile and stationary objects and their positioning etc.

Object control in the system is directive and centralized. The automatic control system of a water supply enterprise has a lot of various communications paths: a local-area network (LAN), half-duplex radio communication with radio modems, GSM channel communication, analog telecommunication and others.

The control environment is defined as trustful. The technological information circulating in the system is not confidential, but there is a need of ensuring its availability, reliability (authenticity) and access isolation within the trusted environment.

The urgency of the modelling problem is caused by occasional and systematic failures of communication links. There is a reduction of personnel and a systematic shortage of shifts at enterprises. As a result there is a need to ensure a quick cross access to technological and information resources in case of difficult and emergency situations.

So we carried out the comparative analysis of describing models of access isolation in different sources. We also consider classical models of access isolation in computer systems: discretionary, mandatory, role-based access. The model of cross access demarcation is also studied [1].

The research objective is developing the model and control methods of access to information and technological resources of an automated control system at water supply enterprises.

Within this objective, the following tasks are solved:

- development of a mathematical model of the research object, including its features;
- development of methods to control the access and ensure access reliability:
  - deposition and preliminary distribution of authenticators,
  - ensuring quick cross access in emergency,
  - control of the period of access authority by means of authenticator evolution;
- development of the automated control subsystem of access to information and technological resources.

2. The formalized control model of access to information and technological resources of automated control systems at water supply enterprises.

Taking into account the peculiarities of this process and the previous works in the field [2,3], we have developed the formalized control model of reliable access to information and technological resources of the automatic control system of a water supply enterprise.

Nominally, it is defined as follows:

$$M = \langle U, T, S, P, R, Z, G, F \rangle, \quad (1)$$

where $U = \{u_1, u_2, ..., u_{uc}\}$ is a set of access subjects (users/subscribers), if uc is a number of access subjects. Each subject has a set of unique identifiers (within the user group). $I = \{i_1, i_2, ..., i_{ik}\}$ is a set of identifiers, if ik is an amount of identifiers. Identifiers serve for unambiguous recognition of subjects among all set members U. Also, each subject has a set of authenticators from set A, by means of which it is possible to confirm the personality and the rights of the subject. Therefore, $A = \{a_1, a_2, ..., a_{ak}\}$ is a set of authenticators, if ak is an amount of authenticators. The following functions were introduced: $user: I \to U$, $id: A \to I$. They join the set of basic functions of system F, i.e., $user, id \subset F$.

$T = \{t_1, t_2, ..., t_{tc}\}$ is a set of information and technological resources (objects), if tc is an amount of resources. Each resource has one or more set members of communication links $C = \{c_1, c_2, ..., c_{ck}\}$, if ck is a number of channels, with the appropriate sets of safety protocols $K = \{k_1, k_2, ..., k_{kc}\}$, if kc is a number of protocols which are also unambiguously connected with the current authenticator of $a_j \in A$. We introduced the following ratio:

$$resourses = \{(t_i, k_j, a_j) \mid (i \le tc) \wedge (c_k \in C)\}. \quad (2)$$

$S = \{s_1, s_2, ..., s_{sc}\}$ is a session identifier, if sc is a number of such session identifiers. The element of this set is created at the first appeal of the subject to the object, when the subject passes identification and authentication. The identifier and an authenticator are stored in the session and do not demand repeated input at the appeal to resources. The following functions are introduced: $sid: S \to I$, $suser: S \to U$, $sauth: S \to A$. They join the set of basic functions of system F, i.e., $\{suser, sid, sauth\} \subset F$.

$P = \{p_1, p_2, ..., p_{pc}\}$ is a set of access rights, if pc is an amount of access rights. This set contains all possible access rights to resources. The following relations are introduced: $UP = U \times P$ is a relation setting correlation between subjects and access rights, and $PH = P \times P$ is a relation of a partial order (hierarchy) in the set of access rights indicated by "$\succeq$". The following functions are introduced: $permission: U \to 2^P$ is a function correlating subject ui with a set of access rights

$permission:(u_i)\subseteq\{p\,|\,(\exists p_0 \succeq p)\wedge((u_i,p_0)\in UP)\}$. They join the set of basic functions of system F, i.e., $permission:(u_i)\subseteq\{p\,|\,(\exists p_0 \succeq p)\wedge((u_i,p_0)\in UP)\}$.

$R=\{r_1,r_2,r_3,r_4,r_5\}$ is a set of roles within system: r_1 - "unauthorized user", r_2 - "authorized user", r_3 - " resource administrator ", r_4 - " user group administrator" and r_5 - "network administrator". The following relations are introduced: $UR=U\times R$ is a relation correlating subjects and roles, $RP=R\times P$ is a relation correlating roles and access rights, $RH\in R\times R$ is a relation of a partial order (hierarchy) in the set of roles indicated by " $\succeq$ ". The following function is introduced: $role:S\rightarrow R$ is a function correlating s_i session, one of the roles acceptable in *R*.

$Z=\{z_1,z_2,...,z_{zc}\}$ is a set of access servers, if *zc* is an amount of access servers included into the network. One access server can serve several groups, subscribers, resources, i.e. $z_i=\langle u_i,t_j,g_k\in G\rangle$.

$G=\{g_1,g_2,...,g_{gc}\}$ is a set of user groups, if *gc* is an amount of user groups. Each user u_j is included into user group g_i, i.e., $u_i\subset g_i$ for $j\in\{1,...,uc\}$ and $i\in\{1,...,gc\}$. The g_i group also includes the following relation:

$serv:Z\rightarrow G; autorization:t_i\times\{permission(u_k)\,|\,u_k\in U\}\times$
$\times U\rightarrow\{ok, access\ denied\}$

They join the set of basic functions of system *F*: $\{serv, autorization\}\subset F$.

$F=\{user, id, suser, sid, sauth, permission, serv, autorization, role\}$ is a set of the system basic functions.

We defined the following rules:

- A rule of interacting users and information resources of various workgroup under the system control. Each information resource correlates with a set of couples like $\langle g, privacy\rangle$. In this case, a user has access to a resource only if the resource belongs to the user's domain, and the level of his access privilege is more or equals the level of resource privileges. So, user *u* has an access to resource *t*, if:

$$\exists\langle g, privacy\rangle\in D_r(t): g\in D_p(u)\wedge privacy\geq access(u),$$

where D_r is an operator which receives a set of couples defining a resource (a characteristic set); D_p is an operator which receives a set of domains (domain hierarchy), to which the user belongs; *access* is a function of the user admission level.

- A rule of providing a quick cross access to technological resources. Each technological resource correlates with a set of couples $\langle a,c\rangle$ and the access is provided if the user is authorized for this session in the communication link with the resource by correlation $resourses$:

$$\exists\langle a,c\rangle\in D_c(t): a\in A, c\in C: a\in sauth(s)\wedge resourses(t,k,a,c)$$

where D_c is an operator which receives a set of couples defining the used channel and the protocol of communication with the resource (a characteristic set).

This model takes into account that there are different communication links with information systems and technological equipment. Besides, it was determined that for increasing control flexibility the model allows introduction of additional restrictions on the combinations of components, for example: restrictions of the subject's access rights on the quantity of sessions or on time and so on.

3. Control methods of access to information and technological resources of automated control systems at water supply enterprises.

The group of methods provides combined use of the control model of access, schemes of preliminary distribution of authenticators, schemes of secret share for development of a temporary access authenticator and an algorithm of authenticator evolution.

In the trusted communication environment, it is recommended to use a unique identifier for communication between each couple of subscribers. This results in a need for a network of *n* subscribers to generate and store *n* (*n* - 1)/2 authenticators. And each of *n* subscribers should store *n* - 1 authenticator.

Application of schemes of preliminary distribution of authenticators allows (Blom, KDP and others) to reduce the number of authenticators generated and stored in the automated control system.

Let us consider Blom's scheme as a basic scheme of preliminary distribution of authenticators [5]. In this scheme, finite field F fixes n different nontrivial elements $r_1, \ldots, r_n \in F$ which are attributed as identifiers to network subscribers. Then a polynomial is chosen over finite field F 2m, $1 \leq m < n$, like

$$f(x, y) = \sum_{i=0}^{m} \sum_{j=0}^{m} a_{ij} x^i y^j, \tag{3}$$

which factors a_{ij} form an invertible square matrix of authenticators symmetric to main diagonal $\Lambda = (a_{ij})_{m \times m}$ over finite field F.

Matrix Λ is confidential and is deposited on the authentication server. Each subscriber A receives set $(a_0^{(A)}, a_1^{(A)}, \ldots, a_m^{(A)})$ consisting of polynomial factors as a universal authenticator:

$$g_A(x) = f(x, r_A) = a_0^{(A)} + a_1^{(A)} x + \ldots + a_m^{(A)} x^m. \tag{4}$$

Then, a unique authenticator of k_{AB} is used for communication between each couple of subscribers A and B:

$$k_{AB} = k_{BA} = f(r_A, r_B) = g_A(r_B) = g_B(r_A), \tag{5}$$

which finally allows to store m authenticators instead of n - 1.

In Blom's classical scheme, identifiers are in open access. When using a combined (matrix and role) access, we determine correlation in the access matrix between the subscriber's rights seeking access and an r_B identifier of the requested information resource or the requested subscriber.

Technological resources often have a fixed access authenticator. Besides, when arranging a cross access, the subscriber's authority can be limited on time or number of sessions. Denial of service by the authentic (protected) communication link is also possible. For emergency access, it is offered to use a combination of the scheme of preliminary authenticators' distribution and the scheme of the full secret share [4]. Let us assume that it is necessary to provide an access to technological resource U having a fixed k_U authenticator. In this case, the authentication server generates accidental bit vector b of length m:

$$b = (b0, b1, \ldots, bm)2. \tag{6}$$

Also temporary authenticator r' is calculated (over GF_2 field.):

$$r' = k_U + \sum_{i=0}^{m} b_i a_i^{(A)}. \tag{7}$$

Temporary authenticator r', calculated in this way, and accidental vector b are transferred to the subscriber by the open reserve channel. Having received them the subscriber can retrieve the k_U authenticator (over the GF_2 field.):

$$k_U = r' + \sum_{i=0}^{m} b_i a_i^{(A)} \tag{8}$$

To increase security, it is possible to use function of hashing of $H(x)$ in addition to summing of authenticators:

$$r' = k_U + H\left(\sum_{i=0}^{m} b_i a_i^{(A)}\right). \tag{9}$$

Carrying out a similar combination is also possible on the basis of scheme KDP [6] founded on intersection of sets. For $n > 2$ subscribers and sets of authenticators of A, $|A| = q$, direct numbering of authenticators 1, 2, ..q is defined. We choose some family $\{S_1, \ldots, S_n\}$of set subcollections $\{1, 2, \ldots, q\}$, which is Shperner's family in which one subcollection does not contain another. In the original KDP scheme, family $\{S_1, \ldots, S_n\}$ represents an unclassified table with numbers of authenticators from set a_{ik}, $k \in S_i$, previously transferred to each subscriber by the secure channel. When using a combined (matrix and role) access, we determine correlation in the access matrix between the subscriber's rights seeking access and S_i element of the requested information resource or the requested subscriber.

To develop a common communication authenticator between subscribers of A and B, we use intersection $S_A \cap S_B$. For example:

$$k_{AB} = k_{BA} = \sum_{i=1}^{|S_A \cap S_B|} a_i \text{ over } GF_2. \quad (10)$$

If there is a need to provide an access to technological resource U having a fixed k_U authenticator, the authentication server generates accidental bit vector b of length $m = |S_A|$, as it was in the previous case:

b = (b0, b1, …, bm)2. (11)

Also temporary authenticator r' is calculated:

$$r' = k_U + H\left(\sum_{i=0}^{m} b_i a_i\right). \quad (12)$$

Temporary authenticator r' calculated in this way and accidental vector b are transferred to the subscriber by the open reserve channel. Having received them, the subscriber can retrieve the k_U authenticator as in the example above.

The drawback of the methods described above is that the k_U authenticator received by subscriber A is constant. To stop temporary emergency access rights of subscriber A to resource U, we offer to use evolution of technology resource authenticators.

We consider the following methods of authenticators' evolution: Lamport's protocol of one-time authentication [7]; dynamic authenticator change in a gamming mode (CTR); combination of two previous methods.

When using Lamport's protocol for each technology resource U having a fixed k_U authenticator, we determine action duration L_U of basic authenticator k_U and estimated frequency of replacement L_i of temporary authenticators k_{Ui}. Then, a required amount of periods $t = L_U/L_i$ is calculated. On the basic step of the protocol, a one-way function of hashing $H(x)$ is used. Temporary authenticator k_{Ui} is defined for the current i - period of action:

$$k_{Ui} = \underbrace{H(H(\ldots(k_U)\ldots))}_{t-i \text{ times}} = H^{t-i}(k_U) \quad (13)$$

Thus to provide temporary access rights to subscriber A, the authentication server transmits by an open communication link: accidental binary vector b, the action period of authenticator L_i, period number i, the amount of periods t and temporary authenticator r' are calculated as:

$$r' = \underbrace{H^{t-i}(k_U)}_{k_{Ui}} + H\left(\sum_{i=0}^{m} b_i a_i^{(A)}\right). \quad (14)$$

Having received them, the subscriber can retrieve the k_{Ui} authenticator operating during period T_i:

$$k_{Ui} = r' + H\left(\sum_{i=0}^{m} b_i a_i^{(A)}\right). \quad (15)$$

The shortcomings of this method are the limited period of the authenticator k_U action, and also an increase of computing loading at the high intensity of information exchange.

When using the gamming mode, authenticator k_U dynamically changes within some period L_i by means of enciphering in the gamming mode. Authenticator k_{Ui} is the recurrent fraction of block gamma:

kUi = Ts (eK(CTRi)), (16)

where $CTR_1 = IV$ is an initiating vector (synchronous packet),
$CTR_i = inc(CTR_{i-1})$ is a register value,
e_K is an encryption block algorithm on secret key K,
T_s - procedure of taking seniors s bits.

In this case, authenticator k_U can act both as key K and as synchronous packet IV.

Two methods described above can be combined as follows. By means of Lamport's protocol temporary administrator of a technological resource, an authenticator, used as a K key, is given, and the administrator in its turn can transfer users temporary authenticators generated in the CTR mode, using the simplified way of generation r':

$$r' = k_{Ui} \oplus H(k_{AB}) \oplus b \qquad (17)$$

where k_{AB} is a common communication authenticator between the user and the administrator.

4. Conclusion

The given model considers the availability of different communication links with information systems and technological equipment. Besides, it has been determined that in order to increase control flexibility, the model allows introduction of additional restrictions on combinations of components, for example: restrictions on the subject's access rights by session amounts or on time and so on.

On the basis of the formalized model and methods of access control, there has been developed a hardware-software complex of rapid access to information and technological resources of automated control systems for water supply enterprises. This complex consists of the administrator's automated workplace and ultimate users.

The program components of the complex implement are:

- the scheme of authenticators' preliminary distribution (Blom's scheme, KDP), in combination with access policies;
- support of authenticators' evolution on the basis of Lamport's protocol or the Counter (CTR) mode;
- generation and transfer of temporary access authenticators using the combination of the scheme of authenticators' preliminary distribution and schemes of the full secret share.

The program part of the complex is implemented with the use of high-level libraries of the PKCS#11standard [8] and the Rutoken CSP crypto service provider, which is the main to provide authentication and integrity (CMAC message authentication code development and other cryptographic transformation). The subsystem is multiplatform and it imposes low system requirements. This provides connectivity with the hardware of technological equipment. The complex hardware is represented by cryptographic identifiers Rutoken ECP Flash and technology equipment controllers.

References

[1] Demidov A V, Ofitserov A I and Afonin S I 2010 *Information technologies in science, education and production* **5** 94-101
[2] Eryomenko V T, Mishin D S, Paramokhina T M, Eryomenko A V and Eryomenko S V 2014 *Information systems and technologies* **3** 51-58
[3] Eremenko V T and Paramokhin V M 2015 *Information systems and technologies* **2(88)** 131-137
[4] Rytov M Yu, Shpichak S A, Pavlinova I V and Pavlinova E I 2014 *Cryptographic providing information protection* 220
[5] Blom R 1983 *Nonpublic key distribution, Advances in Cryptology, Proceeding of EUROCRYPT '82 Plenum New York* 231-236
[6] Dyer M, Fenner T, Frieze A and Thomason A 1995 *On key storage in secure network, J Cryptology* **8** 189-200
[7] Lamport L 1978 *Communications of the ACM* **21(7)** 558-565
[8] PKCS #11 Base Functionality v2.30: Cryptoki - Draft 4, RSA Laboratories 2009

Texture-based forest segmentation in satellite images

S V Sai, E V Mikhailov

Pacific National University, 136, Tikhookeanskaya st., Khabarovsk, 680035, Russia

E-mail: sai1111@rambler.ru

Abstract. NDVI images are often used to segment vegetation areas in satellite images. Unfortunately, NDVI lack information to effectively separate forest regions from grass regions, as it is computed basing solely on spectral characteristics of red and infrared bands, and both forest and grass display similar spectral characteristics. On the other hand, forest areas are visually distinguishable, as they contain a lot of small shadows or dark spots in satellite images, while grass areas look flat. In the paper, we use this observation by extracting simple texture features from the panchromatic band. In our case, it was enough to compute standard deviation for a sliding window to separate forest from grass.

1. Introduction

A normalized difference vegetation index (NDVI) is often used to analyse satellite images to detect vegetation areas. NDVI is based on spectral information from red band and near-infrared band images [1]. It is easy to compute and it produces reliable results even if combined with simple thresholding. Although, NDVI is also applied for quantitative analysis of vegetation areas and is used as an indicator of vegetation density, biomass, etc. [2, 3, 4], NDVI is not sufficient to reliably segment forest from dense grass. NDVI is an old method and does not fully utilize modern satellites, which produce high resolution imagery. For example, WorldView-2 produces 0.5 meter resolution panchromatic images and 2 meter resolution multispectral images. As seen in figure 1, forest and trees are clearly distinguishable from grass even in a panchromatic image, because trees have light and dark spots, and grass looks flat. The problem is that NDVI uses only spectral information, while spatial information is required to distinguish.

To make the proposed method aware of spatial information, computation of texture features is added [5, 6]. Even simple features such as standard deviation for a sliding window combined with NDVI thresholding is enough to reliably segment forest areas.

2. Method overview

General steps for the proposed method are as follows.

1. Compute top of atmosphere reflectance (ToAR) for each band [7, 8, 9].
2. Compute NDVI based on red and near-infrared images.
3. Extract texture features from a panchromatic image. In the paper, standard deviation for a sliding window is used.
4. Combine NDVI thresholding and texture features image thresholding to segment forest areas.

Source images are shown in figure 1. Those are images of a small area in Khabarovsk made by the WorldView-2 satellite. Calculation of NDVI requires using reflectance data, but initial source images values are just digital numbers which do not have physical meaning. More importantly, images for different bands are not properly weighted against each other. First spectral radiance response and scaling factors are taken into account to produce top of atmosphere radiance. This step just requires to weight images of different bands by some fixed band-specific values. Given radiance registered by satellite sensors, known as solar spectral irradiance and current position of the scene and the sun, it is possible to calculate ToAR values. The last step depends on the image generation date and time, as well as on geographical position of the region. The whole procedure is satellite specific, so one usually should refer to satellite technical documentation.

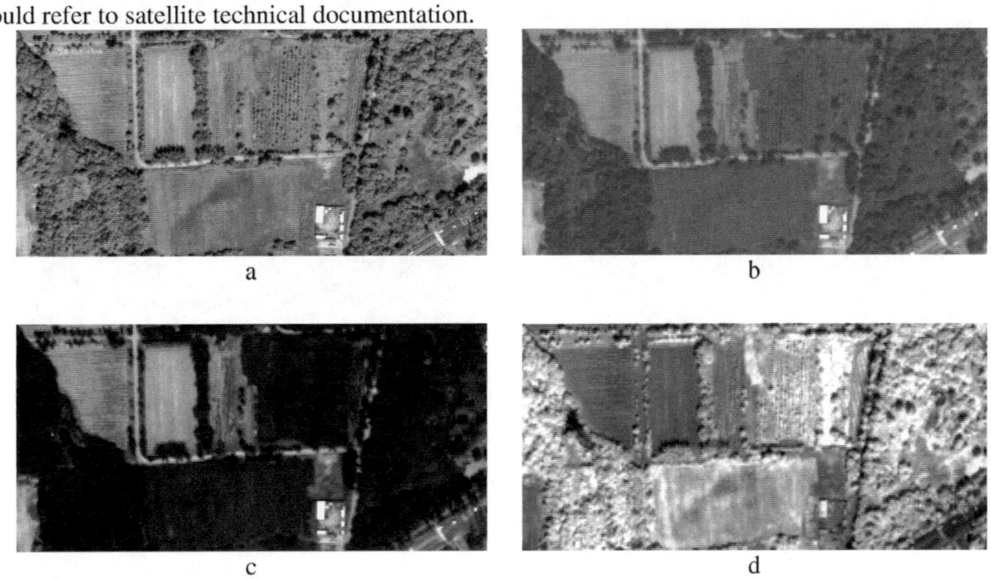

Figure 1. Source images: a – panchromatic image; b – multispectral image (rgb); c – red band; d – near-infrared band.

After ToAR images are computed, it is possible to compute NDVI images. NDVI is derived from red band and near-infrared band images, and then NDVI can be used to segment vegetation areas by simply applying threshold. This segmentation method is based solely on spectral information, so it cannot effectively distinguish grass and forest. To improve this result, an additional step is proposed, so that the method can utilize high resolution of the satellite. This step is based on extracting texture features from the panchromatic image, so that the method becomes somewhat aware of spatial information. More information on NDVI images and texture feature extraction is given in the following sections.

3. NDVI image

The idea of NDVI is based on spectral characteristics of plants, as it is mainly used to analyze satellite images and check whether a region contains green vegetation or not [1, 10]. Plants convert energy of visible light into chemical energy through photosynthesis. The main leaves pigment used in photosynthesis is chlorophyll, and it strongly absorbs red and blue light, so that healthy plants appear to be green in visible light. Photons at infrared wavelengths do not transmit enough energy for photosynthesis, and plants strongly reflect it to prevent overheating. Based on that information one could expect vegetation areas to be bright in the near-infrared spectrum and dark – in the visible red spectrum, as it is clearly seen in figure 1. NDVI exploits that observation and it is computed as follows:

$$\mathrm{NDVI} = \frac{\mathrm{NIR} - \mathrm{VIS}}{\mathrm{NIR} + \mathrm{VIS}}, \tag{1}$$

where NIR and VIS are reflectance in near-infrared and visible red spectrums, respectively. The range of NDVI values is between –1.0 and 1.0. For vegetation areas, VIS should be near zero and an overall value of expression should be near 1.0. Usually, condition NDVI>0.6 is a strong indicator of dense vegetation. Bare land and soils have slightly above zero NDVI values and water areas have negative NDVI values or zero. Buildings, roads and other artificial regions have quite a wide range of NDVI values, but usually they appear to be similar to soils NDVI values or below zero.

Figure 2.a shows the results of computing NDVI for the area of interest. Both grass and forest areas appear to be bright (values near 1.0), while other areas appear to be dark (values near zero or negative values). Using threshold NDVI>0.6, we can easily separate vegetation areas from soils and roads as seen in figure 2.b. NDVI thresholding also identifies correctly a small water body to the left of the region as a non-vegetation area. On the other hand, NDVI is ineffective if one needs to separate forest from grass, as it can only estimate vegetation density and not plants height.

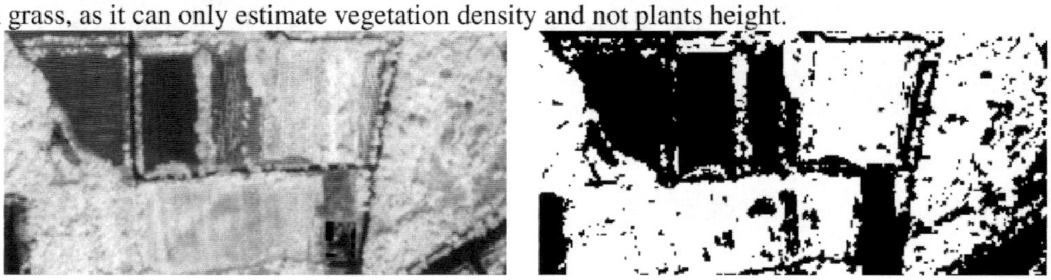

a b

Figure 2. Segmentation of vegetation areas using NDVI: a – NDVI image; b – dense vegetation areas (NDVI>0.6).

4. Texture features and forest segmentation

Although we cannot use NDVI to separate forest from grass, one can visually distinguish forest areas in panchromatic images, without even considering color image. As seen in figure 1, grass areas usually appear to be flat as it contains pixels of almost the same brightness. Forest usually has a lot of height variation, which results in a lot of small dark (tree shadows) and bright (actual trees) spots. Moreover, these spots usually have certain size, which roughly equals to an expected tree crown diameter. As a result, in order to find forest in satellite images, we need to consider spatial information, and NDVI is only aware of spectral information.

To address this issue, we propose extracting simple texture features [5, 6, 11]. These features should be computed for a sliding window big enough to contain at least one tree with a shadow. Small windows observe high frequency spatial information, and they are suitable for highlighting edges, so a sliding window should not be too small. Moreover, it is preferable to remove high frequency noise before feature extraction. A sliding window also should not be too big, as grass areas over big regions also contain some brightness variation and big windows are more suitable for detecting low frequency spatial information. Given a proper window size, it is enough to use even simple texture feature to separate forest from grass, and we use standard deviation for a sliding window. Overall, the extraction of texture features goes as follows.

1. Pre-process the panchromatic image by removing high frequency noise. For that, we use Gaussian blur to remove features smaller than a tree.
2. Compute standard deviation for a sliding window. The window size should be slightly bigger than that of a tree.
3. Post-process the image of texture features with blur. As distance between trees varies, we also want to include small grass area near trees into forest regions. A blur window should be bigger than a tree.
4. Threshold the resulting image to segment forest.

A panchromatic image is used as a source to compute texture features, as it is usually has a higher resolution. The first two steps are used to find regions with a certain spatial structure to highlight areas with spots of a tree size. Generally, we want to avoid small or thin areas of grass or forest, so we apply the third step. It is possible to replace it with more aggressive post-processing, such as morphological image processing of the final result. The final result after texture feature thresholding combined with NDVI thresholding is shown in figure 3. As compared to figure 2, the final result excludes grass in the lower part of the image and represents the forest area more accurately.

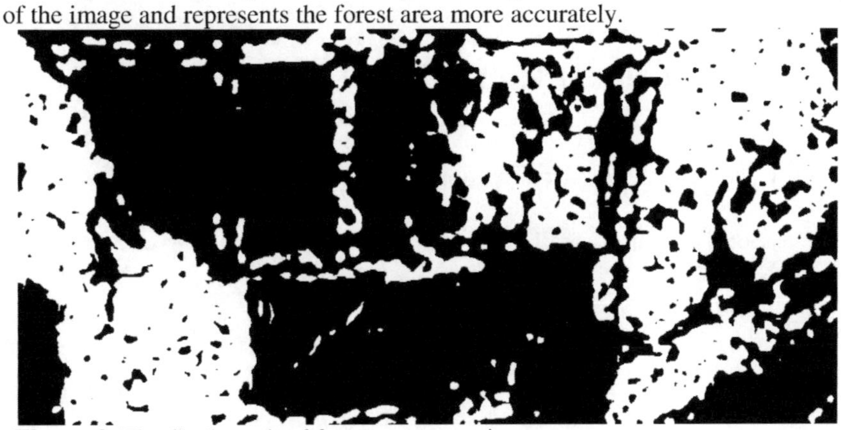

Figure 3. The final result of forest segmentation.

5. Conclusion

NDVI on its own is not sufficient to segment forest areas, as it does not consider spatial information, so the proposed method combines it with texture features extraction and the method is capable to effectively separate forest from grass.

The further work should consider more complex texture features, such as Gabor filters, texture features of the second order and others, as simple standard deviation is not particularly aware of spatial information. These new features might allow for better classification, for example it might be used not only for separating forest, but for estimating tree density, tree height and other parameters.

The proposed method also requires selecting thresholds. Although the process is straightforward for NDVI, it becomes tricky, when it comes to thresholding on more complex features, such as texture features. So it is better to automate thresholding by deploying some machine learning algorithms.

References

[1] Rouse J W, Haas R H, Scheel J A, and Deering D W *Proceedings, 3rd Earth Resource Technology Satellite (ERTS) Symposium* **1** 48–62
[2] Todd S W, Hoffer R M, and Milchunas D G 1998 *International Journal of Remote Sensing* **19(3)** 427–438
[3] Carlson T N and Ripley D A 1997 *Remote Sensing of the Environment* **62(3)** 241–252
[4] Moleele N, Ringrose S, Arnberg W, Lunden B, and Vanderpost C 2001 *International Journal of Remote Sensing* **22(5)** 741–756
[5] Newsam S, Kamath C 2004 *Proceedings of SPIE* **5433** 21–32
[6] Zhang C, Guo X, Wilmshurst J and Sissons R 2006 *Canadian Journal of Remote Sensing* **32(4)** 281-287
[7] Chavez P S 1996 *Programming Engineering and Remote Sensing* **62** 1025–1036
[8] Chandler G, Markham B L, Helder D L 2009 *Remote Sensing of Environment* **113** 893-903
[9] Parente C 2013 *Proceedings in EIIC* 2(1)
[10] Tucker C J 1979 *Remote Sensing of Environment* **8(2)** 127-150
[11] Srinivasan G N, Shobha G 2008 *Proceedings of World Academy of Science, Engineering and Technology* **36** 1264-1269

Adaptive interpretation of gas well deliverability tests with generating data of the IPR curve

V L Sergeev[1], Nguyen T H Phuong[1] and A I Krainov[1]

[1]National Research Tomsk Polytechnic University, Lenina Ave., 30, Tomsk, 634050, Russia

E-mail: SergeevVL@ignd.tpu.ru , nguyenphuongtpu1512@gmail.com

Abstract. The paper considers topical issues of improving accuracy of estimated parameters given by data obtained from gas well deliverability tests, decreasing test time, and reducing gas emissions into the atmosphere. The aim of the research is to develop the method of adaptive interpretation of gas well deliverability tests with a resulting IPR curve and using a technique of generating data, which allows taking into account additional a priori information, improving accuracy of determining formation pressure and flow coefficients, reducing test time. The present research is based on the previous theoretical and practical findings in the spheres of gas well deliverability tests, systems analysis, system identification, function optimization and linear algebra. To test the method, the authors used the field data of deliverability tests of two wells, run in the Urengoy gas and condensate field, Tyumen Oblast. The authors suggest the method of adaptive interpretation of gas well deliverability tests with the resulting IPR curve and the possibility of generating data of bottomhole pressure and a flow rate at different test stages. The suggested method allows defining the estimates of the formation pressure and flow coefficients, optimal in terms of preassigned measures of quality, and setting the adequate number of test stages in the course of well testing. The case study of IPR curve data processing has indicated that adaptive interpretation provides more accurate estimates on the formation pressure and flow coefficients, as well as reduces the number of test stages.

1. Introduction

Deliverability tests with the resulting inflow performance relationship (IPR) curve, run in the gas wells with steady-state filtration, are one of the most informative and common methods well tests to characterize the behavior of well and the bottomhole conditions. Currently, the data obtained via deliverability tests are interpreted using the methods described in [1-3], which are based on Forchheimer binomial equation for gas filtration:

$$p_{\mathrm{r}}^2 - p_{\mathrm{w}}^2 = aq + bq^2 , \qquad (1)$$

where $p_{\mathrm{r}}^2 . p_{\mathrm{w}}^2$ are formation pressure and bottomhole pressure, respectively; a and b - flow coefficients dependent on bottomhole zone parameters and bottomhole structure; q - flow rate. Coefficients a and b for IPR curve model (1) should be estimated using the least square method, with the formation pressure being known [2-4]. IPR interpretation based on model (1) and the least square method is challengeable as a field method, which is attributed to the following facts: the initial formation pressure is difficult to determine, estimates should be robust and accurate, the number of test stages (a number of "cycles" characterized by a stabilized flow when the pressure and the flow rate are recorded) is reduced.

To ensure that the estimates are accurate and robust, in the work [5] we suggest interpreting the IPR curve using integrated IPR curve models with account of additional a priori data on the formation pressure and flow coefficients. However, the question is how to provide additional a priori data on the formation pressure and flow coefficients and to determine the adequate number of test stages to secure preassigned estimate accuracy.

In this research the method of adaptive interpretation of deliverability tests with resulting IPR curve and generating data with variable parameters, additional a priori data on the formation pressure and flow coefficients is suggested and investigated. The method implies that the parameters depend on the number of test stages. To define the formation pressure and to generate data of bottomhole pressure and the flow rate at different test stages, we use the empirical power law for gas filtration, which is widely applied in deliverability test analysis over the years. [6-8]

$$q = \lambda (p_{\mathrm{r}}^2 - p_{\mathrm{w}}^2)^{\gamma} \tag{2}$$

where λ - productivity index; γ - constant factor with theoretical value ranging from 0.5 (turbulent flow) to 1.0 (laminar flow).

2. Models and Algorithms for adaptive interpretation of the IPR curve with generating data

The basis to develop algorithms for gas well deliverability test data interpretation is an integrated system of IPR curve models (1) with variable parameters dependent on the number of the test stage, takes into account of generated values of squared bottomhole pressure $\hat{y}_j$, additional a priori data on the formation pressure $\overline{p}^2_{\mathrm{r},n}$ and flow coefficients $\overline{a}_n, \overline{b}_n$:

$$\begin{cases} y_i^* = p_{\mathrm{r},i}^2 - a_i q_i - b_i q_i^2 + \xi_i = y_i + \xi_i, i = \overline{1,n}, \\ \hat{y}_j = y_j + \eta_j, \overline{p}^2_{\mathrm{r},n} = p_{\mathrm{r},n}^2 + \nu_n, j = \overline{1,l}, \\ \overline{a}_n = a_n + \varepsilon_{1,n}, \overline{b}_n = b_n + \varepsilon_{2,n}, n = \overline{2,nk}, \end{cases} \tag{3}$$

where $y_i^* = p_{\mathrm{w},i}^2, q_i$ are values of squared bottomhole pressures and flow rates obtained at test stage number i; $\overline{p}^2_{\mathrm{r},n}, \hat{y}_j = \hat{p}^2_{\mathrm{w},j}$ are estimates of formation pressure and generated values of squared bottomhole pressures; $y_j = p_{\mathrm{r},j}^2 - a_j \hat{q}_j - b_j \hat{q}_j^2$ are values of squared bottomhole pressures given by model (1) with generated values of flow rate $\hat{q}_j$; nk is unknown parameter defines the number of test stages n appropriate to secure preassigned estimates accuracy for the formation pressure and flow coefficients p_{r}^2, a, b ; $\xi_n, \eta_l, \nu_n, \varepsilon_{1,n}, \varepsilon_{2,n}$ are random variables, i.e. errors in measurements, recovery data, and estimates of flow coefficients, as well as deficiencies of gas filtration models (1),(2) etc.

The additional data on formation pressure $\overline{p}^2_{\mathrm{r},n}$ and parameters estimates $\overline{\lambda}_n$ and $\overline{\gamma}_n$ of model (2) can be obtained by solving the following optimization problem:

$$\overline{\boldsymbol{\alpha}}_n = \arg\min_{\boldsymbol{\alpha}} \sum_{i=1}^{n} f(q_i^* - \alpha_1 (\alpha_2 - y_i^*)^{\alpha_3}) \tag{4}$$

where $\arg\min_x f(x)$ is minimum point x^* of function $f(x)$ $(f(x^*) = \min_x f(x))$; $\overline{\boldsymbol{\alpha}}_n = (\overline{p}^2_{\mathrm{r},n}, \overline{\lambda}_n, \overline{\gamma}_n)$ is a vector of estimates; f is a known function. The additional data on flow coefficients $\overline{a}_n, \overline{b}_n$ can be obtained from the system of linear equations:

$$z_i = a q_i + b q_i^2, i = \overline{1,n}, \tag{5}$$

which is the result of grouping models (1),(2) for depression $p_r^2 - p_w^2$ where $z_i = \sqrt[\overline{\gamma}_n]{q_i/\overline{\lambda}_n}$, q_i is a value of the flow rate obtained at test stage number i; $\overline{\lambda}_n, \overline{\gamma}_n$ are optimal estimates obtained by solving problem (4).

The optimal values of squared formation pressure $p_{r,n}^2$ and flow coefficients a_n, b_n of model (3) represented for convenience as matrix

$$\begin{cases} \tilde{\mathbf{y}}_k = F_k \boldsymbol{\alpha}_k + \boldsymbol{\xi}_k, k = \overline{1, n+l}, \\ \overline{\boldsymbol{\alpha}}_{k(n)} = \boldsymbol{\alpha}_k + \boldsymbol{\eta}_k, n = \overline{3, nk}, \end{cases} \quad (6)$$

are calculated using the method of adaptive identification by solving optimization problems (7),(8)

$$\boldsymbol{\alpha}_k^*(\boldsymbol{\beta}_k, h_k) = \arg\min_{\boldsymbol{\alpha}_k}(J_0(\boldsymbol{\alpha}_k, h_k) + J_a(\boldsymbol{\alpha}_k, \boldsymbol{\beta}_k)), \quad (7)$$

$$\boldsymbol{\beta}_k^*, h_k^* = \arg\min_{\boldsymbol{\beta}_k, h_k} J_0(\boldsymbol{\alpha}_k^*(\boldsymbol{\beta}_k, h_k)), \quad (8)$$

where $\tilde{\mathbf{y}}_k = (\tilde{y}_k, k = \overline{1, n+l}, n = \overline{3, nk}) = (y_i^*, \hat{y}_j, i = \overline{1, n}, j = \overline{1, l}, n = \overline{3, nk})^T$ is a combined vector of initial data and generating data on squared bottomhole pressures; $F_k = (\varphi_n, \varphi_l)^T, k = \overline{1, n+l}, n = \overline{3, nk}$ is a combined matrix of vectors $\varphi_n = (1, q_i, q_i^2), i = \overline{i, n}$ and $\varphi_l = (1, \hat{q}_j, \hat{q}_j^2), j = \overline{1, l}$; $\boldsymbol{\alpha}_k = (\alpha_{1,k} = p_{r,k}^2, \alpha_{2,k} = a_k, \alpha_{3,k} = b_k), k = \overline{1, n+l}$ is a combined vector of parameters $(\boldsymbol{\alpha}_i, \boldsymbol{\alpha}_j, i = \overline{1, n}, j = \overline{1, l})^T$ of initial data and generating data; $\overline{\boldsymbol{\alpha}}_{(n)} = (\overline{\alpha}_{1,k} = \overline{p}_{r,k}^2, \overline{\alpha}_{2,k} = \overline{a}_k, \overline{\alpha}_{3,k} = \overline{b}_k)_{(n)}$ is a vector of additional a priori data obtained at stage number n; $J_0(\boldsymbol{\alpha}_k) = \sum_{k=1}^{n+l} \omega_k(h_k) \cdot \psi_0(\tilde{y}_k - \tilde{\varphi}_k^T \boldsymbol{\alpha}_k)$, $J_a(\boldsymbol{\alpha}_k^*, \boldsymbol{\beta}_k) = \sum_{j=1}^{3} \beta_{j,k} \psi_a(\omega_{j,k} \overline{\alpha}_{j,k} - a_{j,k})$ are measures of model (6); $\boldsymbol{\beta}_k = (\beta_{j,k}, j = \overline{1,3})$ is a vector of control parameters defining the importance (weight) of additional a priori data $\overline{\alpha}_{j,k}, j = \overline{1,3}$; ψ_0, ψ_a are known functions; $\omega_k(((k-i)/h_k), i = \overline{1, n+l-1}, k = \overline{1, n+l})$ is a weighting function with decay parameter h_k to secure adaptive identification and interpretation $(\omega(x_1) < \omega(x_2), x_1 < x_2)$; $kr_{j,k}$ is an adjustment parameter for additional data $\overline{\boldsymbol{\alpha}}_{k(n)}$.

The solution on the time for deliverability test with the resulting IPR curve to be completed can be taken via visual analysis of the graph (see figures 2-4) or using the criterion for estimates stabilization, where nk is such test stage k that

$$\left| (\alpha_{j,k}^*(\beta_{j,k}^*, h_k^*) - \alpha_{j,k-1}^*(\beta_{j,k-1}^*, h_{k-1}^*)) / \alpha_{j,k}^*(\beta_{j,k}^*, h_k^*) \right| \le \varepsilon_j, j = \overline{1,3}, k = k_0, k_0 + 1, .. \quad (9)$$

is a valid inequality, where ε_j is preassigned accuracy.

It is noteworthy that there are many strategies of generating data of IPR curve $y_i^* = p_{r,i}^2, q_i, i = \overline{1, n}$ to define values of additional flow rates $\hat{q}_j$ and bottomhole pressures $\hat{y}_j = \hat{p}_{w,j}^2, j = \overline{1, l}$ (3). In this research we use a simple method to generate data of the IPR curve that the total number of data $n+l$ is doubled $n=l$,

$$\hat{P}_{w,j} = P_{w,i} + \Delta_i = P_{w,i} + (P_{w,i+1} - P_{w,i})/2, \ \hat{q}_j = \overline{\lambda}_n (\overline{p}_{r,n}^2 - \hat{p}_{w,j}^2)^{\overline{\gamma}_n}, j = \overline{2, n} \quad (10)$$

Where $\overline{p}_{r,n}^2, \overline{\lambda}_n, \overline{\gamma}_n$ are the estimates of power law parameters for gas filtration (2) obtained by solving problem (4).

The algorithm given below represents the method of adaptive interpretation of the IPR curve with generating data:

1. Forming combined vector $\tilde{\mathbf{y}}_k$ (3) of initial and generating data $\hat{q}_j, \hat{p}^2_{w,j}$ (10), beginning with the minimum number of test stage n = 2.

2. Defining the vector of additional data $\bar{\boldsymbol{\alpha}}_{k(n)} = (\bar{\alpha}_{1,k} = \bar{p}^2_{r,n}, \bar{\alpha}_{2,k} = \bar{a}_k, \bar{\alpha}_{3,k} = \bar{b}_k)_{(n)}$ by solving problem (4) and system of linear equations (5).

3. Selecting measures of model (6) quality $J_0(\boldsymbol{\alpha}_k, h_k), J_a(\boldsymbol{\alpha}_k, \boldsymbol{\beta}_k)$.

4. Solving problems (7), (8) using the appropriate method of function optimization.

5. Checking condition (9): if the condition is fulfilled, the test is completed; if condition (9) fails to be fulfilled, the next test stage $n+1$ is arranged, and one should start new research with step 1 of the algorithm.

3. Results of IPR curve interpretation for gas wells.

The results of a case study of the deliverability test with the resulting IPR curve run in wells 1 and 2 of the Urengoy gas and condensate field are given in figures 1-4 and tables 1, 2.

For example, figure 1 shows the IPR curves for wells 1 and 2, initial data on five test stages and generating data on bottomhole pressure and the flow rate.

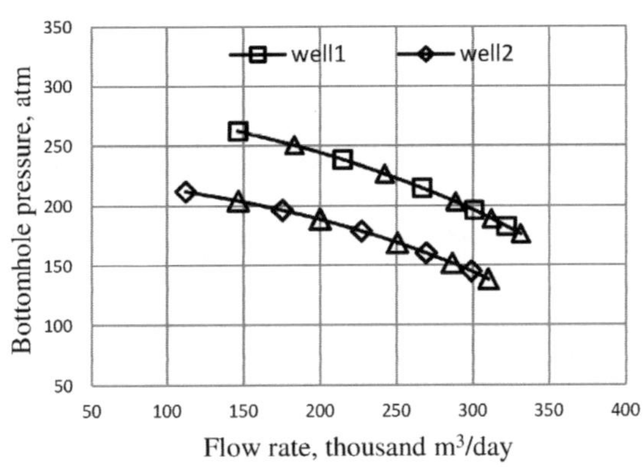

Figure 1. *Initial and generating data (Δ) of IPR curves for wells 1 and 2.*

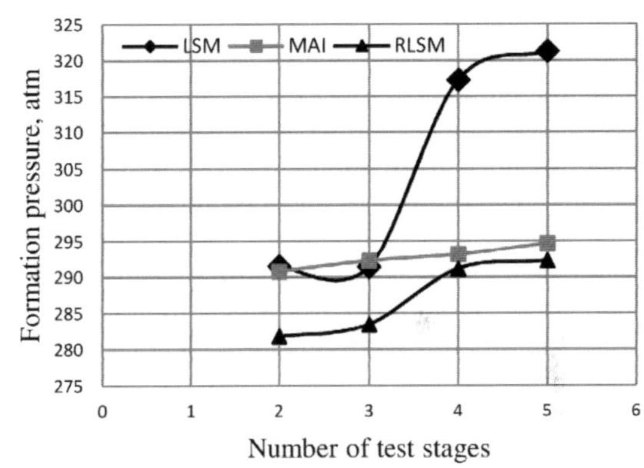

Figure 2. Formation pressure estimates for well 1.

Figures 2-4 show the estimates of formation pressure and flow coefficients of well 1, which are obtained using the following techniques:

1. the method of adaptive interpretation (MAI) (7) with quadratic measures of quality $\psi_0(x) = \psi_a(x) = x^2$ by solving the system of linear equations when $\mathbf{kr}_k = (kr_{j,k}, j = \overline{1,3})$ and $\beta_{j,k} = \beta_k, j = \overline{1,3}$ [9].

$$(F_k^T W_k(h_k^*) F_k + \beta_k \mathbf{I})\, \boldsymbol{\alpha}_k^*(\beta_k^*, h_k^*) = (F_k^T W_k(h_k^*) \tilde{\mathbf{y}}_k + \beta_k^* \cdot \mathbf{kr}_k \cdot \bar{\boldsymbol{\alpha}}_{k(n)}), k(n)=2n,\ k = 4,5,\ldots,nk \quad (11)$$

where the estimates of control parameter β_k^* and decay parameter h_k^* are defined by solving problem (8) using the downhill simplex method [10]; $W(h_k^*) = diag(\exp((k-i)/h_k^*), i = \overline{1,2n-1})$ is a diagonal matrix of weighting function values;

2. the least squares method (LSM) from (11) with $\beta_k^* = 0, \bar{\boldsymbol{\alpha}}_{k(n)} = 0$

3. the regularized least squares method (RLSM) from (11), with $\bar{\boldsymbol{\alpha}}_{k(n)} = 0$.

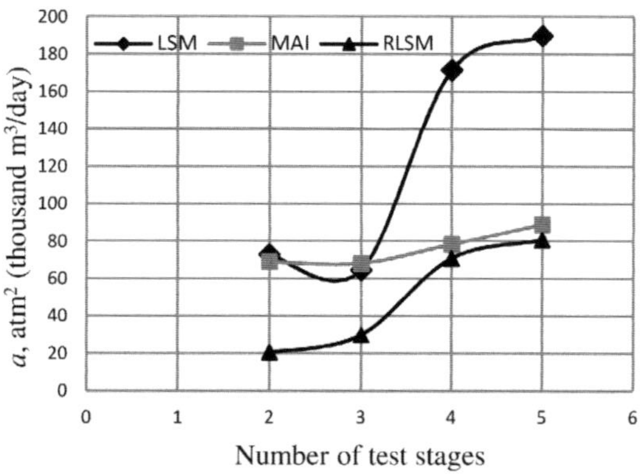

Figure 3. Estimates of flow coefficients a in well 1.

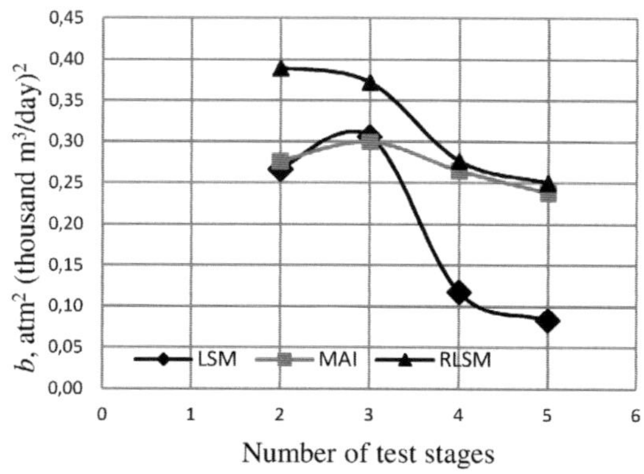

Figure 4. Estimates of flow coefficients b in well 1.

Table 1 shows the estimates of flow coefficients and the formation pressure of well 1 given by different methods.

Table 2 gives the estimates of the formation pressure and flow coefficients $\bar{\alpha}_{k(n)} = (\bar{\alpha}_{1,k} = \bar{p}^2_{r,n}, \bar{\alpha}_{2,k} = \bar{a}_k, \bar{\alpha}_{3,k} = \bar{b}_k)_{(n)}$ of wells 1 and 2, which are used as additional data in (3) and obtained by solving optimization problem (4) using Gauss-Newton method with $f(x)=x^2$ [9] and the system of linear equations (5).

As can be seen in figures 2–4 and table 1, the suggested method of adaptive interpretation of the IPR curve with generating data as well as taking into account additional data allows obtaining more accurate estimates of the formation pressure and flow coefficients with less amount of field data, compared to the method of least squares. For example, for the adaptive interpretation method, three test stages are enough (see figures 2–4 and table 1).

Table 1. Flow coefficients and formation pressure estimates of well 1.

Number of the test stage (n)	The total amount of data ($n+l$)	Method	Flow coefficient estimate $a^*_n = \alpha^*_{2,n}(\beta^*_n, h^*_n)$, atm²/(thousand m³/day)	Flow coefficient estimate $b^*_n = \alpha^*_{3,n}(\beta^*_n, h^*_n)$, atm²/(thousand m³/day)²	Formation pressure estimate $p^*_{r,n} = \alpha^*_{1,n}(\beta^*_n, h^*_n)$, atm
2	4	MAI	68.95	0.2760	290.80
		LSM	73.02	0.2664	291.53
		RLSM	20.44	0.3893	281.85
3	6	MAI	68.21	0.2995	292.27
		LSM	64.58	0.3063	291.43
		RLSM	29.75	0.3721	283.46
4	8	MAI	78.43	0.2646	293.12
		LSM	171.57	0.1177	317.29
		RLSM	70.96	0.2766	291.17
5	10	MAI	88.97	0.2381	294.63
		LSM	189.76	0.0833	321.26
		RLSM	80.81	0.2500	292.25

Table 2. Additional data.

Number of the test stage (n)	Well	Formation pressure $\overline{p}_{r,n}$ atm	Flow coefficient $\overline{a}$, atm^2/(thousand m^3/day)	Flow coefficient $\overline{b}$, atm^2/(thousand m^3/day)^2
2	1	286.8	46.85	0.3277
	2	226.3	29.87	0.2385
3	1	285.2	31.32	0.3807
	2	225.7	29.00	0.2341
4	1	286.8	49.19	0.3205
	2	225.3	23.88	0.2555
5	1	289.0	61.94	0.2904
	2	225.4	24.81	0.2511

4. Conclusion

To overcome the challenges of interpreting deliverability tests with the resulting IPR curve of gas wells, the method of adaptive interpretation has been suggested. This method allows:

1. Generating field data with the resulting IPR curve, using the power law for gas filtration.
2. Accounting additional a priori data on the formation pressure and flow coefficients.
3. Defining estimates of the formation pressure and flow coefficients within the period of test time.
4. Setting the number of test stages adequate for efficient well testing.

The case study of IPR curve interpretation for two wells of the Urengoy gas and condensate field has indicated that adaptive interpretation provides robust and more accurate estimates of the formation pressure and flow coefficients, as well as allows reducing the number of test stages.

References

[1] Houpert A 1959 On the flow of gases in porous media *Revue de L'Institut Francais du* Petrole **14 (11)** pp 1468-1684
[2] Aliev Z S, Gritsenko A I et al 1995 *Guidelines for Well Testing* (Moscow: Nauka) p 523
[3] Aliev Z S and Zotova G A 1980 *Gas reservoir and Well Testing: Handbook* (Moscow: Nedra) p 301
[4] Akhmedov K S, Gasumov R A and Tolpaev V A 2011 Technique of data processing of wells hydrodynamic studies *Petroleum Engineering* **3** pp 8–11
[5] Nguyen T H P and Sergeev V L 2015 Identification of IPR curve for interpreting gas well deliverability tests *Bulletin of the Tomsk Polytechnic University: Georesources Engineering* **326** **12** pp 54–59
[6] Rawlin E L and Schellhardt M A 1936 *Back pressure data on natural gas wells and their application to production practices* (United States Bureau of Mines, Monograph 7) p 210
[7] Al-Subaie A A, Al-Anazi B D and Al-Anazi A F 2009 Learning from modified isochronal test analysis Middle East gas well case study *Nafta Scientific Journal* **60** pp 405–415
[8] AL-Attar H and Al-Zuhair S 2009 A general approach for deliverability calculations of gas wells *Journal of Petroleum Science and Engineering* **67** pp 97–104
[9] Polishchuk V I and Sergeev V L 2015 Adaptive identication method of a signal from stray field magnetic sensor for turbogenerator diagnostics *Journal of Siberian Federal University: Mathematics & Physics* **8 (2)** pp 201–207
[10] Letova T A and Panteleev A V 2002 *Optimization Methods: Examples and Problems* (in Russian) (Moscow: Vysshaya Shkola) p 544